Wind Turbines: Technologies and Applications

Wind Turbines: Technologies and Applications

Edited by Benjamin Wayne

Larsen & Keller
www.larsen-keller.com

Wind Turbines: Technologies and Applications
Edited by Benjamin Wayne
ISBN: 979-8-88836-082-8 (Hardback)

Larsen & Keller
Published by Larsen and Keller Education,
5 Penn Plaza,
19th Floor,
New York, NY 10001, USA

Cataloging-in-Publication Data

Wind turbines : technologies and applications / edited by Benjamin Wayne.
p. cm.
Includes bibliographical references and index.
ISBN 979-8-88836-082-8
1. Wind turbines. 2. Wind power. 3. Green technology.
4. Environmental engineering. I. Wayne, Benjamin.
TJ828 .W56 2023
621.312 136--dc23

For more information regarding Larsen and Keller Education and its products, please visit the publisher's website www.larsen-keller.com

Table of Contents

Preface

A wind turbine refers to a device that produces electricity and is propelled by kinetic energy. It comprises three blades that are mounted on a tubular tower for catching the wind energy, forming a low-pressure air pocket which turns the rotor. Wind turbines have many different applications ranging from producing electricity for a single house to harnessing offshore wind. Due to increased demand, the wind turbine technology is also evolving rapidly. The wind turbines can be broadly categorized into two types, namely, horizontal axis wind turbines (HAWTs) and vertical axis wind turbines (VAWTs). HAWT is the most widely utilized method of generating wind energy. Wind turbines do not emit pollutants into water or air, and they do not need water for cooling. They help in decreasing the use of fossil fuels for electricity generation, resulting in lower carbon dioxide (CO_2) emissions and overall air pollution. This book is a compilation of chapters that discuss the most vital technologies and applications of wind turbines. It is a resource guide for experts as well as students.

After months of intensive research and writing, this book is the end result of all who devoted their time and efforts in the initiation and progress of this book. It will surely be a source of reference in enhancing the required knowledge of the new developments in the area. During the course of developing this book, certain measures such as accuracy, authenticity and research focused analytical studies were given preference in order to produce a comprehensive book in the area of study.

This book would not have been possible without the efforts of the authors and the publisher. I extend my sincere thanks to them. Secondly, I express my gratitude to my family and well-wishers. And most importantly, I thank my students for constantly expressing their willingness and curiosity in enhancing their knowledge in the field, which encourages me to take up further research projects for the advancement of the area.

Editor

1

Modeling of Wind Turbines based on DFIG Generator

Sabra Ahyaten * and Jalal El Bahaoui

Department of Physics, Faculty of science Tetouan, Abdelmalek Essaadi University, 93 030 Tetouan, Morocco; jelbahaoui@yahoo.com

* Correspondence: ahyatensabra@gmail.com or sabra.ahyaten@etu.uae.ac.ma

† Presented at the 14th International Conference on Interdisciplinarity in Engineering—INTER-ENG 2020, Târgu Mureș, Romania, 8–9 October 2020.

Abstract: The role of wind energy is so promising as a source of future energy all over the world. However, whether the unpredictable nature of wind speed fluctuations and the stability of the power systems be affected by a high penetration of wind power remains an unanswered question. Therefore, an accurate analysis study of the effectiveness and robustness performance of the doubly fed induction generator (DFIG) is one of the challenges in wind turbine applications. The present works tackle the issue of grid connection of DFIG modeling and the dynamic operation to evaluate the capabilities and their impact on the pitch angle, current and voltage stability. The model is simulated using MATLAB/Simulink software, and the curves' results are indicated based on the system that was connected to eight wind turbines in the wind farm. The results show the dynamic features of the DFIG response to the necessary output variables. On the other hand, its show us a view of the waveforms obtained, which means that an adequate study for choosing the type of control suitable for the DFIG is necessary.

Keywords: doubly fed induction generator (DFIG); grid; power system; renewable energy; wind turbine

1. Introduction

At the global scale, the improvement of renewable energy in modern production power systems has exponentially increased due to the increase in greenhouse gas concentrations in the atmosphere, which are extremely destructive to our planet. Wind energy has become a suitable answer for producing clean energy and is currently the quickest developing source when correlated with other sustainable power sources. Nonetheless, the use of available energy depends on weather conditions (wind speed) and its integration produces volatility in the power system. Integrating renewable energies with network connection, intelligent control, and storage systems could result in a change in generating electricity and reducing CO_2. The demand for a 100% renewable electricity supply by the global energy storage has already analyzed [1,2]. According to the author of International Electrotechnical Commission, the need for energy storage arises as a result of the increasing use of renewable energy and innovations in smart grids [3]. For 2050, it is foreseen that renewable energy sources (RESs) will be able to meet 70% of the world's energy needs. Wind, solar energy, and storage systems will be the most important. The investment in these energy alternative sources reduces in the electricity sector greenhouse gas emissions and also decreases fossil fuel dependence [4]. Wind energy will probably be the most dominant and will be the first immediate practical power of a global energy system, and the bulk of most nuclear and fossil fuel sources will also be replaced completely. At present, efforts to increase and improve the wind sector's participation in the electricity generation market have aroused the interest of numerous researchers [5]. It is important to mention that the development of power electronic technologies (power semiconductor devices, converters, pulse width modulation (PWM) techniques) has played an enormous role in the improvement of wind turbine efficiency. Therefore, this strong problem of wind energy in the electrical system has been strictly linked to the enhancement of wind turbine conversion and control technology [6]. These systems can be operated in order to encounter the network requirements since they can supply frequency, and active/reactive power. Essentially, there are several types of generators used in the wind power fields such as the squirrel-cage induction generator (SCIG), doubly fed induction generator (DFIG) and a direct-drive synchronous generator (DDSG) [7–9]. However, the doubly fed induction generator (DFIG) is the most frequently chosen. The reason behind this popularity is the fact that DFIG is based on the operation and independent control of active and reactive power. One of the examples of using this kind of generator in power systems application is presented by Raúl in [10], who investigated the DFIG-based hybrid system by modeling the different battery and ultra-capacitor energy storage systems. This modeling has been performed by using intelligent control techniques such as fuzzy logic to generate a compensating active power term that is changed between both energy storage systems (ESSs) to adjust their state of charge (SOC) according to the operating conditions. The impact of the wind turbines on the dynamic behavior of the system experiencing a fault and how the wind turbines behave in the system when it experiences a transient fault have also been analyzed [11]. Furthermore, it has also been found that if a transient fault in the system lead to considerable excursions in voltage and/or frequency the wind turbines were to disconnect and reconnect only once the system has returned to stable operation [12–14]. While increasing wind power penetration leads to the problem that a considerable amount of generation might disconnect in case of a transient fault in the system, this causes the system to become unstable from an otherwise harmless fault situation [15]. According to this, the grid-connected DFIG wind turbines was used in the present work. This paper indicates how the study of the dynamic behavior of the DFIG characteristic analysis and its effectiveness and optimal performance can assess the impact on the output variables of a wind farm encompassing many wind turbines. The purpose of this research was to have an exact view of the distribution and degree of fluctuations presented in the pitch angle, current, and voltage curves. In order to choose a suitable DFIG-t type control based on a series-connected wind farm by the conjunction a hybrid system consisting of several wind turbines, an energy storage system must be selected in future work to avoid such situations.

2. Power System Model Integrated with Wind Farm Using DFIG

A dynamic model of a DFIG wind turbine can be indicated in terms of the equations of a piece system. The stator winding is connected directly to the grid, while a bidirectional power converter feeds the rotor windings, allowing variable frequency operation in the rotor currents. This power converter is made up of two back-to-back insulated gate bipolar transistor (IGBT) bridges (rotor side converter, or RSC; and grid side converter, or GSC) linked by a DC bus. The wind turbine is also equipped with a blade pitch angle controller that limits wind power capture and rotational speed for high winds. Figure 1 represents its configuration.

The output power of the turbine is controlled to follow a determined power–speed characteristic, called tracking characteristic. The electrical output power on the network the wind turbine terminals are added to the power losses is compared to the reference power obtained from the tracking feature. In the converter system on the rotor side, AC voltage and Volt Ampere Reactive (VAR) are regulated. The DC–DC intermediate circuit contains two converters: first converter AC to DC and second converter DC to AC. The intermediate DC to DC circuit consists of current regulation, DC voltage and pitch control system. The last one is adjusted at zero degrees by the regulator of pitch angle to the extent that the speed is required to follow the characteristics of the control system. The DC voltage output from intermediate circuit was applied to the grid side converter which consists of an insulated gate bipolar transistor (IGBT) two-level inverter, generating an AC voltage at 60 Hz. A 12 MW wind farm consisting of eight 1.5 MW wind turbines connected to a 25 kV distribution system exports power to a 120 kV grid through a 30 km, 25 kV feeder. The wind speed is maintained constant at 15 m/s.

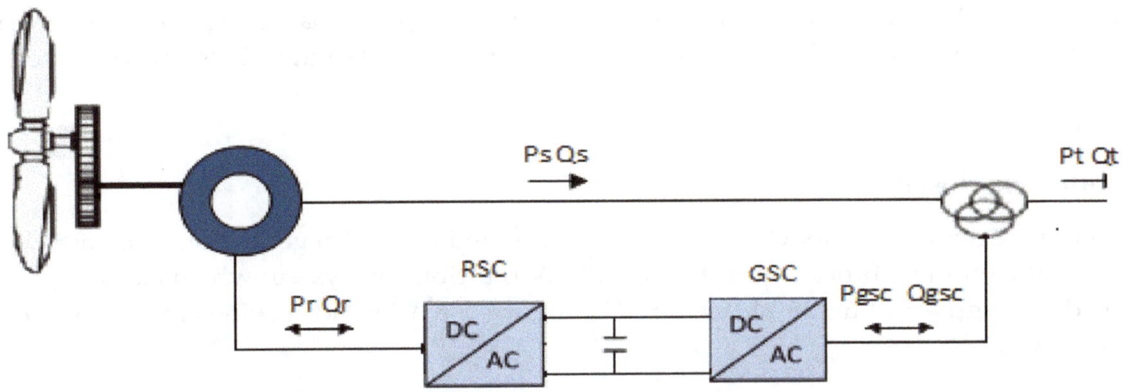

Figure 1. Doubly fed induction generator (DFIG) wind turbine scheme by Rotor side converter (RSC) and grid side converter (GSC).

2.1. *Mechanical System Model*

The system mechanic contains the drive train and aerodynamic rotor. The model of the rotor expresses the mechanical torque extracted from the wind, defined by the actuator disk theory. This torque is transmitted to the DFIG rotor through the drive train, represented by the well known two masses model. In this model, two rotating masses—the aerodynamic rotor and the DFIG rotor—are elastically connected via springs, characterized by the stiffness and damping factor of the coupling [9].

2.2. *Electrical System Model*

The DFIG and its power converter are two important factors that can be indicated in the electrical system model. The DFIG system can be implemented under a fifth-order model to define its behavior. The four electrical equations are indicated in a direct (d)-quadrature (q) for the rotor and stator voltage coordinates, which include the synchronous turning speed in this wind farm. The equations are determined by

$$U_{ds} = R_s i_{ds} + \frac{d}{dt}\varphi_{ds} - \omega\varphi_{qs} \tag{1}$$

$$U_{qs} = R_s i_{qs} + \frac{d}{dt}\varphi_{qs} - \omega\varphi_{ds} \tag{2}$$

$$U_{dr} = R_r i_{dr} + \frac{d}{dt}\varphi_{dr} - (\omega - \omega_r)\varphi_{qr} \tag{3}$$

$$U_{qr} = R_r i_{qr} + \frac{d}{dt}\varphi_{qr} + (\omega - \omega_r)\varphi_{dr} \tag{4}$$

where U represents the voltage, i represents the current, ω is the rotating speed, φ denotes the magnetic flux, R represents resistance and L inductance, indexes d and q stand for the direct and quadrature components, and indexes s and r refer to the stator and rotor, respectively.

Furthermore, the mechanical power available from a wind turbine is typically given by (5) [9]:

$$P_w = 0.5\rho\pi R^2 V_w^3 C_p(\lambda, \beta) \tag{5}$$

$$C_p = \frac{1}{2}\left(\lambda - 0.022\beta^2 - 5.6\right)e^{-0.17\lambda} \tag{6}$$

$$\lambda = \frac{V_w}{\omega\beta} \tag{7}$$

where P_w is the extracted power from the wind, ρ is the air density, R is the blade radius and ω_β is the wind speed. C_p is called the 'power coefficient' and is given as a nonlinear function of the parameters' tip speed ratio λ and blade pitch angle β. The calculation of the performance coefficient requires the use of blade element theory. ω_β is the rotational speed of the turbine. Usually, C_p is approximated as [12]:

$$C_p = \alpha\lambda + \beta\lambda^{2+} \gamma\lambda^3 \quad (8)$$

3. Simulation and Results

The Figure 2 define the model based on the doubly fed induction generator. This model consists of eight 1.5 MW wind turbines connected to a 25 kV distribution system which exports power to a 120 kV grid through a 30 km, 25 kV feeder. The eight 1.5 MW wind turbines form a wind farm of 12 MW of power.

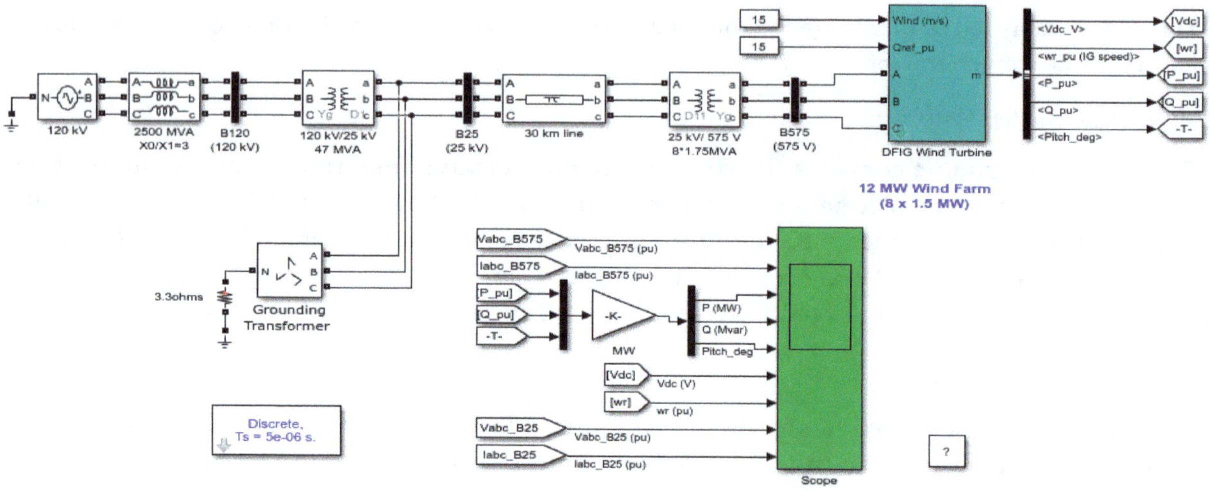

Figure 2. Schematic of the doubly fed induction generator model of MATLAB/Simulink.

In the following, we will represent and analyze simulation 1 over a period of 60 s by showing the graphs of the rotational speed, pitch angle, active power and DC link capacitor voltage and the current of wind turbines. Most of the variables of interest cited above were represented in pu values where the base power is the nominal power of each wind turbine (1.5 MW). The value 1 pu delimits the nominal power of the wind turbine. We will start by representing two important mechanical parameters, which are the rotational speed and the pitch angle of wind turbine. As we can see, the rotational speed of the rotor is regulated through the angle of the blades. From Figure 3a, the pitch angle starts to oscillate right after the disturbance. Nevertheless, it can be seen the oscillation decays after 5 seconds and the system is stable. Moreover, it is clear in Figure 3b that the rotational speed of the wind turbine starts at 1.2 pu during the first 10s of the simulation. However, the rotational speed decreases as well to a value below (1pu), which means the generator rotor of the wind turbine does not rotate at its rated value anymore. This behavior shows that the rotational speed is highly dependent on wind power. From Figure 3c, it can be seen that the DC link capacitor voltage shows small fluctuations on rotor active power and output power caused by wind speed variations. The voltage and currents considerably change the generator's active and reactive powers, which are shown in Figure 3d. It also remarks in Figure 3e,g, that the wind farm DC voltage is the sum of the DC voltages of wind turbines, which means that the GSC is controlling the wind farm DC voltage by maintaining it to a set constant value and also assuring the active power exchange from the generator to the grid . It can also be seen from Figure 3f, h that the active power and reactive power diminution of the wind turbine fluctuates until the end of the simulation. The other parameter that is necessary to verify in order to check that our Simulink model is working perfectly is the DC voltage.

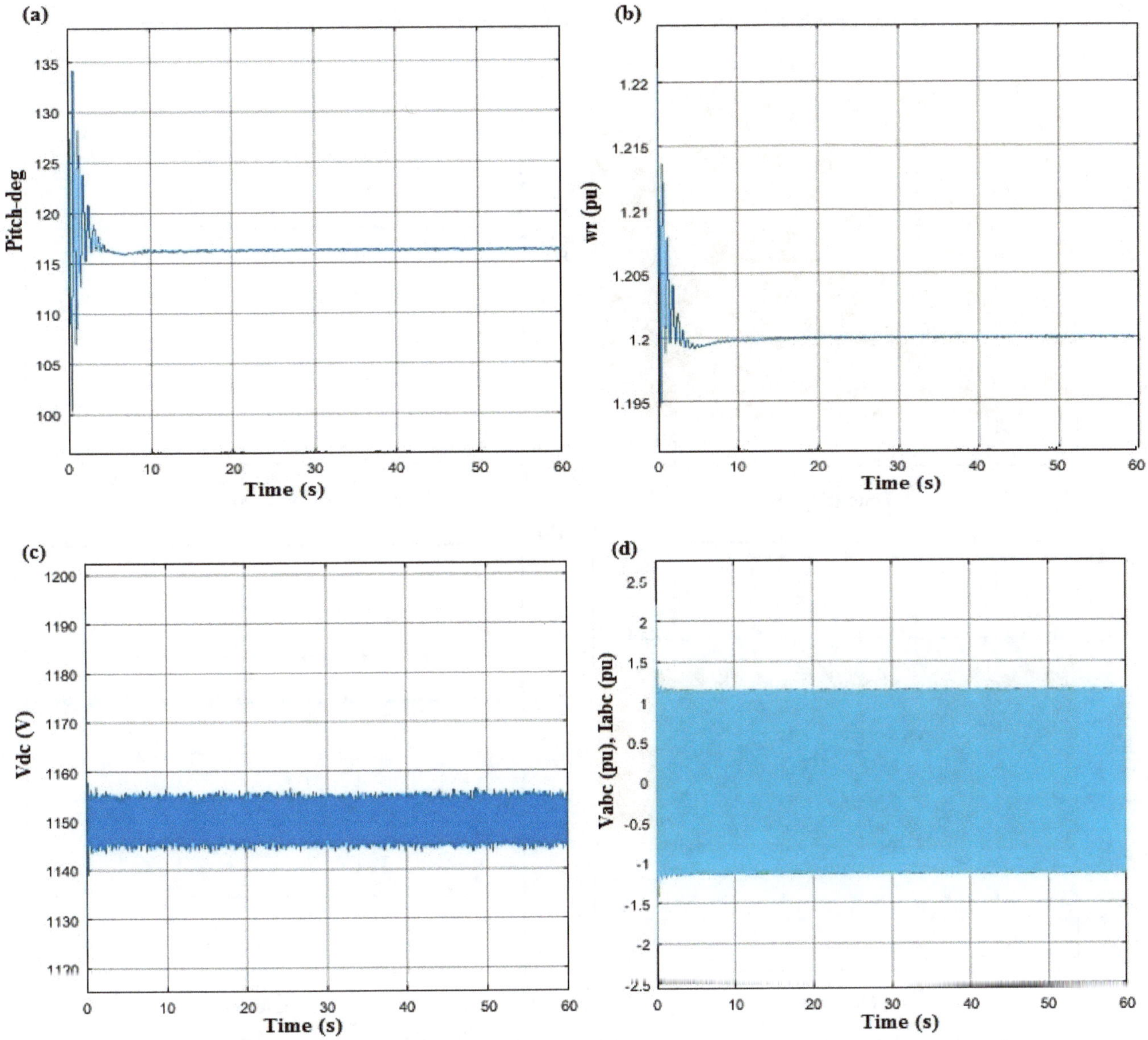

Figure 3. *Cont.*

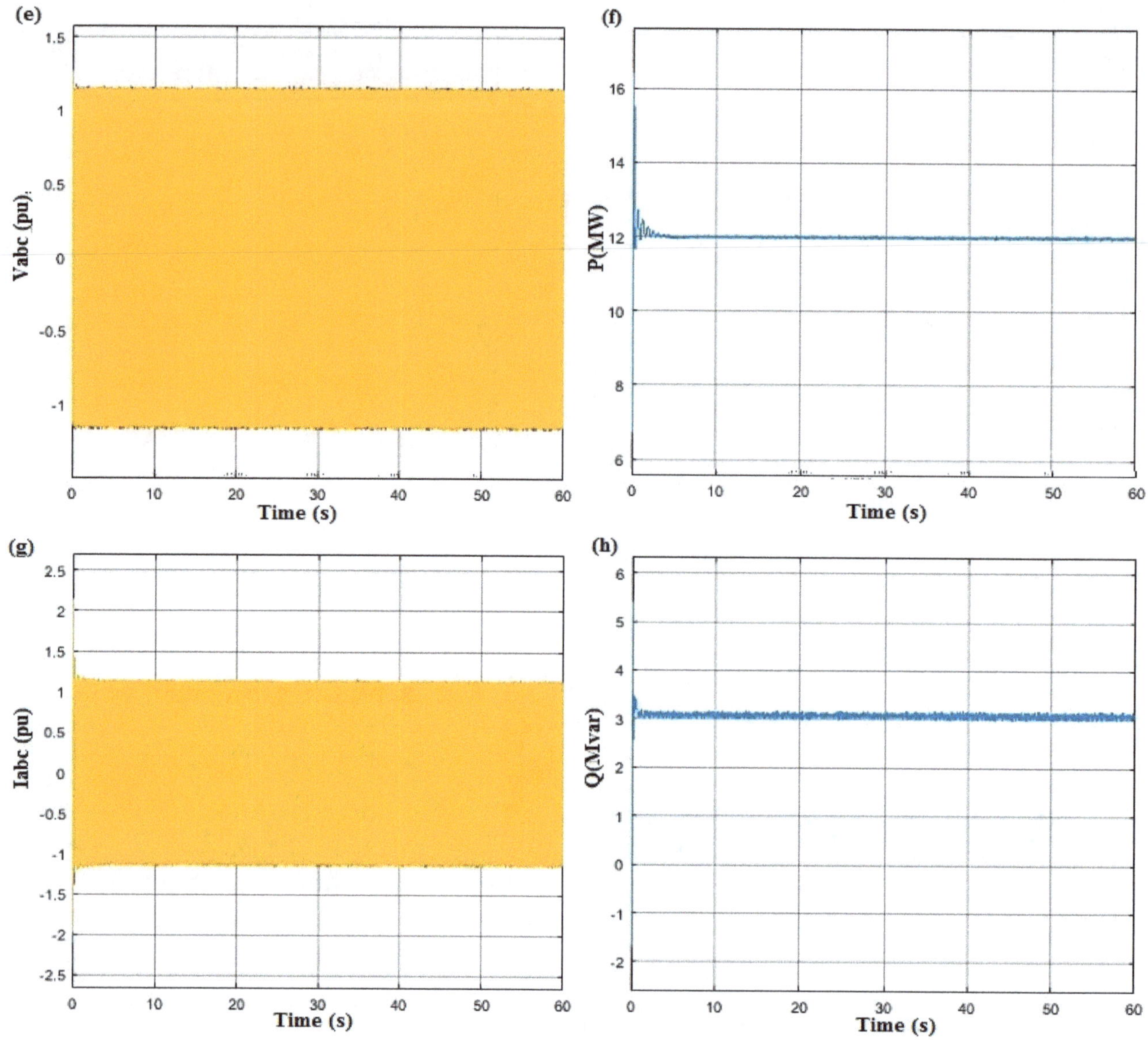

Figure 3. Schematization of the set output variables of DIFG: (**a**) pitch-deg; (**b**) W_r (pu); (**c**) V_{dc} (V); (**d**) V_{abc} (pu) I_{abc} (pu); (**e**) V_{abc} (pu); (**f**) I_{abc} (pu); (**g**) P (MW); and (**h**) Q (Mvar).

4. Conclusions

In this research, the DFIG and the dynamics operation were modeled to evaluate the responses and to assure the electrical distribution with the wind in terms of grid voltage and frequency fluctuations. Moreover, the main objective of this paper is the study on the dynamic behavior of the DFIG characteristic analysis and its effectiveness and optimal performance to assess the impact on the output variables of a wind farm contain many wind turbines. For this reason, the continuation of the proposal of this research is to make a hybrid system consisting of several wind turbines and an energy storage system will be selected to provide a clear idea and exact view of the distribution and degree of fluctuations as presented in the analysis and simulation in the pitch angle, current, and voltage curves to prevent such fluctuation situations and others. Each subsystem will be controlled via non-isolated power converters, and then coupled with grid or local load through the inverter. Simulations of the intelligent control system will also be improved through energy management among all the components of the hybrid system. Therefore, it is not enough to claim that one energy storage device is the best among all the others for each generator, but it is more appropriate to state that each of them has a better performance and is the most suitable for certain applications. Different configurations will be studied to eliminate the weaknesses of DFIG generator wind turbines that will allow the use of electronic power converters. The storage devices comprised in the hybrid system add the flexibility and capacity to the control and regulate the active power generation of the hybrid system, which yield to adapt the changes on the grid demand.

Abbreviations

The following abbreviations and notations are used in this manuscript:

DFIG	Doubly fed induction generator
RES	Renewable energy sources
PWM	Pulse width modulation
WECS	Wind energy conversion systems
ω_β	Rotational speed of turbine
V_w	Wind speed
Λ	Tip speed ratio
β	Blade pitch angle
U	Voltage
R	Resistance
P_w	Power wind turbine
C_P	Polynomial function of λ and β
GSC	Grid source converter

References

1. Pleßmann, G.; Erdmann, M.; Hlusiak, M.; Breyer, C. Global energy storage demand for a 100% renewable electricity supply. *Energy Procedia* **2014**, *46*, 22–31. [CrossRef]
2. Bussar, C.; Moos, M.; Alvarez, R.; Wolf, P.; Thien, T.; Chen, H.; Cai, Z.; Leuthold, M.; Sauer, D.U.; Moser, A. Optimal allocation and capacity of energy storage systems in a future European power system with 100% renewable energy generation. *Energy Procedia* **2014**, *46*, 40–47. [CrossRef]
3. IEC White Paper Energy Challenge: 2010. *Coping with the Energy Challenge The IEC 's Role from 2010 to 2030*; IEC: Geneva, Switzerland, 2010.
4. IRENA. *Global Energy Transformation: A Roadmap to 2050*, 2018th ed.; IRENA: Abu Dahbi, UAE, 2018.
5. Sun, Z.; Wang, H.; Li, Y. Modelling and simulation of doubly-fed induction wind power system based on Matlab/Simulink. *IET Conf. Publ.* **2012**, *2012*. [CrossRef]
6. Jami, H. *World Wind Resource Assessment Report*, 2014th ed.; World Wind Energy Association: Bonn, Germany, 2014.
7. Rolán, A.; Pedra, J.; Córcoles, F. Detailed study of DFIG-based wind turbines to overcome the most severe grid faults. *Int. J. Electr. Power Energy Syst.* **2014**, *62*, 868–878. [CrossRef]
8. Fernández, L.M.; García, C.A.; Saenz, J.R.; Jurado, F. Equivalent models of wind farms by using aggregated wind turbines and equivalent winds. *Energy Convers. Manag.* **2009**, *50*, 691–704. [CrossRef]
9. Krim, Y.; Abbes, D.; Krim, S.; Mimouni, M.F. Intelligent droop control and power management of active generator for ancillary services under grid instability using fuzzy logic technology. *Control Eng. Pract.* **2018**, *81*, 215–230. [CrossRef]
10. Precup, R.; Kamal, T.; Hassan, S.Z. *Advanced Control and Optimization Paradigms for Wind Energy Systems*; Springer: Singapore, 2019; ISBN 978-981-13-5994-1.
11. Shi, G.; Zhang, J.; Cai, X.; Zhu, M. Decoupling control of series-connected DC wind turbines with energy storage system for offshore DC wind farm. In Proceedings of the 2016 IEEE 7th International Symposium on Power Electronics for Distributed Generation Systems (PEDG), Vancouver, BC, Canada, 27–30 June 2016. [CrossRef]
12. Mu, W.; Wang, J.; Feng, W. Fault detection and fault-tolerant control of actuators and sensors in distributed parameter systems. *J. Franklin Inst.* **2017**, *354*, 3341–3363. [CrossRef]

13. Li, W.; Chao, P.; Liang, X.; Sun, Y.; Qi, J.; Chang, X. Modeling of complete fault ride-through processes for DFIG-Based wind turbines. *Renew. Energy* **2018**, *118*, 1001–1014. [CrossRef]
14. Nazir, M.S.; Wu, Q.H.; Li, M.; Luliang, Z. Lagrangian-Based Approach for Non-linear Dynamic Control of an Islanded Power System Short title: Non-linear dynamic control of power system. *Int. J. Comput. Sci. Inf. Secur.* **2017**, *15*, 24–29.
15. Sarrias-Mena, R.; Fernández-Ramírez, L.M.; García-Vázquez, C.A.; Jurado, F. Improving grid integration of wind turbines by using secondary batteries. *Renew. Sustain. Energy Rev.* **2014**, *34*, 194–207. [CrossRef]

2

Demonstration Experiment and Numerical Simulation Analysis of Full-Scale Barge-Type Floating Offshore Wind Turbine

Ko Matias Adrian Kosasih [1,*], Hideyuki Suzuki [2], Hideyuki Niizato [1] and Shigeki Okubo [1]

1 Hitachi Zosen Corporation, Osaka 551-0022, Japan; niizato@hitachizosen.co.jp (H.N.); ohkubo_s@hitachizosen.co.jp (S.O.)
2 Department of Systems Innovation, Graduate School of Engineering, The University of Tokyo, Tokyo 113-8656, Japan; suzukih@sys.t.u-tokyo.ac.jp
* Correspondence: adrian@hitachizosen.co.jp

Abstract: The development of Floating Offshore Wind Turbines (FOWT) has been progressing steadily. To utilize the moderate water depth of 50–100 m ocean space around Japan, a barge-type FOWT was installed in Kitakyushu as part of a demonstration project conducted by the New Energy and Industrial Technology Development Organization (NEDO) of Japan. The FOWT mounts a 3 MW two-bladed wind turbine with blade diameter of 100 m and hub height of 72 m. The barge-type floating support structure is equipped with a moonpool in the center and a skirt at its bottom and is moored with 9 lines of catenary chains. To investigate the dynamic behavior of the barge-type FOWT in extreme condition and the validity of the numerical simulation in modeling the effect of the complex flow around the floating structure to the FOWT's motion response, the FOWT's motion data during typhoon Tapah on 23 September 2019 were measured and compared with the simulation results. As the results, the simulation results showed a good agreement in general to the measurement data. However, some shifts in the peak frequency of the simulation's motion spectrum and a disagreement in waves with shorter wave periods were also observed. The possible causes of these differences are discussed thoroughly in this paper.

Keywords: floating offshore wind turbine; barge-type; full-scale demonstration project; time-domain numerical simulation; floating support structure dynamic behavior

---

1. Introduction

To realize a low-carbon society and promote the utilization of ocean space, the development of offshore wind turbine in the world has been progressing steadily. In Japan itself, the development of both bottom-fixed foundation wind turbine and floating offshore wind turbine (FOWT) have been increasing in recent years. In 2013, the first operational scale FOWT in Japan, a 2 MW spar-type FOWT, was installed in Nagasaki [1]. Then, as part of the Fukushima FORWARD project [2], in 2013, a 2 MW semi-submersible FOWT [3], in 2015, a 7 MW V-shaped semi-submersible [4], and in 2016, a 5 MW advanced spar-type [5] FOWTs were installed in Fukushima. Since then, with aim to verify the feasibility of a low-cost FOWT suitable for moderate water depth of 50–100 m, a demonstration 3 MW barge-type FOWT was installed in Kitakyushu in 2018 by the New Energy and Industrial Technology Development Organization (NEDO) of Japan. The demonstration wind turbine has been in operation and collecting data since 2019.

In this case of moderate water depth of 50 m to 100 m, the barge-type floating foundation is considered advantageous over other types of offshore wind turbine foundations. For water depth of more than 50 m, floating foundations are considered more cost-effective than bottom-fixed foundations, however, for water depth less than 100 m, the selection of the floating foundation types is limited to floating foundations with shallow draught. Floating foundations with deeper draught, such as a spar-type floating foundation, is difficult or impossible to be installed due to the water depth restriction. Then, compared to floating foundations with similar draught, such as semi-submersible foundations, the barge-type floating support structure is easier to design and manufacture resulting in its lower installation cost.

However, modeling a barge-type FOWT is not an easy feat. In 2007, Jonkman and Buhl [6] presented a fully coupled load analysis of a barge-type FOWT and showed the presence of excessive pitching in extreme wave conditions, which prompts the need to decrease the motion response of the barge-type FOWT. To do so, usually moonpool or skirts can be installed on the barge floating support structure. This introduces the need to confirm the effectiveness of the moonpool and the skirts, as well as to accurately model the effect of those additional parts to the FOWT's motion responses. Thus, in 2015, Beyer et al. [7] presented a numerical analysis using coupled Multibody System–Computational Fluid Dynamics (MBS-CFD) method to compute the motion response of a barge-type floating support structure equipped with a moonpool and bottom skirt. The numerical analysis was able to model the motion response of the experiment quite accurately and indicated the presence of vortex shedding of the hull and moving water in the moonpool which caused the heave damping on the floating support structure. However, there is still a need for a numerical simulation method capable of modeling the motion response of such barge-type FOWT coupled with the dynamics of the wind turbine, not only accurately, but also quickly, as usually thousands of environmental conditions are needed to be evaluated in the design of a single FOWT.

In 2016, Vijay et al. [8] investigated the motion responses of several moonpool configurations of barge-type FOWT using Boundary Element Method (BEM) potential flow theory to model the hydrodynamics of the floating structure and FAST codes to model the aerodynamics of the wind turbine. The study which modeled the moonpool as Oscillating Water Column (OWC) wave energy converter analyzed the motion response of the barge-type FOWT in relation to its moonpool configuration. In 2019, Kosasih et al. [9] compared the motion characteristics of the 1/50 model of two types of barge-type FOWTs in a wave tank experiment and the corresponding experiment results as a preliminary study to the NEDO 3 MW barge-type FOWT demonstration project. The numerical simulation was done using the BEM hydrodynamic analysis added with additional viscous damping coefficients to model the moonpool and the bottom skirt. The study showed a good agreement between the experiment and the simulation results, which indicated that the numerical simulation based on BEM potential theory is good enough to model the dynamic behavior of the barge-type FOWT including a moonpool and bottom skirt. As BEM-based numerical simulation can be calculated considerably quicker compared to a full CFD analysis, using BEM potential flow theory to evaluate the sheer number of cases in the design of a FOWT may be more favorable to CFD. However, validation of such numerical simulation method is still needed, especially compared to the motion response of an actual FOWT.

Thus, for this paper, the motion response of the actual NEDO 3 MW barge-type FOWT during typhoon Tapah on 23 September 2019 was investigated and compared with the time domain numerical simulation results based on the BEM potential flow theory coupled with the aerodynamics of the wind turbine. The power spectral density of the 6 degrees of freedom (DOF) motion response of the FOWT during the typhoon was investigated and compared with the simulation. Then, the FOWT motion's statistics from before, during, and after the typhoon were compared with numerical simulation results. Finally, the validity of the numerical simulation and the possible causes of the differences between the measurement and the simulation results shall be discussed in this paper.

2. Demonstration Experiment

2.1. *General Specifications of the FOWT*

The demonstration project's FOWT was installed in Kitakyushu about 15 km from the shoreline at water depth of 53 m, as shown in Figure 1. The main dimensions and parts of the barge-type FOWT are shown in Figure 2, while the general specifications of the FOWT are summarized in Table 1.

(a) (b)

Figure 1. (**a**) Location of the demonstration project's Floating Offshore Wind Turbines (FOWT) in Kitakyushu, Fukuoka Prefecture, Japan; (**b**) photo of the barge-type FOWT (photo: New Energy and Industrial Technology Development Organization (NEDO)).

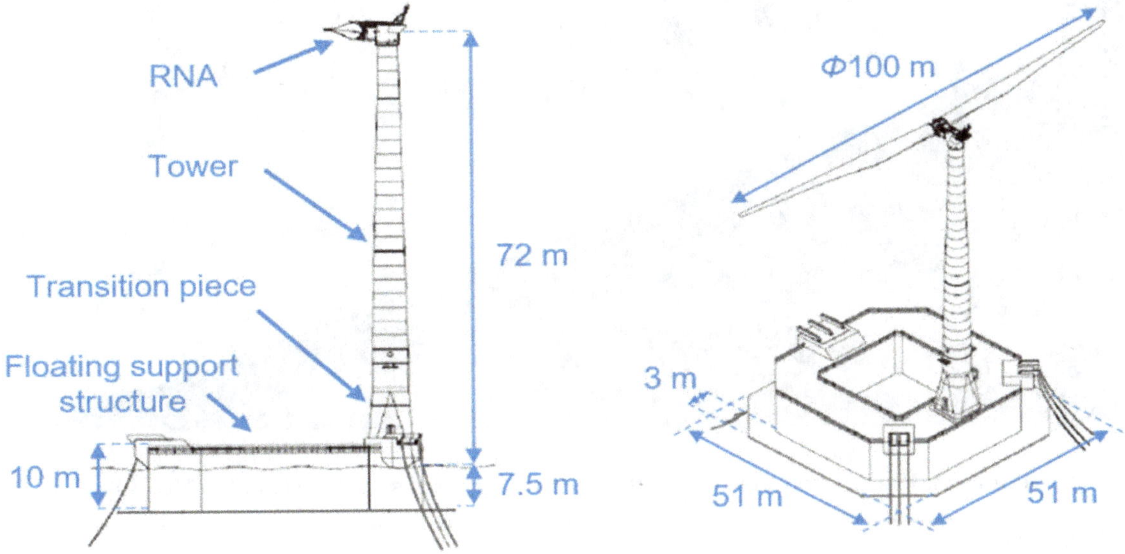

Figure 2. Main dimensions and parts of the barge-type FOWT.

Table 1. General specifications of the barge-type FOWT.

Parameter	Value
Rated power	3 MW
Rotor diameter	100 m
Number of blades	2
Rotor orientation	Upwind
Hub height [1]	72 m
Draught	7.5 m
Floating support structure's dimension (including skirt)	51 m × 51 m ×10 m
Water depth	53 m
Total mass (incl. ballast and mooring lines)	9,858,000 kg

[1] Height is measured from still-water level (SWL).

2.2. Floating Support Structure

The floating support structure of the barge-type FOWT has an octagonal shape of 45 m × 45 m × 10 m size and is equipped with a 3-m wide skirt on the bottom and a moonpool of 26.2 m × 26.2 m in the center of the floating support structure, as shown in Figure 2. Furthermore, several brackets were installed on the skirt to ensure its structural integrity, as shown in Figure 3. The barge-type FOWT's center of gravity during operation is located on the center of the moonpool at the height of 0.225 m below still-water level (SWL).

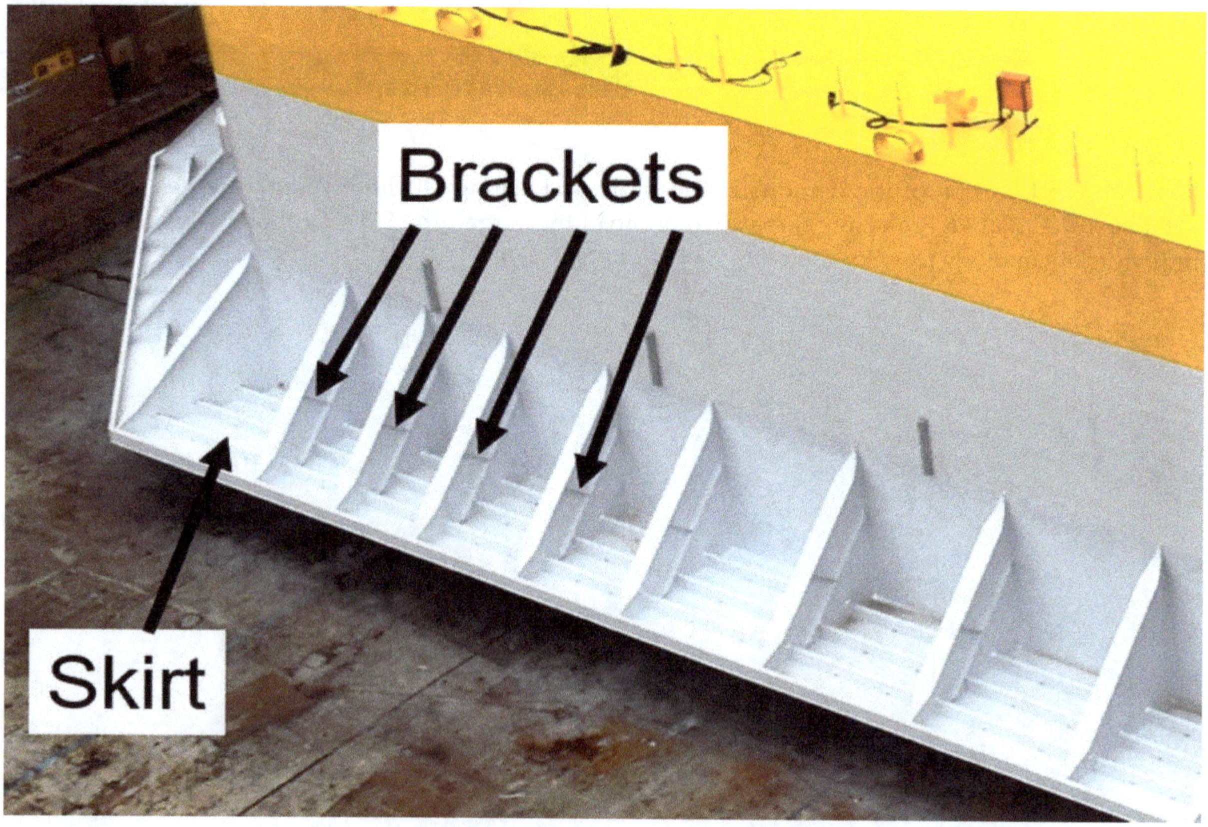

Figure 3. Brackets on the skirt of the floating support structure (photo: NEDO).

2.3. Mooring System

The mooring system of the barge-type FOWT is composed of three groups of three catenary chain mooring lines with 522 m length each, as shown in Figure 4. The mooring lines are made with studless chains with nominal diameter of 132 mm. Properties of the mooring lines are summarized in Table 2.

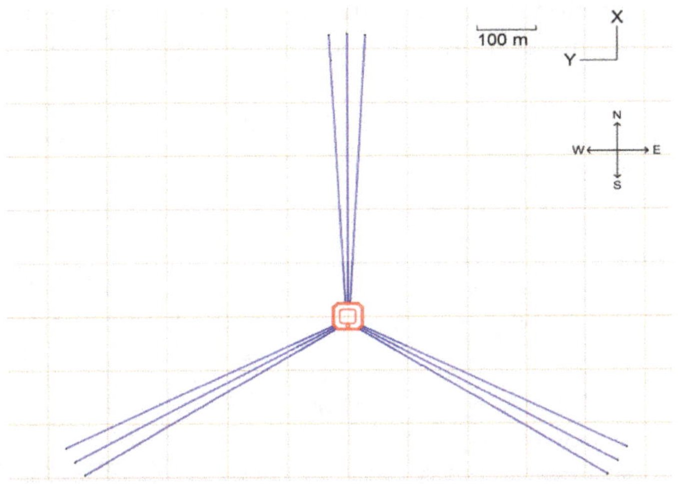

Figure 4. Mooring configuration of the barge-type FOWT.

Table 2. Properties of the mooring lines.

Parameter	Value
Line length	522 m
Line type	Studless chain
Nominal diameter	132 mm
Weight in air	0.349 t/m
Minimum Breaking Load	15,965 kN

2.4. *Wind Turbine*

The wind turbine and its tower are installed on the aft side of the floating support structure. The wind turbine is an upwind two-blades 3 MW wind turbine with a rotor diameter of 100 m and hub height of 72 m. The rotor-nacelle assembly (RNA) is supported by a single tower structure composed of a transition piece and a steel tubular tower as shown in Figure 2. The specifications of the wind turbine are shown in Table 3.

Table 3. Specifications of the wind turbine [10].

Parameter	Value
Wind turbine maker	Aerodyn Engineering GmbH
Wind turbine type	SCD3MW–NEDO
Rated power	3 MW
Rotor diameter	100 m
Hub height [1]	72 m
Number of blades	2 blades
Rotor orientation	Upwind
Tilt angle	3°
Coning angle	5°
Control scheme	Variable speed, individual pitch controlled
Cut-in wind speed	3 m/s
Cut-out wind speed	25 m/s
Rated rotor speed	17.1 rpm
RNA mass	133,000 kg

[1] Height is measured from SWL.

2.5. *Measuring Instruments*

Several measuring instruments are installed on the barge-type FOWT to monitor its structural and motion response during all environment conditions. Some of the sensors used to measure the environment conditions and motion response of the floating support structure are shown in Figure 5. The measuring instruments mainly used in this study are shown in Table 4.

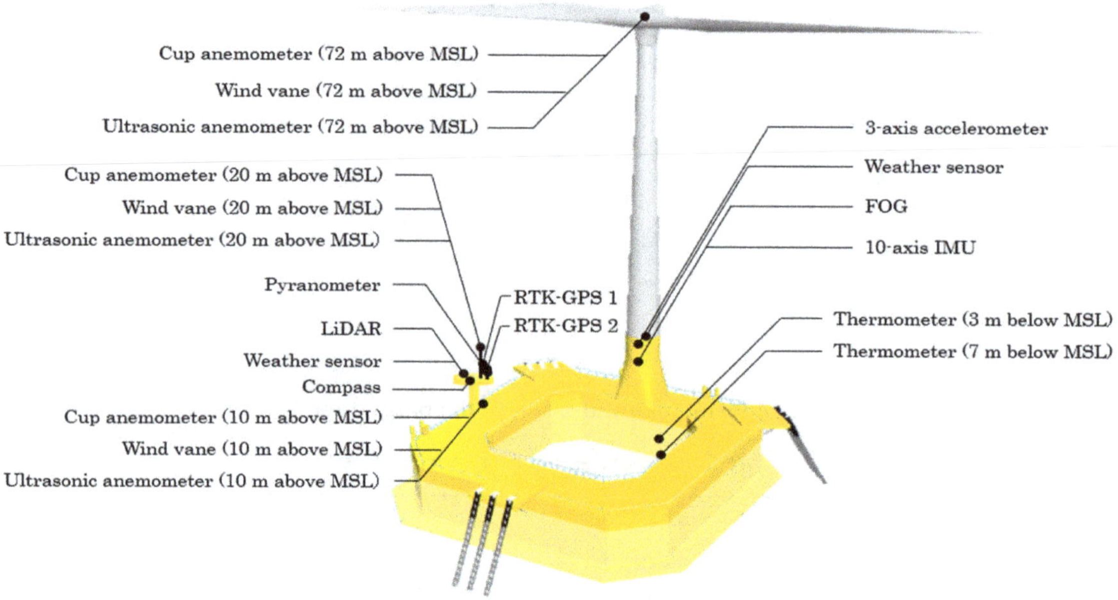

Figure 5. Measuring instruments installed on the barge-type FOWT.

Table 4. Measuring instruments used in this study.

Measurement Item	Instrument Used (Location)
Wind speed	Cup anemometer (72 m above SWL)
Wind direction	Wind vane (72 m above SWL)
Wave height, period, and direction	Ultrasonic wave meter (120 m west of FOWT)
Surge motion	RTK-GPS 1 (FOWT's starboard side deck)
Sway motion	RTK-GPS 1 (FOWT's starboard side deck)
Heave motion	RTK-GPS 1 (FOWT's starboard side deck)
Roll	Fiber-Optic Gyroscope (inside transition piece)
Pitch	Fiber-Optic Gyroscope (inside transition piece)
Yaw	Compass (FOWT's starboard side deck)

The environment conditions and the motion data from the sensors were taken at 1 Hz sampling frequency, except for the wave data which were taken at 0.5 Hz sampling frequency. The hourly statistical parameter, such as mean, maximum, minimum, and standard deviation of the environment conditions, were calculated from these measured time series data and then used as inputs to the numerical simulation analysis. Furthermore, to minimize the coupling effect of the translational motion with the rotational motion due to the difference in the sensor locations, the measured motion data from the real-time kinematic Global Positioning System (RTK-GPS) and the gyroscope were transformed to the FOWT's center of gravity before doing further analysis.

3. Numerical Simulation

A coupled numerical simulation analysis of the floating support structure's hydrodynamics, tower's structural response, and wind turbine's aerodynamics was conducted using the information of the demonstration project's FOWT described in the previous chapter. The simulation can be divided into hydrodynamic analysis and time domain analysis.

3.1. *Hydrodynamic Analysis*

The hydrodynamic database and hydrostatic stiffness of the floating support structure were calculated using Aqwa, a BEM potential flow solver developed by Ansys. The software solves the first-order wave loads, the diffraction wave load, the added mass coefficients, and the radiation damping coefficients of the floating support structure at various wave directions and periods. Then, to consider the second-order wave loads on the FOWT, the wave drift Quadratic Transfer Function (QTF) was also calculated using near-field method. These loads at various wave conditions along with the floating support structure's hydrostatic coefficients were then added to a hydrodynamic database, which will be used to calculate the hydrodynamic loads acting on the FOWT at each time step during the time domain analysis.

The calculation of the hydrodynamic loads acting on the floating support structure is mainly done by firstly assuming the fluid is inviscid and incompressible, the wave amplitude is small, and the motion of the floating support structure is of small amplitude, so that linear hydrodynamic theory can be used. Then, the velocity potential of the fluid flow field surrounding the floating support structure is calculated by using pulsating Green's function, which utilizes the boundary conditions defined by the wetted floating support structure's surface, the free surface, and the seabed condition. The calculated velocity potential around each panel of the floating support structure's mesh are then used to calculate the hydrodynamic pressure acting on it. Finally, the pressure around the whole floating support structure's mesh is integrated to get the hydrodynamic loads acting on the FOWT's center of gravity.

Regarding the analysis' parameters, the wave conditions were set with an interval of 22.5° and the wave periods were set from 2.5 s to 70 s. Then, to avoid standing waves on the moonpool, an external lid with 5% damping factor and 26.2 m gap was placed on the moonpool at the SWL. In the hydrodynamic analysis, the skirt is modeled without brackets. The reason for this is because BEM analysis could not capture the mainly viscous effect caused by the brackets. The viscous damping effect of the skirt and brackets is to be considered with additional damping coefficients in the time domain analysis. The mesh used in the hydrodynamic analysis is shown in Figure 6. The lid elements on the moonpool is indicated by the red arrow inside the figure.

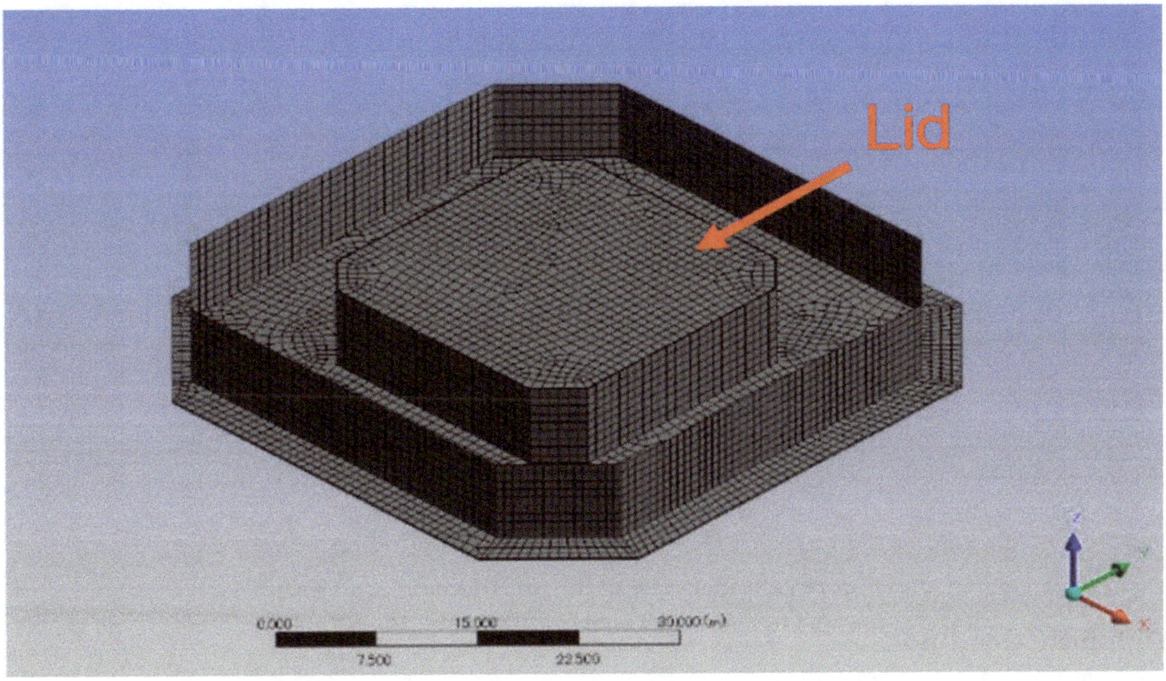

Figure 6. The mesh used in the hydrodynamic analysis.

3.2. Time Domain Analysis

In the time domain analysis, the hydrodynamic and hydrostatic loads on the floating support structure, the aerodynamic loads on the wind turbine's blades, and the structural response of the tower, the blades and the mooring lines were calculated at every time step to solve the motion response of the FOWT by using Orcaflex software developed by Orcina Ltd., Ulverston, UK coupled with FAST software developed by the National Renewable Energy Laboratory (NREL), Golden, CO, US.

To calculate the hydrodynamic loads acting on the FOWT in the time domain analysis, the hydrodynamic load database, the hydrostatic stiffness, and the wave drift QTF, which were calculated by the BEM hydrodynamic analysis, were imported and were used to calculate the hydrodynamic loads at every time step. Then, to consider the effect of the viscous loads on the skirt and its brackets, additional linear and quadratic damping coefficients and several heave plates modeled with Morison drag elements were used. The value of those additional damping coefficients and the heave plates' drag coefficients were obtained from the wave tank experiments conducted in the preliminary design phase of the project [9] and were further adjusted using the actual measurement data during the typhoon Tapah.

The wind load acting on the blades is calculated at each time step by using Blade-Element/Momentum theory. The wind load calculation is coupled with the structural response of the tower and the motion of the FOWT. The instantaneous blade-element's position and velocity and its surrounding wind speed at each time step is considered during the calculation of the wind load. The calculated wind load on each element of the blade is transferred along the blades to the hub, the nacelle, the tower, and ultimately to the floating support structure. Then, considering the hydrodynamic loads on the floating support structure, wind loads on the wind turbine, and the forces from the mooring lines, the FOWT's motion is calculated. Here, the tower and the blade structures were modeled using the dimension and mass of the demonstration project's barge-type FOWT, while the nacelle was modeled with a point mass and inertia matrix attached to the top of the tower. Meanwhile, the mooring system was modeled using analytic catenary lines and the floating support structure was assumed to be rigid. The simulation model created for the time domain analysis is shown in Figure 7.

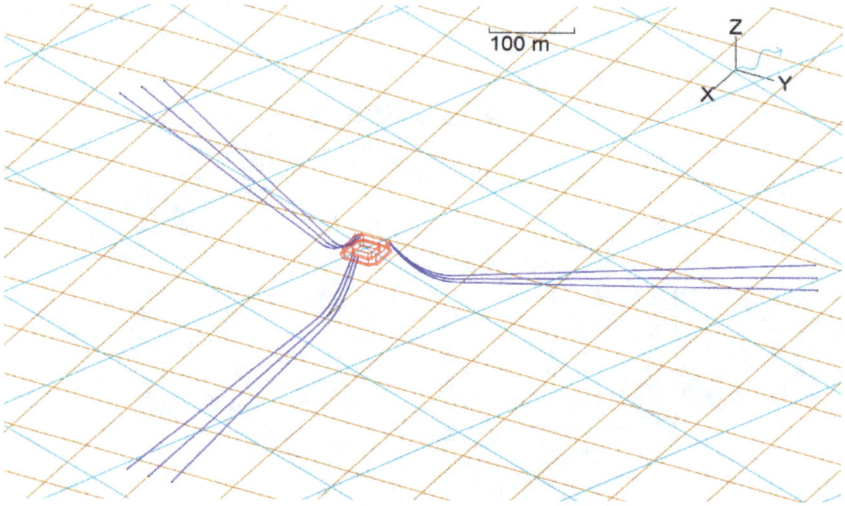

Figure 7. Model view of the time domain analysis.

4. Environment Condition

The environment condition measured by the barge-type FOWT during typhoon Tapah is investigated in this study. The hourly statistical environment data from 22 September 2019 12:00 to 23 September 2019 12:00 GMT + 9 is summarized in Table 5. These hourly statistical environment conditions were used as simulation inputs to the time domain numerical analysis. The high-frequency wind time series were re-generated using Kaimal spectrum using the average wind speed and the 1-h turbulence data, while the waves were re-generated using ISSC spectrum using the hourly significant wave height and significant wave period from the measurement. The wave spreading was modeled using $\cos^n$ spreading function. Here, a wave spreading exponent n of 8.0 is assumed for all cases. The generated wave train in the form of directional wave spectrum was used to recreate the sea surface's time series and shall be compared with the measured wave spectrum to ensure the accuracy of the time domain simulation. During the whole duration of the typhoon, the wind turbine was in idle condition.

For the spectral analysis of the motion response, the data during 23 September 2019 03:00–04:00 shall be used. While for the statistical analysis, only the hourly standard deviation of the motion response data will be used.

Table 5. Environment condition during typhoon Tapah.

Time	Mean Wind Speed [m/s]	Wind Direction [1] [°]	Turbulence Intensity [%]	Significant Wave Height [m]	Significant Wave Period [s]	Wave Direction [1] [°]
22-Sep-19, 12 h	20.13	126.6	8.6%	1.33	6.06	14.3
22-Sep-19, 13 h	21.18	128.6	7.8%	1.31	5.86	30.9
22-Sep-19, 14 h	19.63	132.3	11.4%	1.34	6.11	42.0
22-Sep-19, 15 h	21.45	127.4	13.1%	1.31	6.50	1.0
22-Sep-19, 16 h	23.10	126.3	6.5%	1.42	6.35	42.3
22-Sep-19, 17 h	24.43	129.0	5.4%	1.52	6.42	50.7
22-Sep-19, 18 h	22.67	135.4	7.7%	1.51	7.02	21.3
22-Sep-19, 19 h	21.05	140.8	8.0%	1.66	7.97	4.0
22-Sep-19, 20 h	24.44	145.1	7.9%	1.96	7.61	14.7
22-Sep-19, 21 h	20.42	159.0	12.9%	2.00	7.98	14.7
22-Sep-19, 22 h	23.21	187.1	17.2%	1.98	8.18	2.0
22-Sep-19, 23 h	24.57	200.1	11.0%	2.19	7.05	323.4
23-Sep-19, 00 h	24.44	212.4	11.0%	2.13	6.46	301.2
23-Sep-19, 01 h	20.72	272.9	14.4%	2.21	7.08	318.0
23-Sep-19, 02 h	22.80	325.9	9.9%	2.65	6.92	318.3
23-Sep-19, 03 h	22.76	336.2	9.2%	4.69	8.75	322.3
23-Sep-19, 04 h	21.10	343.1	10.9%	4.93	9.11	332.0
23-Sep-19, 05 h	17.34	347.0	14.6%	4.83	9.43	336.7
23-Sep-19, 06 h	16.28	348.9	14.1%	4.85	9.39	338.3
23-Sep-19, 07 h	14.77	349.2	13.8%	4.20	9.29	340.0
23-Sep-19, 08 h	13.47	358.8	12.8%	3.91	9.59	340.3
23-Sep-19, 09 h	11.91	8.1	12.3%	3.64	9.68	341.3
23-Sep-19, 10 h	11.03	11.6	17.3%	3.28	9.33	340.3
23-Sep-19, 11 h	11.52	13.6	11.4%	3.01	9.20	345.0

[1] Wind and wave direction is defined as the direction where the wind or wave is coming from measured clockwise; 0° is coincident with wind or wave coming from north direction and 90° is coincident with wind or wave coming from east.

5. Results and Discussion

5.1. Power Spectral Density

The power spectral density (PSD) of the wave elevation and 6 DOF the barge-type FOWT motion response during 23 September 2019 03:00–04:00 are shown in Figures 8 and 9, respectively. Although different wave time series was used in the time domain analysis, the spectrum of the wave elevation is roughly the same as the wave elevation during the typhoon. This ensures that the wave train acted on the barge-type FOWT in the simulation and in during the actual operation are statistically similar.

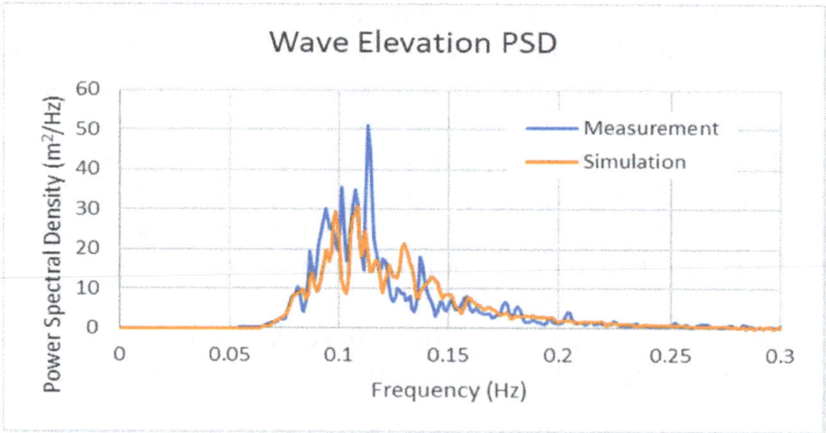

Figure 8. Power spectral density of the wave elevation during 23 September 2019 03:00–04:00.

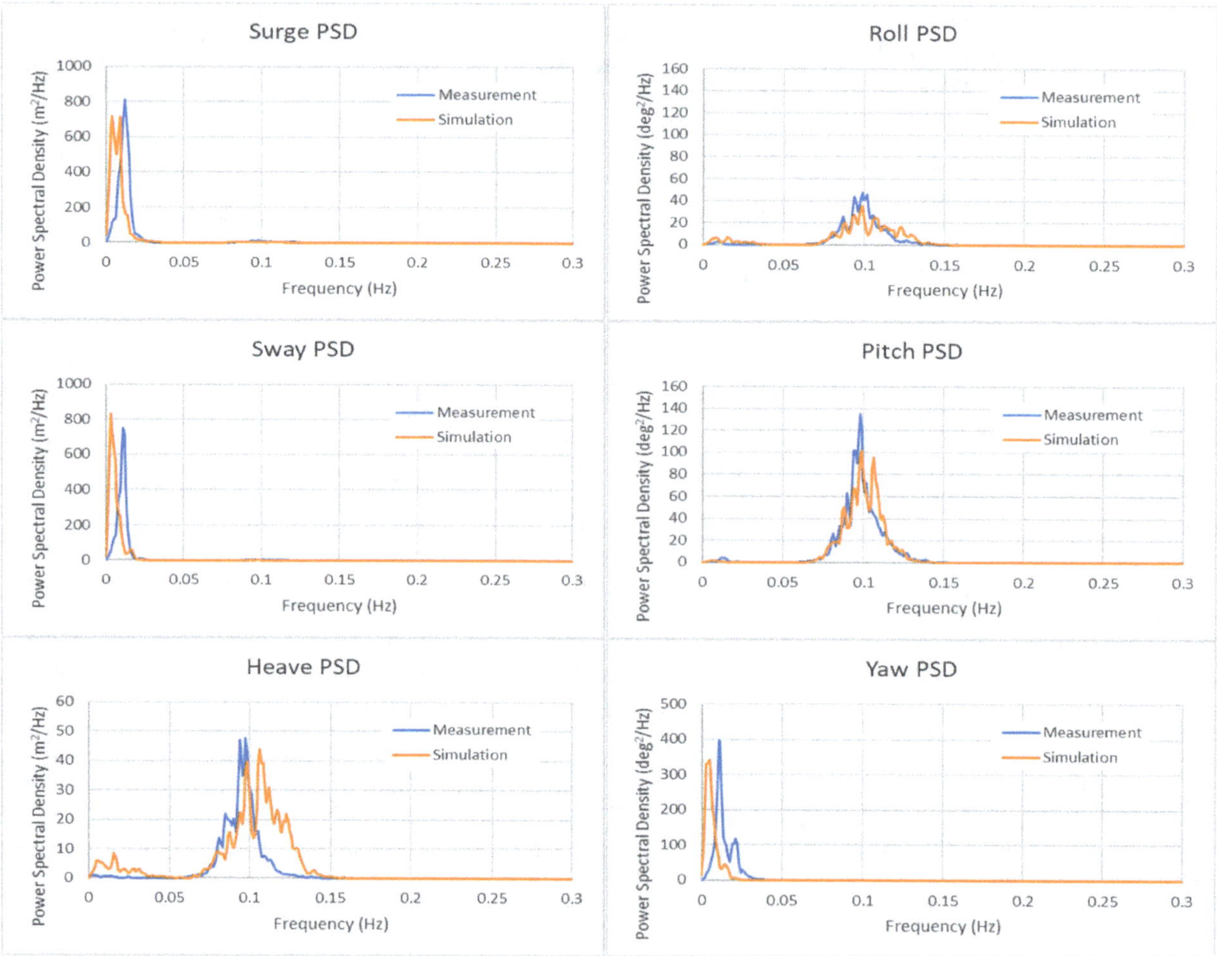

Figure 9. Power spectral density of the 6 degrees of freedom (DOF) motion response during 23 September 2019 03:00–04:00.

Although the best practice is to input the wave spectrum directly into the numerical simulation model, only the significant wave height, the significant wave period, and the wave spreading coefficient were inputted to the numerical simulation due to the availability of the measured wave data and to ensure that the generated wave trains include wave spreading.

The surge and sway PSD graphs show that the measured surge and sway response has much lower frequency than the wave, with peak at 0.011 Hz (90 s). It also can be observed that surge and sway responses are quite identical. However, there is a clear peak frequency shift between the measurement and the simulation. The surge response's peak frequency in the simulation is 0.0088 Hz (113 s) and sway is at 0.0029 Hz (345 s).

Another peak frequency shift of the simulation response is also observed in the heave PSD graph, 0.106 Hz (9.43 s) for the simulation and 0.094 Hz (10.63 s) for the measurement data. The overall distribution of the simulation's heave PSD also shows more response in the higher frequency compared to the measured data. In the simulation's heave PSD, a small bump at around 0.01 Hz (100 s) to 0.03 Hz (33 s) is also observed. This indicates that the heave response is coupled with the surge, sway, or yaw quite strongly in the simulation.

The spectral response of roll and pitch of the simulation and measurement are in a good agreement. Both the power spectral density and the frequency distribution of the simulation match the measurement data. Meanwhile, the yaw response of the simulation shows similar frequency shift to the surge and sway response to the lower frequency compared to the measured one.

The low-frequency responses of the surge, sway, and yaw measurement data indicates that these modes are mainly affected by low-frequency forces, like wave drift force or mooring restoring force. In these modes, high-frequency response is hardly visible, which indicates lower influence from the incoming wave force. On the other hand, the motion response of the roll, pitch, and heave mode showed high-frequency response, which the peak is about the same as the wave peak frequency. This indicates that the roll, pitch, and heave mode of the FOWT are affected mostly by incident wave forces. Although the high-frequency responses of the simulation and the measurement showed good agreements, the low-frequency responses showed large shifts in peak frequencies. This suggests that the high-frequency radiation and diffraction wave forces are well modeled by the BEM hydrodynamic analysis, but the low-frequency wave drift or mooring forces need more calibration.

Then, the heave PSD of the simulation result showed more heave response at higher frequency of more than 0.1 Hz (10 s) and at lower frequency almost coincident with the surge, sway, and yaw's peak frequency compared to the measurement data. The possible cause of these differences is that current modeling of the moonpool and the skirt is not adequate to model the entrapped water in the moonpool and the vortex shedding around the floating support structure which may induce viscous drag as described by Beyer et al. [7]. According to Waris and Ishihara on their study of semi-submersible FOWT [11], the addition of heave plates or skirt may decrease the FOWT's natural frequency in the heave mode. In this study, the moonpool is modeled only as an external damping lid, which functions mainly to prevent the occurrence of standing waves inside the moonpool during the hydrodynamic analysis. Meanwhile, the possible additional damping effect from the moonpool and the skirt is modeled by the additional damping coefficients, as described in the previous chapter. Adding an oscillating water column mechanism into the time series simulation may increase the fidelity of the numerical simulation of a floating support structure with moonpool.

Regarding the peak shifts, compared to the experiment, the surge and sway response's peak frequencies of the simulation are shifted to the lower frequency, while the heave's peak frequency is shifted to the higher frequency. This is interesting because, if those shifts are caused inaccurate mass definition of the FOWT, all translational motion modes should be shifted to the same higher or lower frequency direction instead to different directions. To calibrate such shift caused by mass inaccuracy, different additional masses can be added to each direction. In this case, positive additional mass can be added to the heave direction to shift the peak to lower-frequency and negative additional mass can be added to the surge and sway directions to shift the peak to higher-frequency. However, such addition needs the measurement data of the natural periods of the actual FOWT from the decay test.

Another explanation for the frequency shift is because of the difference in the mooring response of the simulation and the actual FOWT. Mooring response differences can occur from the difference between the actual mooring lengths and the design mooring lengths, although only to a degree. The mooring response difference can also be caused by the difference in the actual conditions of the mooring lines. Those conditions include marine growth condition, which affects buoyancy and drag force of the mooring.

For further studies, the cause of the peak frequency's shift should be confirmed. Calibration of the wave drift forces using wave tank experiment or Computational Fluid Dynamic (CFD) analysis can be investigated to adjust the low-frequency motion responses. Dynamic mooring analysis and inclusion of marine growth may give better mooring response compared to the catenary mooring method due to the difference in the drag force. Then, inclusion of more detailed modeling of the moonpool can also be investigated.

5.2. Sample Time Series During the Typhoon

Time series of the measured and simulated motion response are shown in Figure 10. As the wave train data inputted into the time domain numerical analysis was only based on its spectrum and statistical data, i.e., not an exact wave elevation measured data, the time series of the simulation differs from the measured ones.

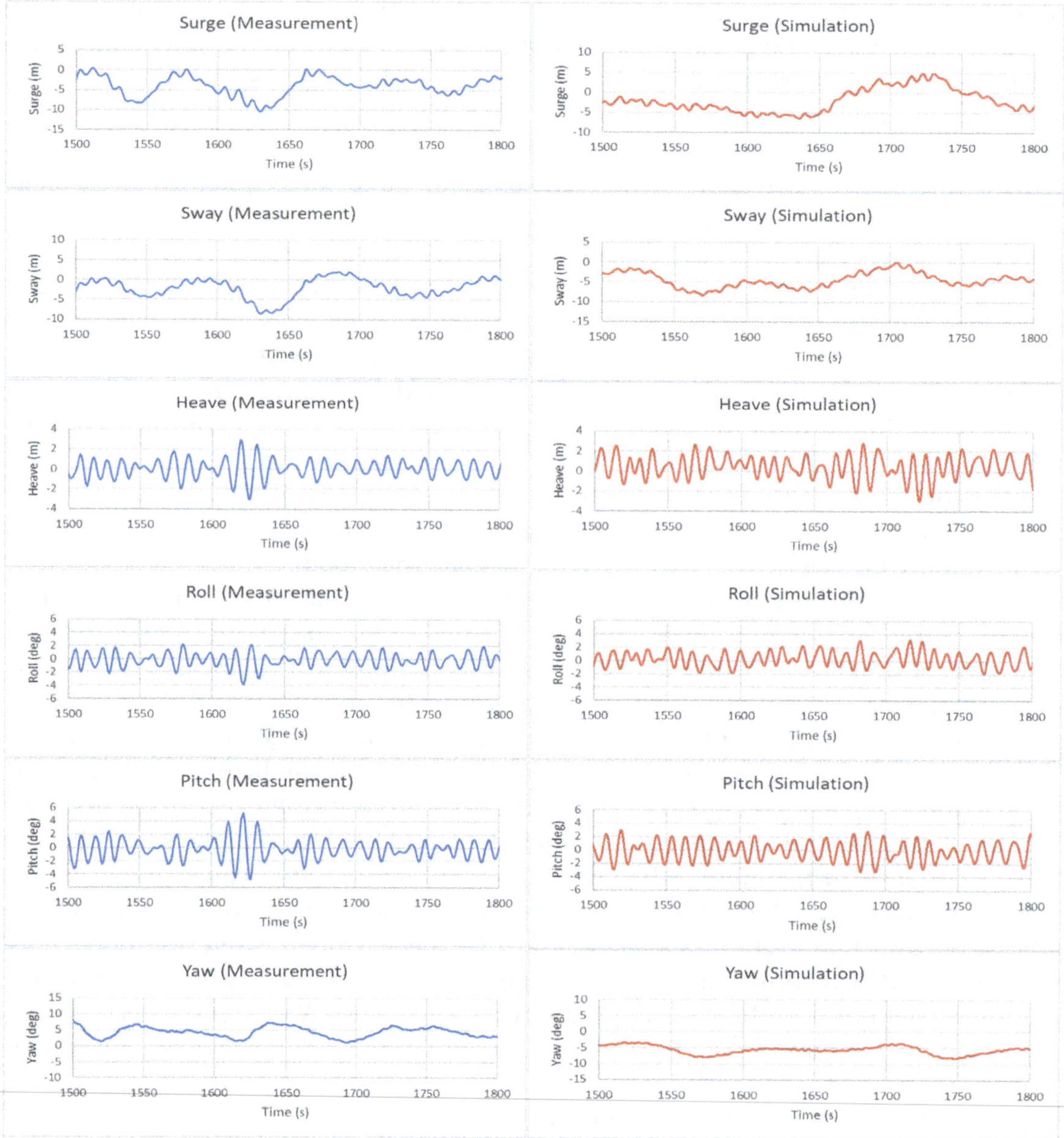

Figure 10. Time series of the barge-type FOWT motion response from 23 September 2019 03:00:00.

Although direct comparison of the time series between the measurement and simulation is quite difficult, the measurement and simulation showed a similar pattern, as indicated by the PSD comparison in the previous sub-chapter.

5.3. Hourly Statistical Motion Response of the Barge-type FOWT during 24 H of the Typhoon

The numerical simulations were also done for the time before the typhoon came to the FOWT and right after. The result of those simulations was summarized as hourly standard deviation of the roll and pitch motion response of the barge-type FOWT which is shown in Figure 11. The horizontal axis of the graph shows the time elapsed since 22 September 2019 12:00 in h. The 0h data represents the statistical data of the 22 September 2019 12:00–13:00 period.

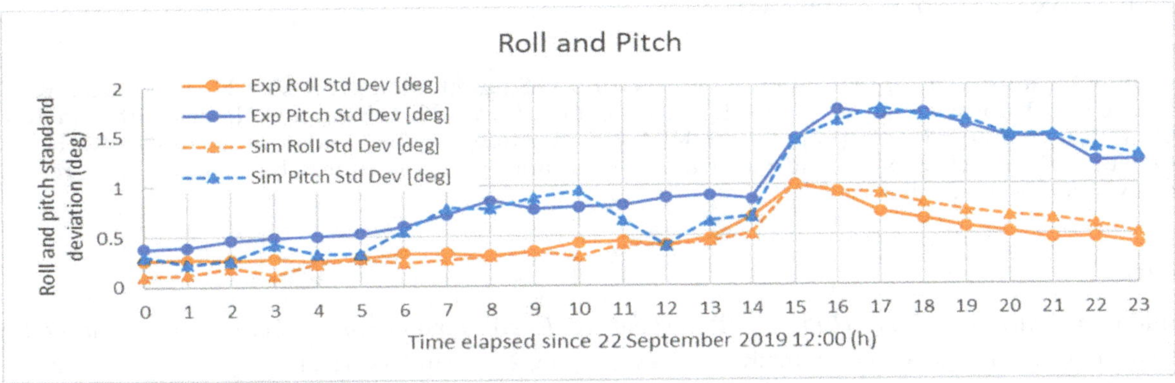

Figure 11. Hourly standard deviation of the barge-type FOWT's roll and pitch response.

Firstly, for the period of the 23 September 2019 03:00–04:00 (15 h marker), which was used in the spectral analysis, the roll and pitch's standard deviation of the simulation and the measurement showed exact match. For the time after 15 h, in which the significant wave periods were quite long at around 9 s, the numerical simulation and the measurement show a good agreement in general. However, there are bigger differences at the time before 15 h, especially before 6 h and at 12 h. For the time before 6 h, the difference is probably caused by the difference of the magnitude of the effect of the additional linear and quadratic damping coefficients to the FOWT motion's attenuation. The additional damping coefficients were added to account the viscous effect of the skirt and brackets; however, they were calibrated to the extreme wave conditions instead of calmer wave condition like in the time before 6 h in the graph, which have significant wave periods of around 6 s. Thus, this suggests that different wave conditions require different additional viscous damping coefficients. This presents a challenge to the method of predicting the value of the damping coefficient suitable for each wave condition and how to model it in the simulation.

As for the 12 h data, it is probably caused by transient phenomenon like sudden wave condition change which can be explained by the wave direction during 12 h (23 September 2019, 00:00–01:00) is the only one different than the adjacent 11 h and 13 h.

Regarding the numerical stability of the time domain analysis, all cases that encompassed the 24 h before and after the Typhoon Tapah were successfully calculated without any error. However, as the hydrodynamic analysis assumed that the floating support structure and the wave amplitude are small, the simulation may become unstable or inaccurate as the FOWT's motion becomes quite large.

6. Conclusions

The power spectral density graphs of the measurement data showed that the surge, sway, and yaw motion response peaked at lower-frequency compared to the waves, while the heave, roll, and pitch peaked around the incoming wave's peak frequency. The comparison between the measurement and simulation spectral analysis result showed a good agreement in the magnitude of the motion response's spectrum especially for the roll and pitch of the FOWT. However, some shifts in the peak frequency of the surge, sway, heave, and yaw were also observed. Possible explanations of the surge and sway's shifts include the difference in wave drift force or the mooring restoring force, while the difference in the heave response may be linked to the effect of the moonpool. Meanwhile, the statistical analysis of the numerical simulation result and the measurement data showed a good agreement at longer significant wave period of around 9 s and a disagreement at shorter significant wave period of around 6 s, which suggests that different wave conditions require different additional viscous damping coefficients. With suitable additional damping coefficients to model the moonpool and skirt's viscous effect, the BEM-based numerical simulation method presented in this paper in general can sufficiently model the dynamic behavior of the barge-type FOWT in extreme wave conditions. However, further studies confirming the cause of the peak frequency shifts and the dynamic behavior in different wave conditions are needed.

Author Contributions: Conceptualization, K.M.A.K. and H.S.; methodology, K.M.A.K.; validation, K.M.A.K., H.S., H.N., and S.O.; formal analysis, K.M.A.K.; investigation, K.M.A.K., H.S., and H.N.; data curation, K.M.A.K. and S.O.; writing—original draft preparation, K.M.A.K.; writing—review and editing, K.M.A.K., H.S., H.N., and S.O.; supervision, H.S.; project administration, H.N. and S.O. All authors have read and agreed to the published version of the manuscript.

References

1. Ishida, M. *Renewable Energy Utilization Report No. 10, Japan's First Commercial Operational Floating Offshore Wind Turbine*; Renewable Energy Institute: Tokyo, Japan, 2017. (In Japanese)
2. Suzuki, H. Fukushima Project and Floating Offshore Wind Turbine System. *Bull. Jpn. Soc. Nav. Archit. Ocean Eng.* **2018**, *81*, 32–33. (In Japanese)
3. Imakita, A. 2 MW Floating Offshore Wind Turbine Facility Fukushima Mirai. *Bull. Jpn. Soc. Nav. Archit. Ocean Eng.* **2018**, *81*, 34–37. (In Japanese)
4. Komatsu, M.; Hitoshi, K.; Ohta, M.; Hiroshi, T.; Mori, H.; Miyazaki, S. Development of Offshore Wind Turbine Floater That Blends Into Japanese Waters; Evaluation of the Validity for Design and Applied Methods for V-shaped Semi-Submersible Floating Structure. *Mitsubishi Heavy Ind. Tech. Rev.* **2016**, *53*, 30–39.
5. Kitakoji, Y. The Construction of Floating Transformer (Fukushima Kizuna) and 5 MW Floating Offshore Wind Turbine (Fukushima Hamakaze) Advanced Spar-type Floating Structure. *Bull. Jpn. Soc. Nav. Archit. Ocean Eng.* **2018**, *81*, 44–47. (In Japanese)
6. Jonkman, J.M.; Buhl, M.L., Jr. Loads Analysis of a Floating Offshore Wind Turbine Using Fully Coupled Simulation. In Proceedings of the Wind Power 2007 Conference and Exhibition, Los Angeles, CA, USA, 3–6 June 2007.
7. Beyer, F.; Choisnet, T.; Kretschmer, M.; Cheng, P.W. Coupled MBS-CFD Simulation of the IDEOL Floating Offshore Wind Turbine Foundation Compared to Wave Tank Model Test Data. In Proceedings of the 25th International Offshore and Polar Engineering Conference (ISOPE-2015), Hawaii, HI, USA, 21–26 June 2015.
8. Vijay, K.G.; Karmakar, D.; Uzunoglu, E.; Soares, C.G. Performance of Barge-Type Floaters for Floating Wind Turbine. In Proceedings of the 2nd International Conference of Renewable Energies Offshore (Renew 2016), Lisbon, Portugal, 24–26 October 2016.
9. Kosasih, K.M.A.; Niizato, H.; Okubo, S.; Mitani, S.; Suzuki, H. Wave Tank Experiment and Coupled Simulation Analysis of Barge-Type Offshore Wind Turbine. In Proceedings of the 29th International Society of Offshore and Polar Engineering Conference (ISOPE-2019), Hawaii, HI, USA, 16–21 June 2019.
10. New Energy and Industrial Technology Development Organization. Floating Offshore Wind Turbine Technology Guidebook. Available online: https://www.nedo.go.jp/content/100891410.pdf (accessed on 25 September 2020). (In Japanese)
11. Waris, M.B.; Ishihara, T. Dynamic Response Analysis of Floating Offshore Wind Turbine with Different Types of Heave Plates and Mooring Systems by Using a Fully Nonlinear Model. *Coupled Syst. Mech.* **2012**, *1*, 247–268. [CrossRef]

3

Prediction and Validation of the Annual Energy Production of a Wind Turbine using WindSim and a Dynamic Wind Turbine Model

Yuan Song [1] **and Insu Paek** [2,3,*]

1 Department of Advanced Mechanical Engineering, Kangwon National University, Chuncheon 24341, Gangwon, Korea; songwon@kangwon.ac.kr

2 Department of Integrated Energy & Infra System, Kangwon National University, Chuncheon 24341, Gangwon, Korea

3 Department of Mechatronics Engineering, Kangwon National University, Chuncheon 24341, Gangwon, Korea

* Correspondence: paek@kangwon.ac.kr

Abstract: In this study, dynamic simulations of a wind turbine were performed to predict its dynamic performance, and the results were experimentally validated. The dynamic simulation received time-domain wind speed and direction data and predicted the power output by applying control algorithms. The target wind turbine for the simulation was a 2 MW wind turbine installed in an onshore wind farm. The wind speed and direction data for the simulation were obtained from WindSim, which is a commercial computational fluid dynamics (CFD) code for wind farm design, and measured wind speed and direction data with a mast were used for WindSim. For the simulation, the wind turbine controller was tuned to match the power curve of the target wind turbine. The dynamic simulation was performed for a period of one year, and the results were compared with the results from WindSim and the measurement. It was found from the comparison that the annual energy production (AEP) of a wind turbine can be accurately predicted using a dynamic wind turbine model with a controller that takes into account both power regulations and yaw actions with wind speed and direction data obtained from WindSim.

Keywords: onshore wind farm; flow analysis; annual energy production; dynamic wind turbine model; yaw motion; peak shaver

---

1. Introduction

South Korea has recently been actively promoting the development of power plants using renewable energy. At the end of 2017, the Ministry of Trade, Industry, and Energy of South Korea issued the "Renewable Energy 3020 Plan", which aims to achieve 20% of the total power generation in South Korea from renewable energy by 2030. The 20% target is approximately 63.8 GW in capacity, about 28% of which, approximately 17.7 GW, corresponds to wind power. Prior to 2019, South Korea's cumulative total of wind power was approximately 1.49 GW, which is about 8.4% of the anticipated renewable energy production in the 3020 plan [1,2]. To achieve the target of the "Renewable Energy 3020 Plan", it is necessary to substantially develop more wind farms and investigate more wind farm sites based on accurate prediction of the annual energy production (AEP), which must be performed in advance.

To predict the AEP of a wind farm at a specific site, the wind data representing that site are needed. The wind data can be either time-based or frequency-based (Weibull distribution). An appropriate program is needed to predict the wind speeds at different wind turbine locations in the wind farm, including the wake effect, and also to predict the AEP from the wind turbines. Programs that are used to predict the AEP of a wind farm are already commercialized and are used worldwide. The programs are classified as linear and nonlinear (CFD) programs based on the equation of motion. The linear programs include WAsP, WindPRO, and WindFarmer, and the nonlinear programs include WindSim, WAsP-CFD, Meteodyn, and Wind Station.

Although the linear programs are known to predict the AEPs of a wind farm well for flat terrains, they have limitations when used for complex terrains. With the increase in computing speed and storage capacity, analysis with CFD is being used more frequently than before to obtain more accurate wind resources.

Some studies have predicted wind resources of various regions to find suitable wind farm sites using commercial CFD codes [3–6]. Dhunny et al. used WindSim to predict wind resources for the island of Mauritius [3]. Similar studies on finding suitable offshore wind farm sites using the CFD simulation WindSim can be found in Yue et al. [4] and Song et al. [5]. Park et al. used WindSim to predict wind resources of wind farms in complex terrains and to find suitable offshore wind farm sites [6]. However, no experimental validations of the simulation results were made in these studies.

There are also studies predicting the AEPs of operating wind farms using commercial software (such as WindPRO, WindSim, WindFarm, and Metodyn) with either measured or reanalyzed wind data and validating the results with actual energy productions. Tabas et al. used WindSim to predict the power of wind by using different turbulence and wake models and compared the results with the measured power. In their study, the root mean square errors were between 0.09% and 0.25% [7]. Kim et al. used WindSim to predict the AEPs of two wind farms on complex terrain in South Korea, and the errors compared with the actual AEPs for three years were between 1.7% and 10.9% [8]. Song et al. used WindSim and WindPRO to predict the AEPs of an offshore wind farm in Denmark, and the errors compared with the actual AEPs for three years were between 0.56% and 4.64% and between 0.09% and 5.71%, respectively [9]. Kim et al. used WindPRO to predict the AEP of a small wind farm with two turbines by considering nearby obstacles. The errors between the prediction and the measured AEP for three years were between 0.9% and 2.4% [10]. Results from these studies indicate that the errors in the AEP prediction of wind farms were mostly within 10% [7–10].

The commercial software used to predict the electrical power produced by a wind turbine is generally based on the Weibull distribution of the wind data and the steady power curve of the wind turbine, so the wind turbine orientation is always matched with the wind direction, and the power output is based on the steady power output. Therefore, the wind turbine yaw dynamics and the control algorithm to face the wind during wind direction changes are not considered, and the transient power variation in actual operating conditions is not simulated. The steady power curve is also based on the wind turbine control, which is applied when the power curve of the wind turbine is made, and therefore, any changes in the operating points of the wind turbine from the steady power curve cannot be simulated.

However, there is increasing interest in research on wind farm control, which actively adjusts the power demand distribution or yaw demand distribution to individual wind turbines in a wind farm to maximize the total power output from the wind farm [11–14]. For this kind of simulation, the wind turbine must be modeled for dynamic simulation, and it must have suitable power control algorithms so that the power output from the wind turbine varies with the commands from wind farm controllers.

A recent study proposed using a dynamic wind turbine simulation model to predict the AEP of a wind turbine [15]. It used a Simulink version of the NREL 5 MW paper wind turbine including a controller to predict the AEP of a wind turbine at four different sites and compared the results with the results from a commercial code, WindPRO. The predictions from the dynamic simulation code were found to be a few percent smaller than those from the commercial code due to yaw motion. However, the results were not experimentally validated in that study.

Therefore, in this study, the previous study by Kim et al. was revisited to experimentally validate the AEP predicted by the model [15]. For this, the dynamic Simulink model was used to model an actual 2 MW wind turbine that is implemented in a wind farm in Korea. The wind turbine control was also tuned to produce the power curve of the wind turbine, and the revised wind turbine model was used to predict the annual energy production of the wind turbine. The wind data for the wind turbine were obtained from a commercial wind farm simulation code, WindSim, with the wind data measured from a meteorological mast near the wind turbine. Finally, the prediction results from the dynamic simulation code were compared with the results from the commercial code, WindSim, and the measurement.

2. Methodology

2.1. Target Wind Turbine

In this study, a wind turbine in a commercial wind farm D in South Korea was used as a target wind turbine. As shown in Figure 1, wind farm D consists of 15 wind turbines. The met mast is at the north of the target wind farm. The target wind turbine is about 2.5 RD (Rotor Diameter) away from the met mast, which corresponds to 220 m in distance. The met mast and the target wind turbine are located in a relatively flat terrain, and they are about 1.7 km away from the sea.

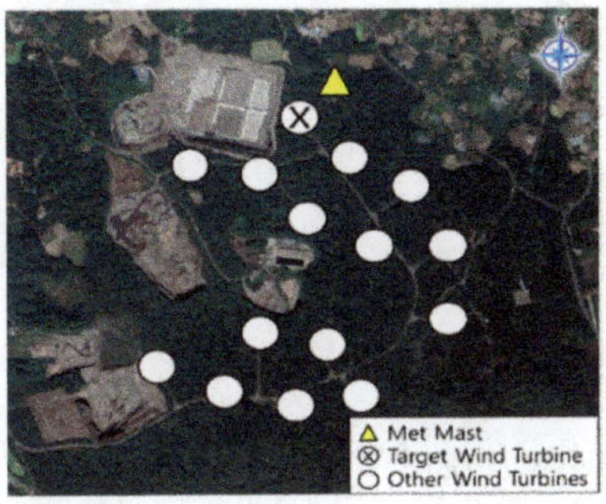

Figure 1. The layout of wind farm D (Source: Google Maps).

Table 1 shows the general specification of the target wind turbine. As shown in the table, the hub height of the wind turbine is 80 m. The cut-in, rated, cut-out wind speeds are 3.5 m/s, 12 m/s, and 25 m/s, respectively. Figure 2 shows the power curve showing the power output with respect to the wind speed for the standard air density, which is 1.225 kg/m^3.

Table 1. General specification of the target wind turbine.

Properties	Value
Rated power	2 MW
Cut-in, rated, cut-out Wind speed	3.5 m/s, 12 m/s, 25 m/s
Hub height	80 m
Power control	Pitch Control

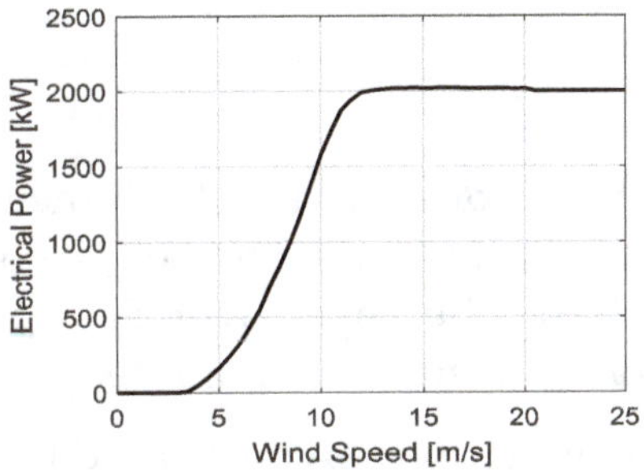

Figure 2. The power curve of the target wind turbine.

2.2. Wind Data

In this study, the 10-min averaged data measured from 1 January 2017 to 31 December 2017 at the meteorological mast were used. The meteorological mast sensors comprise anemometers, wind direction vanes, a temperature sensor, a barometric pressure sensor, and a relative humidity sensor. The anemometers were installed at heights of 80 m, 78 m, and 40 m. The wind direction vanes were installed at heights of 78 m and 40 m. The temperature sensor, barometric pressure sensor, and relative humidity sensor were installed at a height of 75 m. The data measured with the 80 m anemometer and 78 m wind direction vane were used to predict the electrical power production of the target wind turbine in the wind farm. Figure 3a shows the Weibull distribution, and Figure 3b is the wind energy rose obtained using the met mast data. As shown in Figure 3, the average wind speed at a height of 80 m is 5.75 m/s, and the prevailing wind direction is 315 degrees in 16 sector-wise angles.

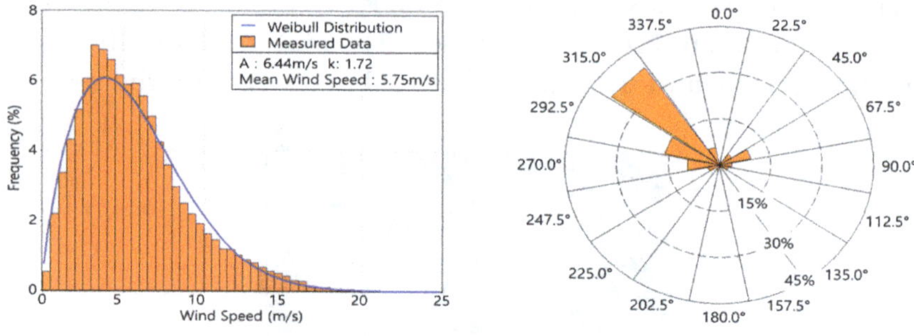

(a) (b)

Figure 3. Measured wind data for the target wind turbine: (**a**) Weibull distribution; (**b**) total wind energy rose.

According to the equation for data normalization provided by the International Electrotechnical Commission (IEC) standards 61400-12-1 2nd edition [16], the air density at the target wind turbine can be calculated using the measured 10-min average data for temperature and air pressure of the mast at a height of 75 m. The 10-min average air density was calculated using Equation (1):

$$\rho_{10min} = \frac{B_{10min}}{R_0 \times T_{10min}}. \tag{1}$$

In the equation, ρ_{10min} is the calculated 10-min average air density, B_{10min} is the measured 10-min average air pressure, T_{10min} is the measured 10-min average air temperature, and R_0 is the gas constant of dry air, which is 287.05 J/(kg·K). The air density calculated at a height of 75 m at the met mast position is 1.076 kg/m^3. The air density for the target wind turbine was assumed to be the same as the air density at the met mast.

2.3. Terrain Modeling

A terrain model that includes the information on both the topology and roughness was used for the CFD simulation analysis of the wind farm. The height contour map provided by the Korea National Spatial Information Portal and the land cover data provided by the Korea Ministry of Environment [17,18] were used to construct the terrain model.

The terrain model of the target wind farm was constructed by the geographic information system (GIS) software Global Mapper based on the height contour and the land cover data and implemented with the CFD simulation program, WindSim. Figure 4a,b shows the elevation and the roughness length information of the terrain model.

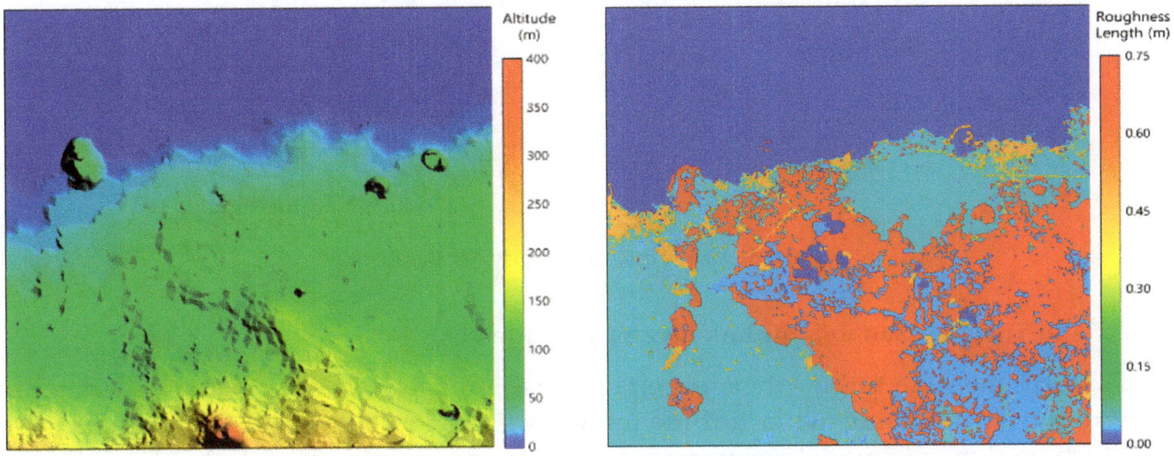

Figure 4. Terrain model of the target wind farm L (**a**) elevation; (**b**) roughness length.

The size and grid spacing of the terrain model has an impact on the prediction results of the power output [19]. Therefore, in this study, the size of the terrain model was determined to be 10 km by 10 km, and the horizontal grid spacing was 25 m by 25 m. The number of vertical grids was 40, and the number of grids below the wind turbine hub height was six, to accurately simulate the wind characteristics in the wind farm.

The wind field analysis was carried out using WindSim based on the Reynolds-Averaged Navier–Stokes (RANS) equation. Because the equation is nonlinear, the solution is solved with the use of multiple iterations until convergence is obtained [20]. This study uses the GCV (Convergence control with new solver) solver in the CFD simulation WindSim to solve the RANS equation. The GCV method uses a pressure-based separation solver strategy to control the mass flux on the control-volume faces, providing faster convergence [21]. It was assumed that the atmospheric boundary layer had a height of 500 m, that the atmospheric flow was stable, and that the wind speed above 500 m was constant with height [22]. The boundary condition at the top was "fixed pressure". The turbulence model used was the standard k-ε eddy viscosity turbulence model without considering temperature changes, where k is the turbulent flow energy and ε is the rate of turbulent dissipation.

Based on the results of the wind field analysis and the N.O. Jensen wake model [23], the 10-min averaged wind speed, wind direction, and electrical power of the target wind turbine were obtained from WindSim.

2.4. *The Wind Turbine Matlab/Simulink Dynamic Model with a Wind Turbine Controller*

In this study, the dynamic simulation model of a wind turbine was used for the prediction of the electrical power production by the target wind turbine. The model was developed in the previous study and modified for this study to be used to predict the power of an actual wind turbine [15].

Figure 5 shows the schematic of the dynamic model used for the target wind turbine. The dynamic model of the wind turbine is composed of four parts: the wind turbulence model, the yaw system model, the wind turbine controller, and the wind turbine model. The model requires 10-min average wind speed and direction data together with their standard deviations as inputs and calculates the electrical power output of the wind turbine with the power and yaw control algorithms. Detailed information on the model is available in Ref. [15].

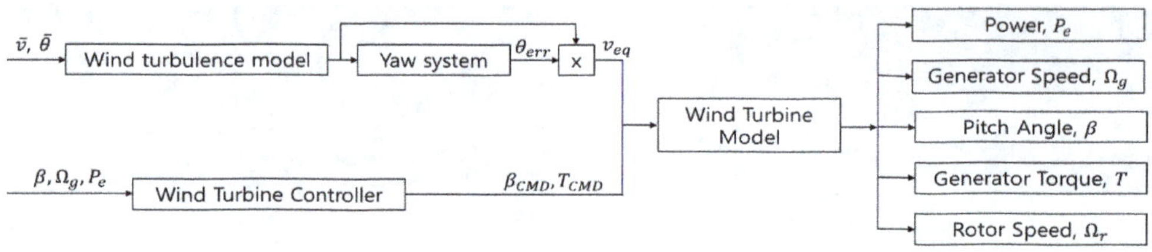

Figure 5. The wind turbine dynamic model.

2.4.1. The Wind Turbine Matlab/Simulink Dynamic Model

In this study, the wind turbine dynamic model was simulated by using Matlab/Simulink, and the block diagram of the dynamic wind turbine model is shown in Figure 6. As shown in the figure, the wind turbine dynamic model is composed of the aerodynamics model, the drive train model, the generator model, the pitch actuator model, and the control system model.

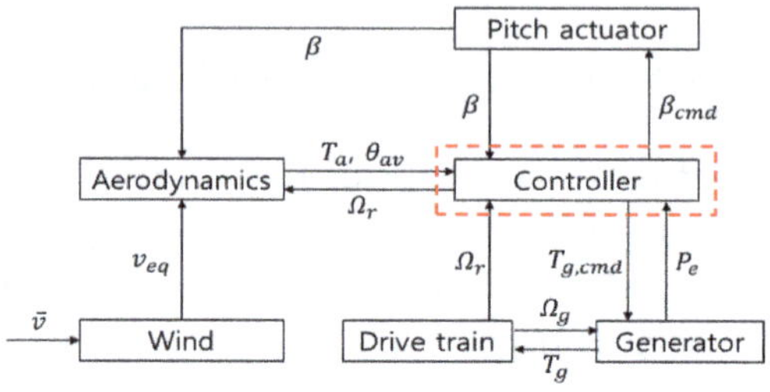

Figure 6. The block diagram of the wind turbine model.

The aerodynamic power can be simply modeled using Equation (2).

$$P_{aero} = \frac{1}{2}\rho\pi R^2 C_P(\lambda, \beta) v^3, \tag{2}$$

where P_{aero} is the aerodynamics power of the wind turbine, ρ is the air density, R is the wind turbine radius, v is the wind speed, and C_P is the power coefficient of the wind turbine which depends on the tip speed ratio (λ) and the blade pitch angle (β). The power coefficients for various tip speed ratios and blade pitch angles are calculated from a BEMT (Blade Element Momentum Theory) software such as Bladed and used as a look-up table in the aerodynamics model.

The drive train model is composed of a low-speed shaft, a gearbox, and a high-speed shaft. The aerodynamics torque and generator torque are the inputs of the drive train model. The low-speed shaft and high-speed shaft can be modeled by the inertia, the stiffness, and the damping. The drive train model was simulated using Equations (3) and (4). The relationship between the low-speed shaft and the high-speed shaft is shown in Equation (5) [24].

$$J_r \frac{d\Omega_r}{dt} = T_a - k_s\left(\theta_r - \frac{1}{N}\theta_g\right) - c_s\left(\Omega_r - \frac{1}{N}\Omega_g\right) - B_r\Omega_g, \tag{3}$$

$$J_g \frac{d\Omega_g}{dt} = \frac{k_s}{N}\left(\theta_r - \frac{1}{N}\theta_g\right) + \frac{c_s}{N}\left(\Omega_r - \frac{1}{N}\Omega_g\right) - B_g\Omega_g - T_g, \tag{4}$$

$$\Omega_g = N\Omega_r, \tag{5}$$

where J is the inertia, Ω is the rotation speed, T is the torque, k_s is the torsional stiffness, c_s is the torsional damping, θ is the rotation angle, B is the damping, and N is the gear ratio. Moreover, a represents the aerodynamic, r and g mean the rotor and the generator, respectively.

The pitch actuator and the generator were modeled as 1st-order systems, which are shown in Equations (6) and (7), respectively [25].

$$\frac{\beta}{\beta_{cmd}} = \frac{1}{1+\tau_\beta s}, \tag{6}$$

$$\frac{T_g}{T_{g,cmd}} = \frac{1}{1+\tau_g s}. \tag{7}$$

In Equations (6) and (7), β is the pitch angle, β_{cmd} is the pitch angle command, and τ_β is the time constant of the pitch actuator. T_g is the generator torque, $T_{g,cmd}$ is the generator torque command, and τ_g is the time constant of the generator torque.

2.4.2. The Wind Turbine Control System

Figure 7 shows the wind turbine controller used in the dynamic wind turbine model. It consists of a basic torque and blade pitch control for power maximization and regulation and a peak shaving to use blade pitching to reduce thrust force by sacrificing power slightly near the rated wind speed region [26].

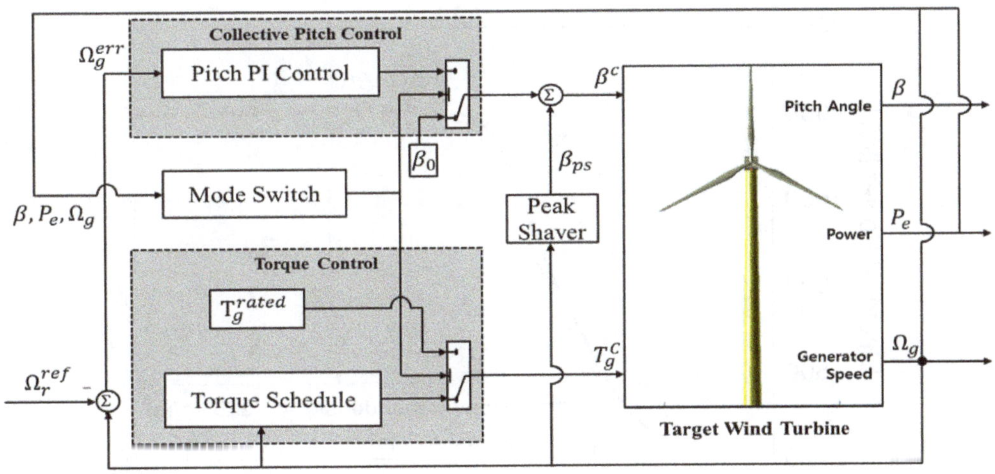

Figure 7. The wind turbine model control system.

To maximize the electrical power of the wind turbine in regions where the wind speed is lower than the rated wind speed, the pitch angle of the blade is fixed to be the fine pitch angle, and the optimal torque command is sent to the generator to maximize the power coefficient of the rotor [27]. To maintain the rated electrical power of the wind turbine in regions where the wind speed is higher than the rated wind speed, the torque is fixed to be the rated torque, and the blade pitch angle of the wind turbine is adjusted to keep the rated rotational speed of the rotor. The torque control algorithm in the control system used was a simple open-loop control using a look-up table with an input of measured generator speed and output of optimal generator torque. Moreover, the blade pitch control in the control system used the classical proportional–integral (PI) algorithm to minimize the error between the measured generator speed and the rated value.

2.5. Conversion of Power Curve to Include Air Density Effect

The power coefficient curve and thrust coefficient curve used for the wind turbine dynamic model are derived from Bladed under the condition of a standard air density of 1.225 kg/m^3. However, the actual air density of a wind farm is different from the standard value, which causes a discrepancy between the prediction and the measured power output.

In this study, a novel method is proposed to match the power of a dynamic wind turbine model with that of a wind turbine with actual air density by tuning the controller. The first step of this method is to convert the power curve of the wind turbine based on the standard air density to that based on the actual mean air density of the site. This conversion is possible by using the correction method for wind turbines with active power control described in the standard, IEC 61400-12-1 [16].

Equation (8) represents the wind speed correction equation calculated based on the IEC standard.

$$V_{10min} = V_n / \left(\frac{\rho_{10min}}{\rho_0} \right)^{1/3}. \quad (8)$$

In Equation (8), V_{10min} is the measured wind speed averaged over 10 min, V_n is the normalized wind speed, ρ_{10min} is the derived 10-min average air density, and ρ_0 is the reference air density, known to be 1.225 kg/m^3. Through Equation (8), the wind speed of a power curve is converted to consider the mean air density at the site.

Figure 8 shows the comparison between the power curve with the standard air density and the power curve with the actual mean air density. It can be seen from the figure that when the power is lower than the rated power, the power with the standard air density is slightly higher than that with the actual mean air density.

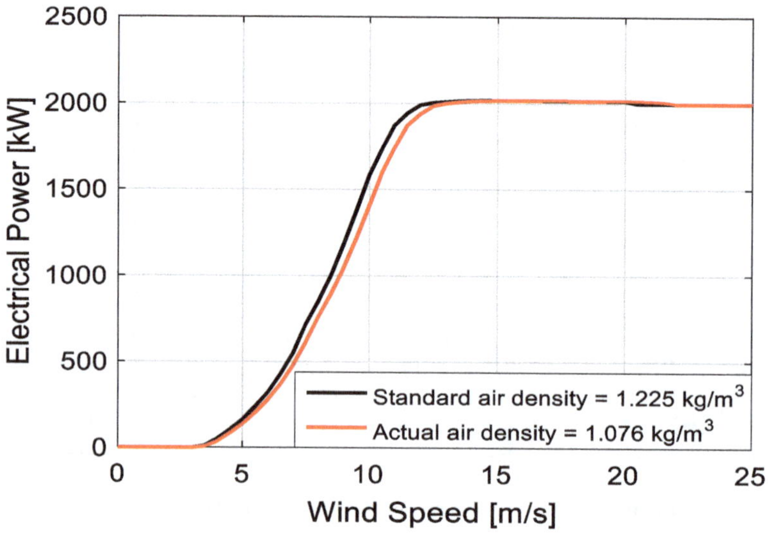

Figure 8. Comparison of the power curve of standard air density and the power curve of actual air density.

2.6. *Tuning of the Wind Turbine Controller to Consider Actual Mean Air Density*

To consider the actual mean air density of the site in the dynamic model, the wind turbine controller needs to be tuned by considering the power curve with the actual mean air density as a target.

In the torque control region where the wind speed is lower than 10 m/s, the look-up table of the torque scheduling was tuned so that the power curve of the dynamic model is matched with the power curve with the actual mean air density. In the transition region where the wind speed is higher than 10 m/s and lower than the rated wind speed, the peak shaving algorithm and torque scheduling were used to match the target power curve. For the rated power region, the closed-loop PI control strategy was used, and no tuning was necessary because the algorithm is not affected by the change of air density.

Figure 9 compares the target power curve with the power curve obtained from the dynamic model after tuning the control algorithms. The two power curves are well matched.

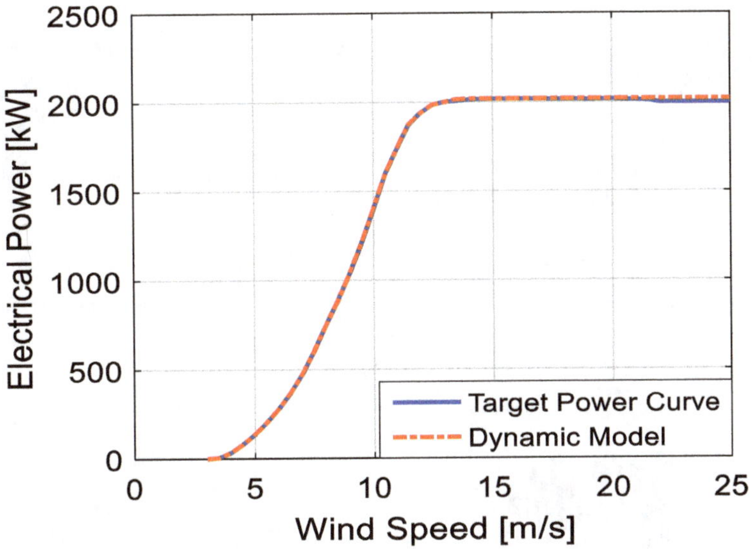

Figure 9. Comparison of the measured power curve report and the power curve of different controllers at 1.076 kg/m^3.

3. Results

Figure 10a–d shows the results of the comparison between the measured electrical power, the electrical power predicted by WindSim, and the electrical power predicted by the dynamic model for four days in January, April, August, and November, respectively. The wind direction data measured from the met mast are also shown in the figure. The measured data and the data obtained from WindSim were 10-min averaged data, and therefore, the data from the dynamic model, which was originally in 1-s intervals, were converted to 10 min averaged data.

As shown in the figure, both predictions by WindSim and the dynamic model were close to the measured power; however, there are regions where relatively large discrepancies exist in Figure 10b,d. For these regions, the wind directions are between 110 degrees and 260 degrees. The reason for these relatively large discrepancies in these regions is that the met mast is located within the wake region of the wind turbines around the target wind turbine for those wind directions, and the wind speed implemented with WindSim to predict the wind speed for the target wind turbine is the wind speed reduced by the wake effect. To obtain the wind speed for the target wind turbine, WindSim again applies a wake model in the case where the wind direction corresponds to the direction causing the wake effect on the target wind turbine. Therefore, the wind speed obtained for the target wind turbine from WindSim in those wind directions becomes lower than the actual wind speed of the target wind turbine and the power output predicted by WindSim and the dynamic model becomes lower than the measured power.

To compare the results in more detail, small portions of Figure 10c,d corresponding to three hours are replotted in Figures 11 and 12, respectively. Figure 11 shows a mean wind speed lower than the rated wind speed, and Figure 12 shows a mean wind speed higher than the rated value. Figures 11a and 12a show the wind speed measured from the met mast and the wind speed of the target wind turbine predicted by WindSim. For both cases, the wind speed of the target wind turbine predicted by WindSim is slightly higher than the wind speed measured from a nearby met mast. This is partly due to the fact that the altitude of the target wind turbine location is slightly higher than that of the met mast location.

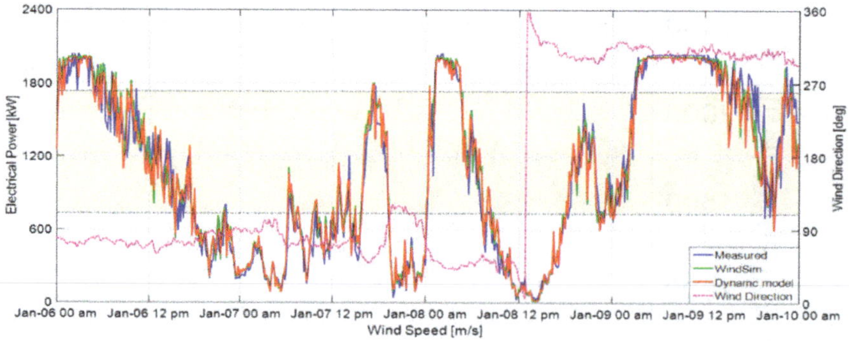

(**a**)

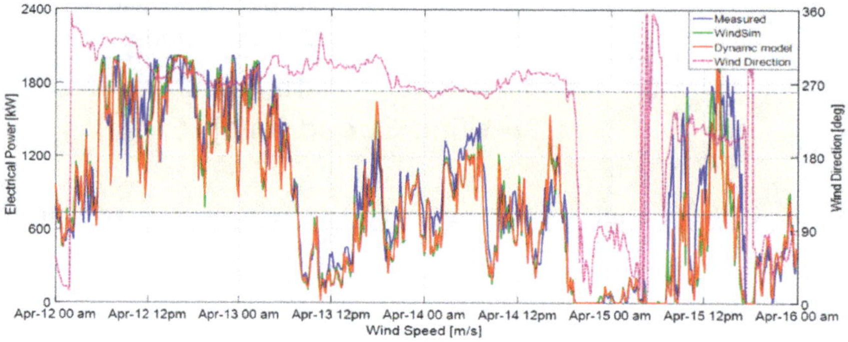

(**b**)

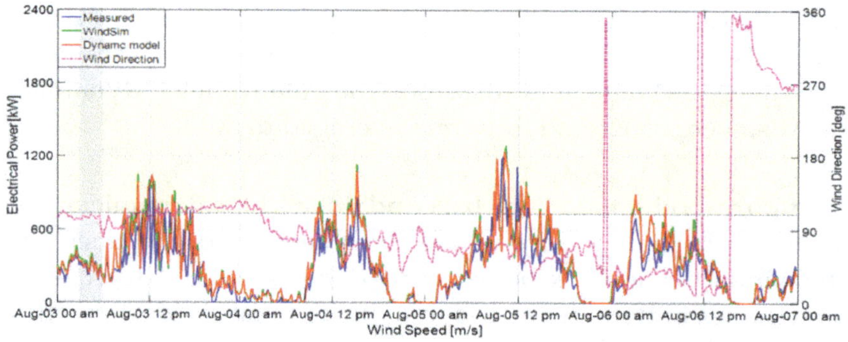

(**c**)

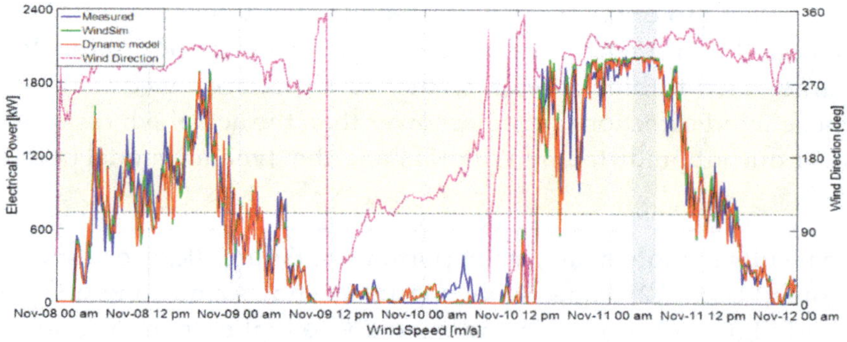

(**d**)

Figure 10. Comparison of the measured and predicted electrical power: (**a**) January, (**b**) April, (**c**) August, (**d**) November.

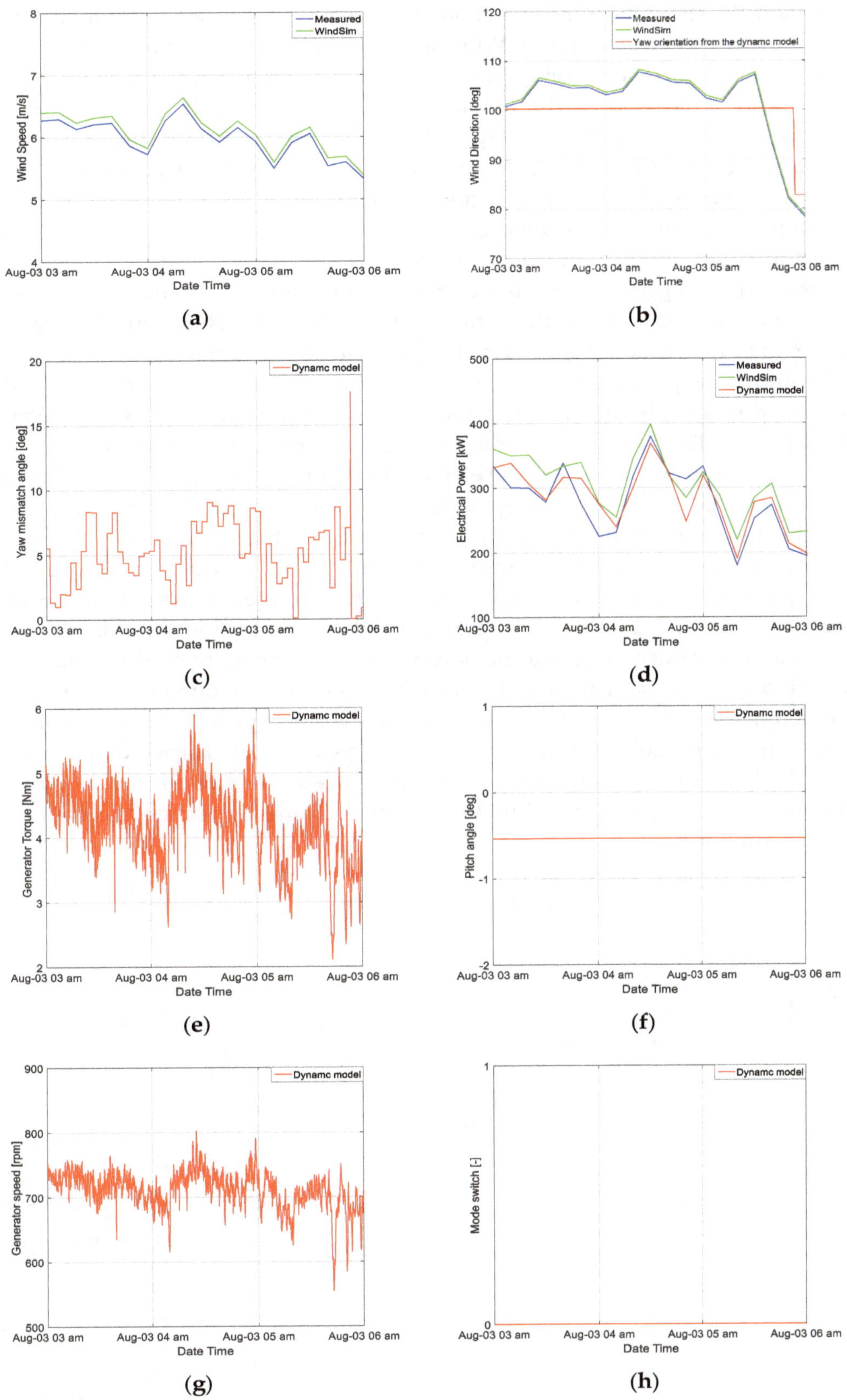

Figure 11. Comparison of the measured and predicted data for 3 h in August: (**a**) wind speed, (**b**) wind direction, (**c**) yaw mismatch, (**d**) electrical power, (**e**) generator torque, (**f**) pitch angle, (**g**) generator speed, (**h**) model switch.

Figures 11b and 12b show the wind direction measured from the met mast and the wind direction at the wind turbine position predicted by WindSim. The wind direction of the target wind turbine predicted by WindSim is basically consistent with the measured wind direction. They also show the turbine orientation predicted by the dynamic model. In Figure 11b, the yaw mismatch angle exceeded 11 degrees in the end, and the yaw system was activated to reduce the mismatch angle. In Figure 12b, the yaw orientation did not change because of the small yaw mismatch angles compared with the threshold. These are also shown in Figures 11c and 12c, which show the yaw mismatch angles. The mismatch angles are three-minute averaged values.

Figures 11d and 12d show the measured electrical power, the electrical power predicted by WindSim, and the electrical power predicted by the dynamic model. Figure 11d shows the electrical power when the wind speed is lower than the rated wind speed. Therefore, the blade pitch angle is fixed to be the fine pitch, and the generator torque is adjusted to maximize the power coefficient. As shown in the figure, the power prediction from WindSim consistently overestimates the measured power. The power predicted by the dynamic model looks similar to the WindSim results but closer to the measured power. The generator torque and the blade pitch angle from the dynamic model is shown in Figure 11e,f. Figure 12d shows the electrical power when the wind speed is mostly higher than the rated wind speed. Therefore, the generator torque is fixed to the rated value, and the blade pitch angle is adjusted to maintain the rated wind speed and the rated power. Figure 12e,f shows the generator torque and the blade pitch angle from the dynamic model. In Figure 12d, there are a few downward spikes in the measured power. The dynamic model shows similar downward spikes but the power from WindSim does not. The reason for this is that WindSim does not consider wind turbine dynamics and yaw mismatch, and the power is solely dependent on the wind speed. Based on Figure 12a, the wind speed is maintained to be mostly higher than the rated wind speed, and therefore, the power prediction from WindSim is mostly close to the rated power. However, if we look at the results from the dynamic model in Figure 12e,h, the mode switch was turned off a few times between 4:00 and 5:00 o'clock, and the generator torque was reduced. As a result, the power decreased as a downward spike. The dynamic model includes wind turbine dynamics and could show similar measured power variations.

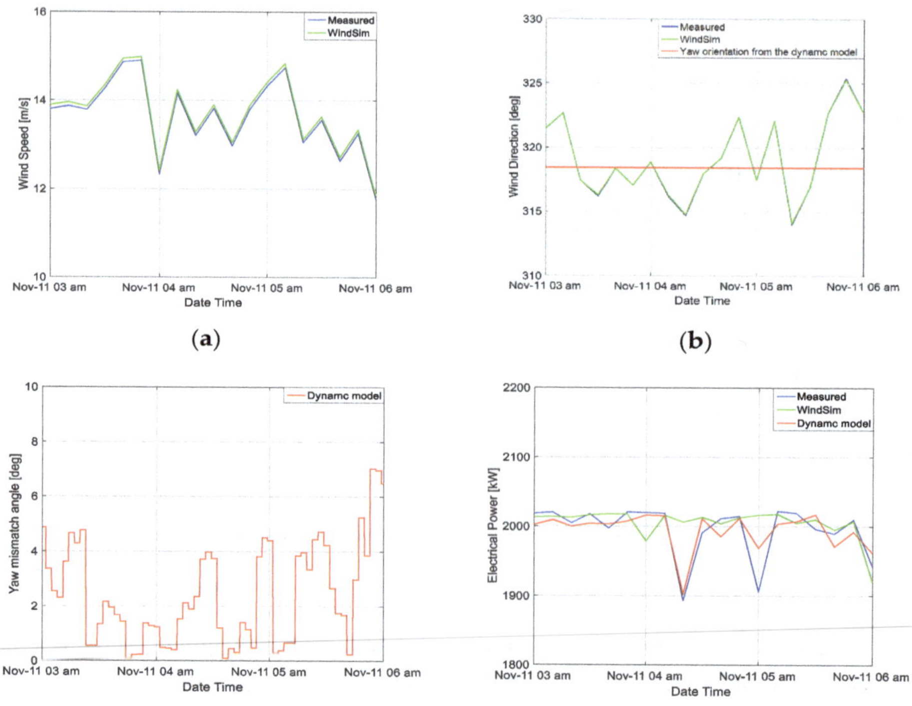

(a) (b) (c) (d)

Figure 12. *Cont.*

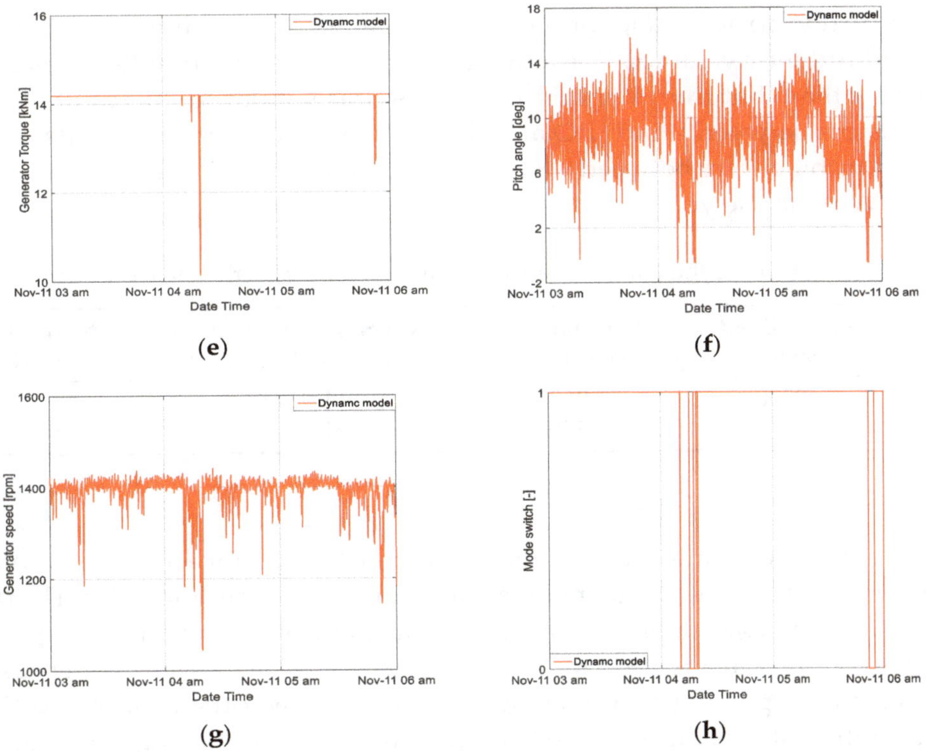

Figure 12. Comparison of the measured and predicted data for 3 h in November (**a**) wind speed, (**b**) wind direction, (**c**) yaw mismatch, (**d**) electrical power, (**e**) generator torque, (**f**) pitch angle, (**g**) generator speed, (**h**) model switch.

Figures 11g and 12g show the generator speed obtained from the dynamic model. For Figure 11g, the generator speed varies between 550 rpm and 800 rpm to maximize the power coefficient; on the other hand, for Figure 12g, the generator speed is generally close to the rated value, which is about 1400 rpm.

Table 2 shows the relative percentage errors of the electrical energy production with respect to the measure values given in Figures 11d and 12d. As shown in the table, the relative errors of the energy production predicted by WindSim and the dynamic model for Figure 11d are 9.45% and 1.12%, respectively, which is the case when the wind speed is lower than the rated wind speed. The errors are reduced to 0.48% and −0.06%, respectively, for Figure 12d in which the wind speed is higher than the rated wind speed. For both cases, the errors from the dynamic model are smaller than those from WindSim, because the dynamic model considers the wind turbine dynamics and the yaw motion. In Table 2, the electrical energy production was normalized by the measured electrical energy production (MWh) of the corresponding period, respectively.

Table 2. Relative errors of electrical energy production prediction for the data shown in Figures 10d and 11d.

	Normalized Measured	Normalized WindSim	Normalized Dynamic Model
Electrical energy production for Figure 10d	1.0000	1.0945	1.0112
Electrical energy production for Figure 11d	1.0000	1.0048	0.9994
Relative errors for Figure 10d	-	9.45%	1.12%
Relative errors for Figure 11d	-	0.48%	−0.06%

Table 3 shows the comparison between the measured AEP, the AEP predicted by the WindSim, and the AEP predicted by the dynamic model with and without yaw motion for the target wind turbine. As shown in the table, the relative error of AEP predicted by WindSim is 2.75%, and those of the dynamic model without and with the yaw system are 2.11% and 0.16%, respectively. Based on the results, the prediction error from ignoring yaw mismatch errors is larger than the prediction error from ignoring wind turbine dynamics. The annual energy production from the dynamic model is also about 2.52% smaller than that from WindSim. This result is similar to the result in Ref. [15], where the relative differences between WindPRO and the dynamic model for four different sites with the NREL 5 MW wind turbine were between 0.83% and 4.91%. In Table 3, the AEP is normalized by the measured AEP (MWh) for one year.

Table 3. The relative errors of the annual energy production (AEP) by the measured and other models.

	Normalized Measured	Normalized WindSim	Normalized Dynamic Model without Yaw	Normalized Dynamic Model with Yaw
AEP	1.0000	1.0275	1.0211	1.0016
Relative error	-	2.75%	2.11%	0.16%

4. Conclusions

In this study, a dynamic wind turbine model was constructed for an actual 2 MW wind turbine, and the wind turbine performance was predicted and validated experimentally. The dynamic model was constructed using Matlab/Simulink to consider the wind turbine dynamics, including yaw systems and a wind turbine controller using basic torque and pitch control algorithms with peak shaving in the transition from maximum power point tracking to the rated power regions. In addition, to consider the density effect of the test site, the torque scheduling and the peak shaving were tuned to reproduce the power curve with the actual mean air density.

For experimental validation, the AEP predicted by the dynamic model was compared with that by WindSim and the measured power. As the result, the relative error of WindSim was 2.75%, and those of the dynamic model without and with yaw were 2.11% and 0.16%, respectively. Although the results are only for a single wind turbine case, the prediction of AEP from the dynamic model was found to be accurate. However, further research is needed for different wind turbines and sites to better understand the performance of the dynamic simulation model.

Author Contributions: Y.S. performed the simulation of the wind turbine, analyzed the results, and wrote the paper. I.P. supervised the research, analyzed the results, and revised the paper. All authors have read and agreed to the published version of the manuscript.

Acknowledgments: The authors would like to acknowledge the support from Han Jin Ind. Co. Ltd. for the wind turbine model construction. The authors would also like to acknowledge the support from Jeju Energy Corp. for the climatology data of the site.

References

1. "Renewable energy 3020" Implementation Plan. Available online: http://www.motie.go.kr/motiee/presse/press2/bbs/bbsView.do?bbs_seq_n=159996&bbs_cd_n=81 (accessed on 3 August 2020). (In Korean)
2. South Korea Wind Turbine Installation Status. Available online: http://kweia.or.kr/bbs/board.php?bo_table=sub03_03 (accessed on 3 August 2020). (In Korean).
3. Yue, C.; Liu, C.; Tu, C.; Lin, T. Prediction of Power Generation by Offshore Wind Farms Using Multiple Data Sources. *Energies* **2019**, *12*, 700. [CrossRef]
4. Song, Y.; Kwon, I.; Paek, I. Investigation on promising offshore wind farm sites and their wind farm capacity factors considering wake losses in Korea. *J. Wind Energy* **2018**, *9*, 27–36. [CrossRef]
5. Park, U.; Yoo, N.; Kim, J.; Kim, K.; Min, D.; Lee, S.; Paek, I.; Kim, H. The selection of promising wind farm sites in Gangwon province using multi exclusion analysis. *J. Korean Sol. Energy Soc.* **2015**, *35*, 1–10. [CrossRef]
6. Dhunny, A.; Lollchund, M.; Rughooputh, S. Wind energy evaluation for a highly complex terrain using Computational Fluid Dynamics (CFD). *Renew. Energy* **2017**, *101*, 1–9. [CrossRef]
7. Song, Y.; Kim, H.; Byeon, J.; Paek, I.; Yoo, N. A Feasibility Study on Annual Energy Production of the Offshore Wind Farm using MERRA Reanalysis Data. *J. Korean Sol. Energy Soc.* **2015**, *35*, 33–41. [CrossRef]

8. Kim, J.; Kwon, I.; Park, U.; Paek, I.; Yoo, N. Prediction of Annual Energy Production of Wind Farms in Complex Terrain using MERRA Reanalysis Data. *J. Korean Sol. Energy Soc.* **2014**, *34*, 82–90. [CrossRef]
9. Tabas, D.; Fang, J.; Porté-Agel, F. Wind Energy Prediction in Highly Complex Terrain by Computational Fluid Dynamics. *Energies* **2019**, *12*, 1311. [CrossRef]
10. Kim, H.; Song, Y.; Paek, I. Predicted and Validation of Annual Energy Production of Garyeok-do Wind Farm in Saemangeu Area. *J. Wind Energy* **2018**, *9*, 19–24.
11. Kim, H.; Kim, K.; Paek, I. Power regulation of upstream wind turbines for power increase in a wind farm. *Int. J. Precis. Eng. Manuf.* **2016**, *17*, 665–670. [CrossRef]
12. Kim, H.; Kim, K.; Paek, I. Model Based Open-Loop Wind Farm Control Using Active Power for Power Increase and Load Reduction. *Appl. Sci.* **2017**, *7*, 1068. [CrossRef]
13. Kim, H.; Kim, K.; Paek, I. A Study on the Effect of Closed-Loop Wind Farm Control on Power and Tower Load in Derating the TSO Command Condition. *Energies* **2019**, *12*, 2004. [CrossRef]
14. Campagnolo, F.; Weber, R.; Schreiber, J.; Bottasso, C. Wind tunnel testing of wake steering with dynamic wind direction changes. *Wind Energy Sci.* **2020**, *5*, 1273–1295. [CrossRef]
15. Kim, K.; Lim, C.; Oh, Y.; Kwon, I.; Yoo, N.; Paek, I. Time-domain dynamic simulation of a wind turbine including yaw motion for power prediction. *Int. J. Precis. Eng. Manuf.* **2014**, *15*, 2199–2203. [CrossRef]
16. International Electrotechnical Commission (IEC). *Wind Turbine Generator Systems Part 12-1: Power Performance Measurements of Electricity Producing Wind Turbines*, 2nd ed.; International Electrotechnical Commission (IEC): Geneva, Switzerland, 2017.
17. Ministry of Land. Infrastructure and Transport, National Spatial Data Infrastructure Portal. Available online: http://www.nsdi.go.kr/lxportal/?menuno=2679 (accessed on 3 August 2020).
18. Ministry of Environment. Environmental Spatial Information Service. Available online: http://egis.me.go.kr/map/map.do?type=land (accessed on 3 August 2020).
19. Hwang, Y.; Paek, I.; Yoon, K.; Lee, W.; Yoo, N.; Nam, Y. Application of wind data from automated weather stations to wind resources estimation in Korea. *J. Mech. Sci. Technol.* **2010**, *24*, 2017–2023. [CrossRef]
20. WindSim AS. Available online: https://www.windsim.com (accessed on 3 August 2020).
21. Cham. The Phoenics Encyclopedia. Available online: http://www.cham.co.uk/phoenics/d_polis/d_enc/enc_gcv.htm (accessed on 3 August 2020).
22. Fallo, D. Wind Energy Resource Evaluation in a Site of Central Italy by CFD Simulations. Ph.D. Thesis, University of Cagliari, Cagliari, Italy, 2007.
23. Katic, I.; Højstrup, J.; Jensen, N. A simple model for cluster efficiency. In Proceedings of the European Wind Energy Association Conference and Exhibition, Roma, Italy, 7–9 October 1986; Volume 1, pp. 407–410.
24. Nam, Y.; Kim, J.; Paek, I.; Moon, Y.; Kim, S.; Kim, D. Feedforward Pitch Control Using Wind Speed Estimation *J. Power Electron.* **2011**, *11*, 211–217. [CrossRef]
25. Bianchi, F.; de Battista, H.; Mantz, R. Wind Turbine Control Systems, Principles, Modelling and Gain Scheduling Design. Available online: https://www.springer.com/gp/book/9781846284922 (accessed on 3 August 2020).
26. Kim, K.; Kim, H.; Paek, I. Application and Validation of Peak Shaving to Improve Performance of a 100 kW Wind Turbine. *Int. J. Precis. Eng. Manuf.-Green Tech.* **2020**, *7*, 411–421. [CrossRef]
27. Jonkman, J.; Butterfield, S.; Musial, W.; Scott, G. *Definition of a 5-MW Reference Wind Turbine for Offshore System Development*; National Renewable Energy Lab. (NREL): Golden, CO, USA, 2009; p. 947422.

4

Free and Forced Vibration Analysis of H-Type and Hybrid Vertical-Axis Wind Turbines

Minhui Tong [1,2,*], Weidong Zhu [2,*], Xiang Zhao [2,3], Meilin Yu [2], Kan Liu [2] and Gang Li [2]

1 Logistics Engineering College, Shanghai Maritime University, Shanghai 201306, China
2 Department of Mechanical Engineering, University of Maryland, Baltimore County, Baltimore, MD 21250, USA; zx1982@umbc.edu (X.Z.); mlyu@umbc.edu (M.Y.); kan7@umbc.edu (K.L.); gangli@umbc.edu (G.L.)
3 School of Civil Engineering and Geomatics, Southwest Petroleum University, Chengdu 610599, China
* Correspondence: mtong@umbc.edu (M.T.); wzhu@umbc.edu (W.Z.)

Abstract: Vertical-axis wind turbines (VAWTs) are compact and efficient and have become increasingly popular for wind energy harvesting. This paper mainly focuses on free and forced vibration analysis of two different types of VAWTs, i.e., an H-type VAWT and a new hybrid VAWT. The H-type VAWT has a lower cost, while the hybrid VAWT has a better self-starting capability at a low wind velocity. Both of them can be used for wind energy harvesting. By using the assumed modes method, the two VAWTs are simplified by a single degree-of-freedom (SDOF) model. By utilizing the method of structural mechanics, a multi-degree-of-freedom (MDOF) model is developed for the two VAWTs and the turbines in them are reasonably simplified. Natural frequency analyses for the SDOF and MDOF models of the two VAWTs are conducted. A beam element model (BEM) of the two VAWTs is created to calculate their natural frequencies and mode shapes and to verify natural frequency results from the SDOF and MDOF models. By using the BEM of the two VAWTs, their amplitude-frequency responses are obtained from harmonic response analysis. To analyze forced vibrations of the two VAWTs, aerodynamic loads on the two VAWTs are obtained from computational fluid dynamics (CFD) simulation. By using solid element models of the two VAWTs, forced transient responses of the two VAWTs are calculated by using the aerodynamic loads from CFD simulation. Steady-state forced response amplitudes of the 1 m-mast hybrid VAWT are 23.8% and 20.5% smaller in X- and Y-directions than those of the 1 m-mast H-type VAWT, respectively. Frequency contents of the aerodynamic loads from CFD simulation are calculated, which confirm that they are periodic, and the power efficiency of the H-type VAWT is about 2.6% higher that of the hybrid VAWT.

Keywords: vertical-axis wind turbine (VAWT); H-type VAWT; hybrid VAWT; natural frequency; mode shape; harmonic response analysis; aerodynamic load; forced transient response; power efficiency

1. Introduction

Wind energy has attracted widespread attention. According to the new Wind Powers America Annual Report released in 2019, wind power has reached over the 105 GW capacity in the U.S., and 7% of the total electricity is provided by wind energy, which is the largest renewable energy in the U.S. so far [1]. Most of 0.8 through 1.5 MW turbines came into production during 1995–2005. As for horizontal-axis wind turbines (HAWTs), the most widely used onshore turbines are 1.5 MW turbines, whose towers are 80 m tall, and this type of turbines is very popular in the U.S. [2]. Vertical-axis wind turbines (VAWTs) are widely used in urban and rural areas for residential, commercial, and industrial applications. In the U.S., Alcoa built several VAWTs whose diameters range from 12.8 to 25 m, and their powers range from 60 to 500 kW. However, high costs of these utility-scale VAWTs prohibit their commercialization. Currently, rapidly developed turbines are smaller types of VAWTs whose capacities are from 20 to 65 kW, which are supported by cantilevered masts. Costs of power generated by VAWTs are reduced by about 18–39% compared with conventional HAWT counterparts [3]. It is reported that VAWTs have fewer moving parts; their reliabilities are increased and their maintenance costs are decreased compared with those of HAWTs. VAWTs have more advantages than HAWTs in both costs and distribution in wind farms when their powers are smaller than or equal to 100 kW.

Use of VAWTs for electricity generation has attracted much interest. VAWTs have developed quickly in the last decade. They are well suited for urban environments, where wind has more turbulence. Many small-scale VAWTs have been recently installed. They can be installed on roofs of tall buildings that have easy access to wind energy, without any acoustic emission. The electricity generated directly provides energy to users in the building without transportation loss. The CO_2 emission rate of new buildings is an important indicator for their environmental friendliness evaluation. Wind turbines provide a successful way to reduce CO_2 emission compared with other sustainable energy sources, such as solar panels. New buildings often take the energy efficiency into account. Installing VAWTs is an effective part of an energy-saving design for new buildings. However, complex aerodynamics of VAWTs pose grand challenges on their structural stability. Vibration analysis and mitigation are an indispensable component to ensure the system reliability [4].

Previous works on technology development of VAWTs are reviewed here. Numerical and experimental approaches are used to study complex aerodynamics and turbulence related to the VAWT technology and performance. Much research focuses on the aerodynamics, which is most important for the VAWT performance. Computational fluid dynamics (CFD) is a typical numerical method to study the aerodynamics of VAWTs [5]. Bianchini et al. [6] used CFD to simulate three different airfoils of blades of VAWTs. Liu et al. [7] used CFD to obtain the hybrid turbine design to achieve the better self-starting capability of a VAWT at small tip-speed ratios (TSRs), which are ratios between tangential speeds of outer blades and the wind speed. Computational methods can be classified into two-dimensional (2D) and three-dimensional (3D) CFD. Zanforlin and Nishino used 2D CFD to analyze VAWTs [8,9]. Chen et al. [10] studied two straight-bladed VAWT power outputs via a CFD method. Other researchers used 3D unsteady numerical analysis to study the aerodynamics of VAWTs. For example, Joo et al. [11], Bhargav et al. [12], and Dessoky et al. [13] used 3D CFD to predict aerodynamic and aeroacoustic behaviors of VAWTs. Lositaño and Danao [14] used 3D CFD to study the performance of H-type VAWTs. Molina et al. [15], ref. [16] combined 3D CFD with wind tunnel testing to study the aerodynamics of a VAWT. Peng et al. [17], ref. [18] used wind tunnel experiments to study the power performance of H-type VAWTs. Research objects of VAWT studies can be classified into five aspects: angle of attack, airfoil section, power efficiency (or coefficient), dynamics and vibration, and VAWT design. Buchner et al. [19] studied effects of the angle of attack on aerodynamic flow by changing its values. Guo et al. [20], ref. [21] studied a method to improve the power efficiency of VAWTs by controlling their pitch angles. Sengupta et al. [22] studied effects of different airfoil sections of blades to improve the performance of H-type VAWTs. Yao et al. [23], ref. [24] studied the power coefficient to obtain the best performance of VAWTs.

Vibration analysis of VAWTs is an important research topic [25]. To investigate the performance of VAWTs, much research discusses VAWT vibration analysis, isolation, and control [26]. Research interest in reducing vibrations of VAWTs has increased in the last decade [27]. Verkinderen and Imam [28] developed an analytical method to analyze the vibration of an H-type VAWT with a simplified model. Wang et al. [29] studied modal responses of a VAWT on a building by experimental methods. Rahman et al. [30] provided vibration control strategies and techniques for VAWTs. Mabrouk et al. [31] studied the effect of the rotational speed of the turbine on the dynamic behavior of a VAWT. Yan and Wiercigroch [32] focused on vibrations of VAWT-like structures induced by high mast support.

The above studies have presented some results that are significant in engineering applications. Most studies just tackle aerodynamic characteristics of VAWTs and numerical methods are usually used. Vibration characteristics of VAWTs have not always been taken into account. According to the literature survey, apart from Refs. [28–31], research on free and forced vibration analysis of VAWTs is quite limited. To the best of authors' knowledge, comparative studies on free and forced vibration analysis of H-type and hybrid VAWTs have not been reported in the literature, which is the objective of the present work. Results from this work show that natural frequencies of the hybrid VAWT are slightly higher than those of the H-type VAWT for a low-mast VAWT. Steady-state forced response amplitudes of the 1 m-mast hybrid VAWT are 23.8% and 20.5% smaller in X- and Y-directions than those of the 1 m-mast H-type VAWT, respectively.

This paper is organized as follows: Section 2 introduces the H-type and hybrid VAWTs and illustrates their main differences. Section 3 presents free vibration analysis of the two VAWTs. The single-degree-of-freedom (SDOF) model of the two VAWTs is introduced in Section 3.1 by using the assumed modes method. The multi-degree-of-freedom (MDOF) model of the two VAWTs is developed in Section 3.2. Based on pipe-16 beam elements in the commercial software ANSYS, the beam element model (BEM) of the two VAWTs is built by ANSYS in Section 3.3 to calculate their natural frequencies and mode shapes. Solid element models (SEMs) of the two VAWTs are built by ANSYS in Section 3.4 to calculate their natural frequencies and mode shapes for comparison purposes. Section 4 calculates forced vibration responses of the two VAWTs. Harmonic response analysis of the two VAWTs is shown in Section 4.1. Based on harmonic response analysis in ANSYS, amplitude-frequency responses of the two VAWTs are obtained there. Forced transient response analysis of the two VAWTs is shown in Section 4.2. Aerodynamic loads due to wind on the two VAWTs are calculated in Section 4.3 via CFD simulation. Numerical results are presented in Section 5. Sections 5.1 and 5.2 show numerical results from the SDOF and MDOF models, respectively. By using the BEM of the two VAWTs, their amplitude-frequency responses are obtained from harmonic response analysis and shown in Section 5.1. Free vibration results of the two VAWTs from their BEM and SEMs are shown in Sections 5.3 and 5.4, respectively. By applying the aerodynamic loads calculated in Section 4.3 on the two VAWTs and using transient dynamic analysis in ANSYS, forced transient responses of the two VAWTs that include both their transient and steady-state responses are obtained using their BEM and SEMs. Frequency contents of the aerodynamic loads calculated in Section 4.3 are shown in Section 5.5. Power efficiency analysis results of the two VAWTs are shown in Section 5.6. Section 6 lists some conclusions from this study.

2. Description of VAWTs

A VAWT consists of a support mast, a turbine shaft, and turbine blades. Vibrations of the H-type and hybrid VAWTs are analyzed in this work by several methods: SDOF analysis, MDOF analysis, and beam element and solid element analyses.

Consider the two VAWTs that have three blades, whose mill diameter is 1.4 m, height is 1 m, and mill weight is 80 kg; the mast height and weight of the turbines are 4 m and 83.6 kg, respectively. The cross-section of the mast is a circle, the diameter of the mast is 0.2 m, and the wall thickness of the mast is 5 mm. The rated power generated by the two VAWTs is 300 W, the rated normal operational wind speed is 10 m/s, and the survival wind speed is 50 m/s. Shapes of the H-type and hybrid VAWTs are shown in Figure 1a,b, respectively.

Most parts of the two VAWTs are similar to each other, which include turbine blades, a turbine shaft, a generator, and the support mast. The main difference between the H-type and hybrid VAWTs is that there are inner blades installed on the shaft of the hybrid VAWT, which can provide a better self-starting capability of the VAWT [7]; the H-type VAWT has no inner blades. Figure 2 shows details of this difference.

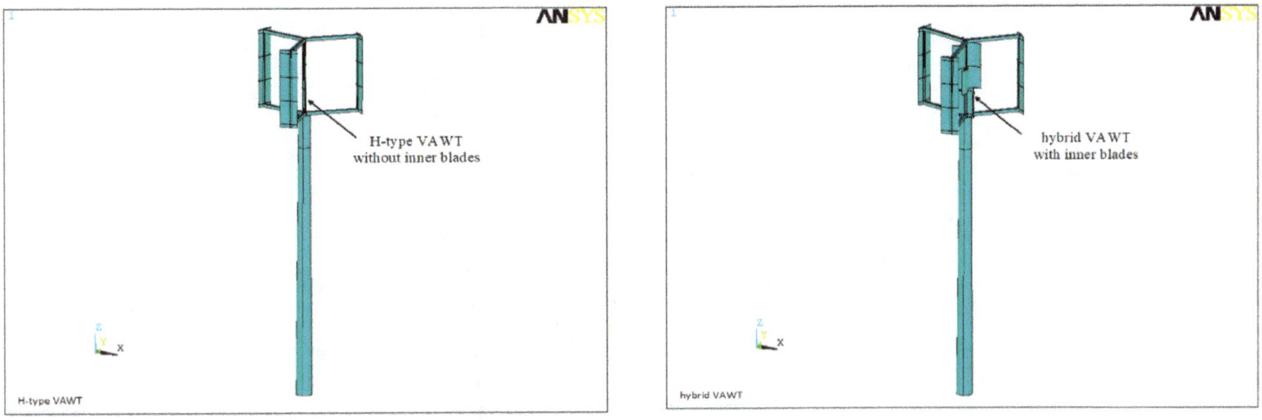

(**a**) (**b**)

Figure 1. Shapes of the (**a**) H-type and (**b**) hybrid VAWTs, where the hybrid VAWT has inner blades and the H-type VAWT does not have them.

(**a**) (**b**)

Figure 2. Details of blades of the (**a**) H-type and (**b**) hybrid VAWTs.

3. Free Vibration Analysis of the Two VAWTs

3.1. SDOF Model

Consider an undamped continuous VAWT system that includes a cantilevered mast with the mass per unit length $m(x)$, where x is the spatial coordinate, the height L, and a turbine with the concentrated mass $m_{turbine}$ at its top. Due to the inner blades in the hybrid VAWT, the mass of the hybrid VAWT is larger than that of the H-type VAWT. In this work, the inner blades are assumed to be made of aluminum alloy; the mass of the inner blades of the hybrid VAWT is 10.5 kg, which is 1.49% of its total mass $m_{turbine} = 704.9$ kg. Therefore, the mass difference between the H-type and hybrid VAWTs is small. The H-type and hybrid VAWTs can be simplified as a cantilever beam with a concentrated mass at its top, as shown in Figure 3.

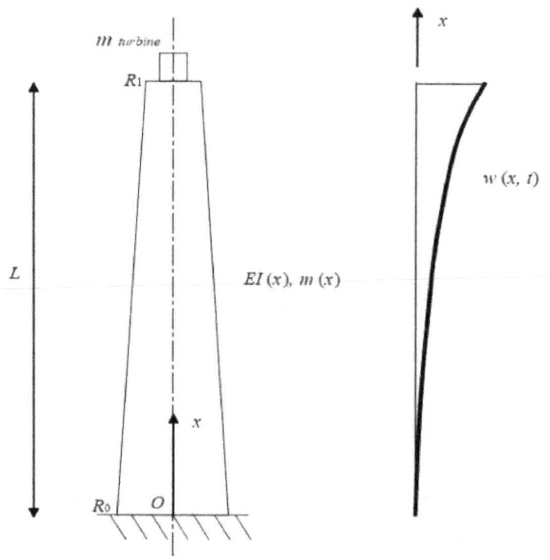

Figure 3. Continuous VAWT system that is modeled as a cantilever beam with a concentrated mass at its top.

The taper mast mass can be expressed as $m(x) = m_0\left(1 - \frac{Kx}{L}\right)$, where m_0 is the mass per unit length of the mast at the bottom, and K is the slope coefficient that is given by $K = (R_0 - R_1)/R_0$, where R_0 and R_1 are radii of bottom and top cross-sections of the mast, respectively. The governing equation of the continuous VAWT system is

$$\frac{\partial^2}{\partial x^2}\left[EI(x)\frac{\partial w^2(x,t)}{\partial x^2}\right] + m(x)\frac{\partial w^2(x,t)}{\partial t^2} = 0, \tag{1}$$

where E and $I(x)$ are Young's modulus and the area moment of inertia of a cross-section of the mast, respectively, and $w(x,\ t)$ is the transverse displacement of the system at the position x at time t. Assume that the cross-section of the mast is annular with a uniform thickness h. Consider $h = 5\ \text{mm}$, which is small compared with the diameter of the cross-section of the mast; the area moment of inertia of the cross-section of the mast is

$$I(x) = I_0 \cdot \left(1 - K\frac{x}{L}\right)^3, \tag{2}$$

where $I_0 = \pi \cdot h \cdot R_0^3$. Boundary conditions of the system are

$$\begin{gathered} w(0,t) = 0,\ w'(0,t) = 0,\ w''(L,t) = 0, \\ EI(L)w'''(L,t) = m_{turbine}\frac{\partial w^2(L,t)}{\partial t^2}. \end{gathered} \tag{3}$$

where a prime denotes partial differentiation with respect to x. By assuming that $w(x,\ t)$ is a product of a generalized coordinate $z(t)$ of a SDOF model of the continuous VAWT system and a shape function $\psi(x)$:

$$w(x,t) = \psi(x) \cdot z(t), \tag{4}$$

the assumed modes method [33] is used to derive the free vibration equation of the SDOF model of the continuous system. Substituting Equation (4) into Equation (1) yields

$$\frac{\partial^2}{\partial x^2}\left[EI(x)\frac{\partial^2}{\partial x^2}(\psi(x)z(t))\right] + m(x)\frac{\partial^2}{\partial t^2}[\psi(x)z(t)] = 0. \tag{5}$$

Multiplying Equation (5) by $\psi(x)$ and integrating the resulting equation over $[0, L]$ yield

$$z(t)\int_0^L \psi(x)\frac{\partial^2}{\partial x^2}\left[\left(EI(x)\frac{\partial^2}{\partial x^2}\psi(x)\right)\right]dx + \ddot{z}(t)\int_0^L \psi(x)m(x)\psi(x)dx = 0. \tag{6}$$

Integrating the integral in the first term in Equation (6) by parts and using the first three equations in Equation (3) yield

$$\begin{aligned}
&\int_0^L \psi(x)\frac{\partial^2}{\partial x^2}[(EI\ (x)\frac{\partial^2}{\partial x^2}\psi(x)\Big)]dx \\
&= \psi(x)\frac{\partial}{\partial x}\left[\left(EI(x)\frac{\partial^2}{\partial x^2}\psi(x)\right)\right]\Bigg|_0^L - \int_0^L \frac{\partial\psi(x)}{\partial x}\frac{\partial}{\partial x}\left[EI(x)\frac{\partial^2\psi(x)}{\partial x^2}\right]dx \\
&= \psi(x)\frac{\partial}{\partial x}\left[\left(EI(x)\frac{\partial^2}{\partial x^2}\psi(x)\right)\right]\Bigg|_0^L - \frac{\partial\psi(x)}{\partial x}\left[EI(x)\frac{\partial^2\psi(x)}{\partial x^2}\right]\Bigg|_0^L + \int_0^L EI(x)[\psi''(x)]^2dx \\
&= \psi(L)EI(L)\frac{\partial^3\psi(L)}{\partial x^3} + \int_0^L EI(x)[\psi''(x)]^2dx.
\end{aligned} \tag{7}$$

Substituting Equation (7) into Equation (6) yields

$$z(t)\left[\psi(L)EI(L)\frac{\partial^3\psi(L)}{\partial x^3} + \int_0^L EI(x)[\psi''(x)]^2dx\right] + \ddot{z}(t)\int_0^L \psi(x)m(x)\psi(x)dx = 0. \tag{8}$$

Based on the fourth equation in Equation (3), one has

$$\ddot{z}(t)\left[\int_0^L m(x)\psi^2(x)dx + m_{turbine}\psi(L)\right] + z(t)\int_0^L EI(x)[\psi''(x)]^2dx = 0. \tag{9}$$

The free vibration equation of the SDOF model is

$$\widetilde{m}\cdot\ddot{z} + \widetilde{k}\cdot z = 0, \tag{10}$$

where

$$\widetilde{m} = \int_0^L m(x)[\psi(x)]^2dx + m_{turbine}\psi(L) \tag{11}$$

is the equivalent mass of the SDOF model of the continuous VAWT system, and

$$\widetilde{k} = \int_0^L EI(x)[\psi''(x)]^2dx \tag{12}$$

is the equivalent stiffness of the SDOF model. In order to select an accurate shape function $\psi(x)$, one assumes that the shape function $\psi(x)$ is in direct proportion to the static deflection of a uniform cantilevered mast with the same length and any flexible rigidity EI, subjected to a unit lateral concentrated force at its top: $u(x) = \left(3Lx^2 - x^3\right)/6EI$ [34]. Hence, one assumes that

$$\psi(x) = u(x)/z_1, \tag{13}$$

where z_1 is the maximum displacement of the uniform mast $u(L)$, which yields

$$\psi(x) = \frac{3}{2}\left(\frac{x}{L}\right)^2 - \frac{1}{2}\left(\frac{x}{L}\right)^3. \tag{14}$$

If one considers the mast tapering from the 200 mm diameter at the bottom to the 150 mm diameter at the top, the slope coefficient is $K = 0.25$ and the equivalent mass of the mast can be calculated from Equation (11):

$$\widetilde{m} = 0.188 \cdot m_0 \cdot L + m_{turbine}. \quad (15)$$

Substituting Equations (2) and (14) into Equation (12) yields

$$\widetilde{k} = 2.491 \cdot \frac{EI_0}{L^3}. \quad (16)$$

The natural frequency of the SDOF model of the continuous VAWT system is given by

$$\omega_n = \sqrt{\frac{\widetilde{k}}{\widetilde{m}}} = \sqrt{\frac{2.491 \cdot EI_0}{(0.188 \cdot m_0 \cdot L + m_{turbine}) \cdot L^3}}. \quad (17)$$

3.2. MDOF Model

As shown in Figure 4a,b, the H-type and hybrid VAWTs mainly include three blades, support spokes, the turbine shaft, the generator, and bearings [35] that are small and not shown there. Blades are made of carbon fiber material or aluminum alloy, depending on their required strength and cost. The blades connect to the turbine shaft with several support spokes and bearings. The blades, support spokes, and the turbine shaft are rotating parts of either VAWT. The generator includes a rotor that is a rotating part of either VAWT and a stator that is a non-rotating part, and the mast is a non-rotating part of either VAWT. The rotating and non-rotating parts of either VAWT are connected by bearings.

The turbines are complex mechanical systems that need to be modeled to study their effects on the VAWTs. Use of a simplified MDOF turbine model with a few DOFs can be adequate to describe their low vibration modes. Due to low rotation speeds of the two VAWTs, rotational effects of their rotating parts are not considered in this work; rotational effects can be considered for high rotation speeds [36].

Table 1 summarizes mass and stiffness contributions of each part in the turbines to the simplified MDOF turbine model in vibration analysis. The mass of the upper support spokes and half of the mass of all the blades and the turbine shaft, which are denoted by m_1, are gathered at the top node 1 in Figure 4c. Similarly, the mass of the lower support spokes and half of the mass of all the blades and the turbine shaft, which is denoted by m_2, is gathered at node 2 in Figure 4c. The mass of the generator denoted by m_3 is assigned to node 3 in Figure 4c. The stiffness of the turbines comes from the bending stiffness of the turbine shaft; the bending stiffness of beam 1 is that of the turbine shaft between nodes 1 and 2, and the bending stiffness of beam 2 is that of the turbine shaft between nodes 2 and 3. Mass and stiffness values of the simplified MDOF turbine model are listed in Table 1. As indicated earlier, the mass difference between the H-type and hybrid VAWTs is 10.5 kg, which is small, but there is much stiffness difference between the two VAWTs. The MDOF model of the VAWTs is shown in Figure 5. Based on the above discussions, the continuous VAWT system that consists of the taper mast and turbine can be discretized into a number of elements, which can be seen as short beams [28]. There are ten nodes in the MDOF model for the two VAWTs in this study: seven elements for the mast and two elements for the turbines, as shown in Figure 5.

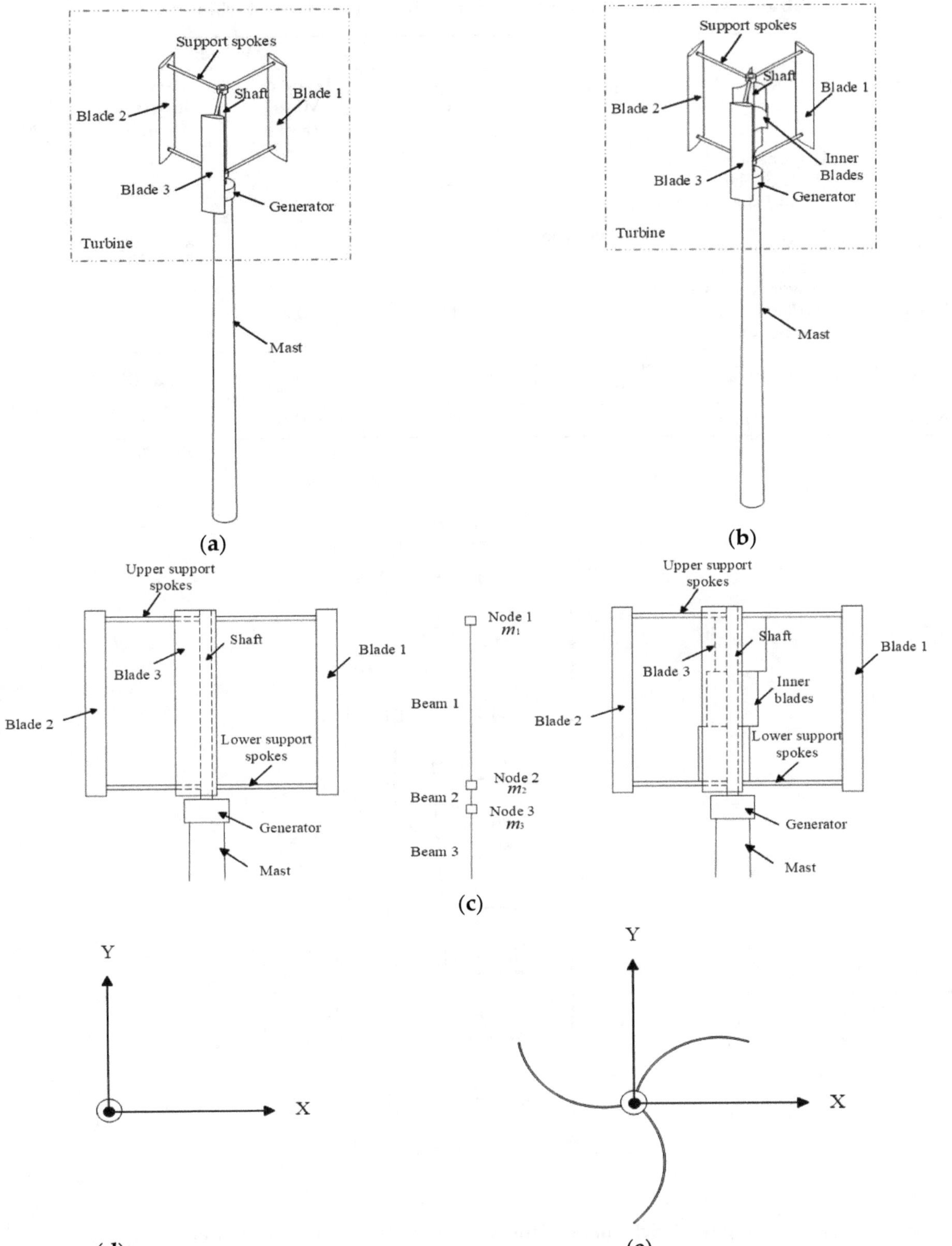

Figure 4. Main parts of the (**a**) H-type and (**b**) hybrid VAWTs and (**c**) their MDOF model; (**d**) an H-type VAWT shaft cross-section and (**e**) hybrid VAWT shaft and inner blade cross-sections.

Table 1. Mass and stiffness values of the simplified MDOF turbine model.

Model Part	Mass	Stiffness	Value			
			H-type Mass (Kg)	**Hybrid Mass (Kg)**	**H-type Stiffness (N·m^2)**	**Hybrid Stiffness (N·m^2)**
Node 1	Upper spokes + 50% of blades and shaft	/	44.9	50.2	0	0
Beam 1	No mass	Shaft between nodes 1 and 2	0	0	2.6×10^4	2.3×10^7
Node 2	Lower spokes + 50% of blades and shaft	/	44.9	50.2	0	0
Beam 2	No mass	Shaft between nodes 2 and 3	0	0	2.6×10^4	2.3×10^7
Node 3	Generator	/	604.5	604.5	0	0

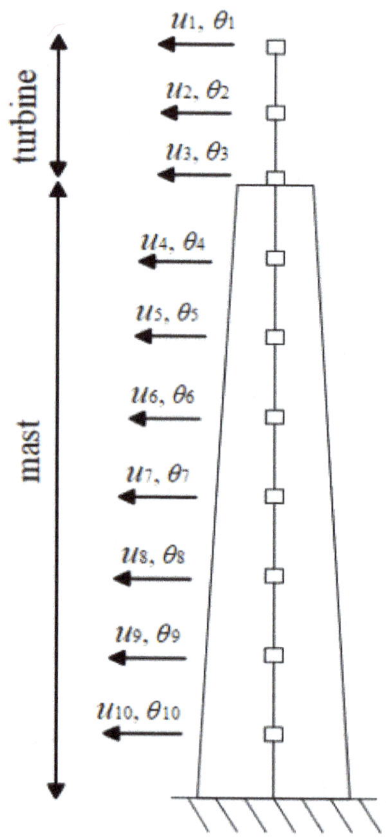

Figure 5. Schematic of the MDOF model of the two VAWTs.

Masses of the two VAWTs are distributed throughout their heights, but are modeled to be concentrated at their ten nodes. The turbines have been modeled by their simplified model shown in Figure 5. There are two DOFs at each node, i.e., the lateral displacement u and rotation angle θ. Stiffness matrices of the VAWTs can be established by applying unit displacements and rotations to their DOFs. For convenience, a simpler model that has only three nodes is shown in Figure 6, and the associated stiffness matrix is calculated below.

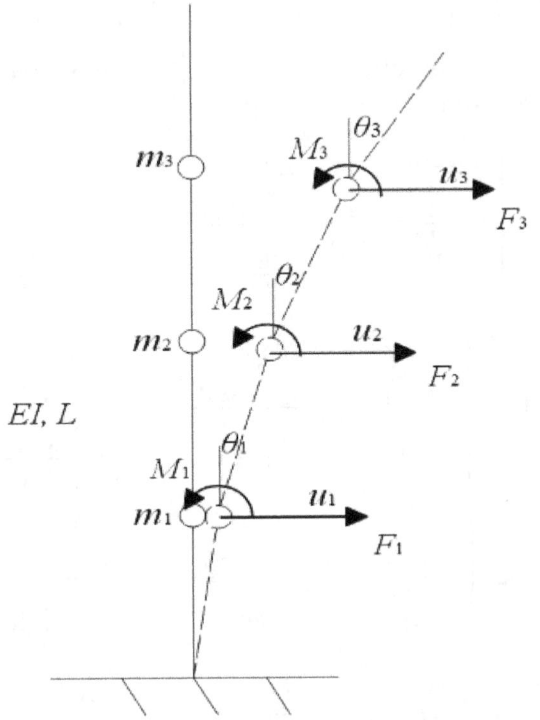

Figure 6. Simpler MDOF model of the VAWTs with only three nodes.

By applying unit displacements and rotations, force and moment balance equations for all the elements are established:

$$F_1 = \frac{12E_1I_1}{L_1^3}u_1 - \frac{12E_1I_1}{L_1^3}u_2 - \frac{6E_1I_1}{L_1^2}\theta_1 - \frac{6E_1I_1}{L_1^2}\theta_2, \tag{18}$$

$$F_2 = -\frac{12E_1I_1}{L_1^3}u_1 + \left(\frac{12E_1I_1}{L_1^3} + \frac{12E_2I_2}{L_2^3}\right)u_2 - \frac{12E_2I_2}{L_2^3}u_3 + \frac{6E_1I_1}{L_1^2}\theta_1 + \left(\frac{6E_1I_1}{L_1^2} - \frac{6E_2I_2}{L_2^2}\right)\theta_2 - \frac{6E_2I_2}{L_2^2}\theta_3, \tag{19}$$

$$F_3 = -\frac{12E_2I_2}{L_2^3}u_2 + \left(\frac{12E_2I_2}{L_2^3} + \frac{12E_1I_1}{L_3^3}\right)u_3 + \frac{6E_2I_2}{L_2^3}\theta_2 + \left(\frac{6E_2I_2}{L_2^2} - \frac{6E_3I_3}{L_3^2}\right)\theta_3, \tag{20}$$

$$M_1 = -\frac{6E_1I_1}{L_1^2}u_1 + \frac{6E_1I_1}{L_1^2}u_2 + \frac{4E_1I_1}{L_1}\theta_2 + \frac{2E_1I_1}{L_1}\theta_3, \tag{21}$$

$$M_2 = -\frac{6E_1I_1}{L_1^2}u_1 + \left(\frac{6E_1I_1}{L_1^2} - \frac{6E_2I_2}{L_2^2}\right)u_2 + \frac{6E_2I_2}{L_2^2}u_3 + \frac{2E_1I_1}{L_1}\theta_1 + \left(\frac{4E_1I_1}{L_1} + \frac{4E_2I_2}{L_2}\right)\theta_2 + \frac{2E_2I_2}{L_2}\theta_3, \tag{22}$$

$$M_3 = -\frac{E_2I_2}{L_2^3}u_1 + \left(\frac{6E_2I_2}{L_2^2} - \frac{6E_3I_3}{L_3^2}\right)u_2 + \frac{2E_2I_2}{L_2}\theta_1 + \left(\frac{4E_2I_2}{L_2} + \frac{4E_3I_3}{L_3}\right)\theta_3, \tag{23}$$

where E_i $(i = 1, 2, 3)$ are Young's moduli of all the elements, I_i $(i = 1, 2, 3)$ are area moments of inertia of cross-sections of all the elements, L_i $(i = 1, 2, 3)$ are lengths of all the elements, u_i $(i = 1, 2, 3)$ are displacements of all the nodes, θ_i $(i = 1, 2, 3)$ are rotation angles of all the nodes, F_i $(i = 1, 2, 3)$ are internal forces at all the nodes, and M_i $(i = 1, 2, 3)$ are internal moments at all the nodes. Equations (18)–(23) can be written in the matrix form:

$$\begin{bmatrix} \mathbf{f} \\ \mathbf{m} \end{bmatrix} = \begin{bmatrix} \mathbf{a} & \mathbf{b} \\ \mathbf{c} & \mathbf{d} \end{bmatrix} \begin{bmatrix} \mathbf{u} \\ \boldsymbol{\theta} \end{bmatrix} \tag{24}$$

where $\mathbf{f} = \begin{bmatrix} F_1 & F_2 & F_3 \end{bmatrix}^T$ is the force vector, $\mathbf{m} = \begin{bmatrix} M_1 & M_2 & M_3 \end{bmatrix}^T$ is the moment vector, $\mathbf{u} = \begin{bmatrix} u_1 & u_2 & u_3 \end{bmatrix}^T$ is the displacement vector, $\boldsymbol{\theta} = \begin{bmatrix} \theta_1 & \theta_2 & \theta_3 \end{bmatrix}^T$ is the rotating angle vector, and four block matrices in the stiffness matrix are given by

$$\mathbf{a} = \begin{bmatrix} \frac{12E_1I_1}{L_1^3} & -\frac{12E_1I_1}{L_1^3} & 0 \\ -\frac{12E_1I_1}{L_1^3} & \frac{12E_1I_1}{L_1^3} + \frac{12E_2I_2}{L_2^3} & -\frac{12E_2I_2}{L_2^3} \\ -\frac{12E_2I_2}{L_2^3} & -\frac{12E_2I_2}{L_2^3} & \frac{12E_2I_2}{L_2^3} + \frac{12E_1I_1}{L_3^3} \end{bmatrix} \tag{25}$$

$$\mathbf{b} = \begin{bmatrix} -\frac{6E_1I_1}{L_1^2} & -\frac{6E_1I_1}{L_1^2} & 0 \\ \frac{6E_1I_1}{L_1^2} & \frac{6E_1I_1}{L_1^2} - \frac{6E_2I_2}{L_2^2} & -\frac{6E_2I_2}{L_2^2} \\ 0 & \frac{6E_2I_2}{L_2^3} & \frac{6E_2I_2}{L_2^2} - \frac{6E_3I_3}{L_3^2} \end{bmatrix} \tag{26}$$

$$\mathbf{c} = \begin{bmatrix} -\frac{6E_1I_1}{L_1^2} & \frac{6E_1I_1}{L_1^2} & 0 \\ -\frac{6E_1I_1}{L_1^2} & \frac{6E_1I_1}{L_1^2} - \frac{6E_2I_2}{L_2^2} & \frac{6E_2I_2}{L_2^2} \\ 0 & -\frac{E_2I_2}{L_2^3} & \frac{6E_2I_2}{L_2^2} - \frac{6E_3I_3}{L_3^2} \end{bmatrix} \tag{27}$$

$$\mathbf{d} = \begin{bmatrix} \frac{4E_1I_1}{L_1} & \frac{2E_1I_1}{L_1} & 0 \\ \frac{2E_1I_1}{L_1} & \frac{4E_1I_1}{L_1} + \frac{4E_2I_2}{L_2} & \frac{2E_2I_2}{L_2} \\ 0 & \frac{2E_2I_2}{L_2} & \frac{4E_2I_2}{L_2} + \frac{4E_3I_3}{L_3} \end{bmatrix} \tag{28}$$

For the MDOF model of the two VAWTs with 10 nodes and 20 DOFs, one can use the same method to derive the force-displacement relationship and the stiffness matrix of the MDOF model [28]:

$$\begin{bmatrix} \mathbf{F} \\ \mathbf{M} \end{bmatrix} = \begin{bmatrix} \mathbf{A} & \mathbf{B} \\ \mathbf{C} & \mathbf{D} \end{bmatrix} \begin{bmatrix} \mathbf{U} \\ \boldsymbol{\Theta} \end{bmatrix}, \tag{29}$$

where

$$\mathbf{F} = \begin{bmatrix} F_1 \\ F_2 \\ \vdots \\ F_9 \\ F_{10} \end{bmatrix}, \ \mathbf{M} = \begin{bmatrix} M_1 \\ M_2 \\ \vdots \\ M_9 \\ M_{10} \end{bmatrix}, \ \mathbf{U} = \begin{bmatrix} u_1 \\ u_2 \\ \vdots \\ u_9 \\ u_{10} \end{bmatrix}, \ \boldsymbol{\Theta} = \begin{bmatrix} \theta_1 \\ \theta_2 \\ \vdots \\ \theta_9 \\ \theta_{10} \end{bmatrix}, \tag{30}$$

$$\mathbf{A} = \begin{bmatrix} \frac{12E_1I_1}{L_1^3} & -\frac{12E_1I_1}{L_1^3} & 0 & \cdots & 0 \\ -\frac{12E_1I_1}{L_1^3} & \frac{12E_1I_1}{L_1^3} + \frac{12E_2I_2}{L_2^3} & -\frac{12E_2I_2}{L_2^3} & 0 & \vdots \\ 0 & -\frac{12E_2I_2}{L_2^3} & \frac{12E_2I_2}{L_2^3} + \frac{12E_1I_1}{L_3^3} & \ddots & 0 \\ \vdots & 0 & \ddots & \ddots & -\frac{12E_9I_9}{L_9^3} \\ 0 & \cdots & 0 & -\frac{12E_9I_9}{L_9^3} & \frac{12E_9I_9}{L_9^3} + \frac{12E_{10}I_{10}}{L_{10}^3} \end{bmatrix}, \tag{31}$$

$$\mathbf{B}=\begin{bmatrix} -\frac{6E_1I_1}{L_1^2} & -\frac{6E_1I_1}{L_1^2} & 0 & \cdots & 0 \\ \frac{6E_1I_1}{L_1^2} & \frac{6E_1I_1}{L_1^2}-\frac{6E_2I_2}{L_2^2} & -\frac{6E_2I_2}{L_2^2} & 0 & \vdots \\ 0 & \frac{6E_2I_2}{L_2^3} & \frac{6E_2I_2}{L_2^2}-\frac{6E_3I_3}{L_3^2} & \ddots & 0 \\ \vdots & 0 & \ddots & \ddots & -\frac{6E_9I_9}{L_9^2} \\ 0 & \cdots & 0 & \frac{6E_9I_9}{L_9^2} & \frac{6E_9I_9}{L_9^2}-\frac{6E_{10}I_{10}}{L_{10}^2} \end{bmatrix}, \tag{32}$$

$$\mathbf{C}=\begin{bmatrix} -\frac{6E_1I_1}{L_1^2} & \frac{6E_1I_1}{L_1^2} & 0 & \cdots & 0 \\ -\frac{6E_1I_1}{L_1^2} & \frac{6E_1I_1}{L_1^2}-\frac{6E_2I_2}{L_2^2} & \frac{6E_2I_2}{L_2^2} & 0 & \vdots \\ 0 & -\frac{E_2I_2}{L_2^3} & \frac{6E_2I_2}{L_2^2}-\frac{6E_3I_3}{L_3^2} & \ddots & 0 \\ \vdots & 0 & \ddots & \ddots & \frac{6E_9I_9}{L_9^2} \\ 0 & \cdots & 0 & -\frac{6E_9I_9}{L_9^2} & \frac{6E_9I_9}{L_9^2}-\frac{6E_{10}I_{10}}{L_{10}^2} \end{bmatrix}, \tag{33}$$

$$\mathbf{D}=\begin{bmatrix} \frac{4E_1I_1}{L_1} & \frac{2E_1I_1}{L_1} & 0 & \cdots & 0 \\ \frac{2E_1I_1}{L_1} & \frac{4E_1I_1}{L_1}+\frac{4E_2I_2}{L_2} & \frac{2E_2I_2}{L_2} & 0 & \vdots \\ 0 & \frac{2E_2I_2}{L_2} & \frac{4E_2I_2}{L_2}+\frac{4E_3I_3}{L_3} & \ddots & 0 \\ \vdots & 0 & \ddots & \ddots & \frac{2E_9I_9}{L_9} \\ 0 & \cdots & 0 & \frac{2E_9I_9}{L_9} & \frac{4E_9I_9}{L_9}+\frac{4E_{10}I_{10}}{L_{10}} \end{bmatrix}, \tag{34}$$

in which E_i $(i=1, 2, \cdots, 10)$ are Young's moduli of all the elements, I_i $(i=1, 2, \cdots, 10)$ are moments of inertia of all the elements, L_i $(i=1, 2, \cdots, 10)$ are length of all the elements, u_i $(i=1, 2, \cdots, 10)$ are displacements of all the nodes, θ_i $(i=1, 2, \cdots, 10)$ are rotating angles of all the nodes, F_i $(i=1, 2, \cdots, 10)$ are forces at all the nodes, and M_i $(i=1, 2, \cdots, 10)$ are moments at all the nodes.

Based on Ref. [33], undamped free vibration equations of the MDOF can be written as

$$\mathbf{m}\cdot\ddot{u}(t)+\mathbf{k}\cdot u(t)=\mathbf{0}, \tag{35}$$

where $\mathbf{m}$ is the mass matrix that is given by

$$m=\begin{bmatrix} m_1 & 0 & \cdots & 0 & 0 & 0 & \cdots & 0 \\ 0 & m_2 & \ddots & 0 & 0 & 0 & \ddots & 0 \\ \vdots & & \ddots & \vdots & \vdots & \ddots & \ddots & \vdots \\ 0 & 0 & \cdots & m_{10} & 0 & 0 & \cdots & 0 \\ 0 & 0 & \cdots & 0 & m_1 & 0 & \cdots & 0 \\ 0 & 0 & \ddots & 0 & 0 & m_2 & \ddots & 0 \\ \vdots & \ddots & \ddots & \vdots & \vdots & \ddots & \ddots & \vdots \\ 0 & 0 & \cdots & 0 & 0 & 0 & \cdots & m_{10} \end{bmatrix}, \tag{36}$$

in which $m_1, m_2, \cdots, m_{10}$ are masses of the 10 elements in the MDOF model, $\mathbf{k}$ is the stiffness matrix whose block matrices are given earlier, and $u(t)$ is the displacement vector.

Natural frequencies and mode shapes of the MDOF model of the two VAWTs for different mast heights can be obtained by solving the eigenvalue problem associated with Equation (35). Natural frequencies are obtained from

$$\left|k-\omega_n^2 m\right|=0, \tag{37}$$

where ω_n is the nth natural frequency.

3.3. BEM

By employing the commercial software ANSYS, the BEM of the two VAWTs is built to analyze their free vibrations. Pipe-16 beam elements are used in the BEM, and there are 20 elements and 21 nodes in the model. Each node has two DOFs, which are the horizontal displacement and rotation angle. The fixed constraint of the BEM is located at the bottom node, which is consistent with the SDOF and MDOF models of a cantilever beam. Hence, the BEM has 42 DOFs.

A beam model of the mast has two types of modeling methods. In the first method, the mass of the mast is not considered in beam elements, but is assigned at nodes of the beam model, as shown in Figure 7a, while in the second method, the mass of the mast is assigned on beam elements, as shown in Figure 7b; the second method is used in this work. To obtain natural frequencies and mode shapes of the two VAWTs, seven methods can be used for modal analysis in ANSYS: Block Lanczos, Subspace, Power dynamics, Reduced, Unsymmetric, Damped, and Quick response damped. The method of Block Lanczos has fast convergence and is suitable for symmetrical mass and stiffness matrices of large structures, which is adopted in this work. The results are shown in Section 5.3.1.

(a) (b)

Figure 7. Schematic of the two BEMs of the two VAWTs: (**a**) the mass of the mast is assigned at nodes of beam elements; and (**b**) the mass of the mast is assigned on beam elements.

3.4. SEM

In order to compare with the BEM, SEMs of the two VAWTs are built. The element type of the SEMs is solid-45 in ANSYS. The SEMs of the two VAWTs are built with two steps. The first step is to build 3D geometric models of the two VAWTs. The second step is to mesh the geometric models with the solid-45 element type. A 3D geometric model consists of the mast, blades, support spokes, and the shaft. The mast is a tapered cylindrical tube. The shape of the blades is defined by the airfoil NACA0015 cross-section. The blades are supported by two straight spokes, which are connected to the central shaft. The shaft combines the mast and blades to become a whole VAWT 3D geometric model. The generator is modeled as a mass point located at the joint of the mast and shaft. After the 3D geometric model is built, the second step is taken by meshing the 3D geometric models into the SEMs.

For the H-type VAWT, there are 131,546 nodes and 112,000 elements in its SEM. For the hybrid VAWT, there are 145,154 nodes and 118,615 elements in its SEM. Differences between the two SEMs for the H-type and hybrid VAWTs are at their inner blades. The hybrid VAWT model has inner blades and has 13,608 more nodes and 6165 more elements than the H-type VAWT model. Two types of materials, i.e., aluminum alloy and steel, are used in the SEMs to model different parts of the two VAWTs. For the blades and support spokes in the two VAWTs, aluminum alloy is used because it has a light density and high strength. Hence, it is suitable to make blades since blades should be light and easily rotated by wind. The density of this material is 2.7×10^3 kg/m^3 and its Young's modulus is 7.2×10^{10} N/m^2. Steel is used for the mast and shaft in the two VAWTs because it has a very high strength, which can endure bending moments produced by wind loads. The density of steel is 7.85×10^3 kg/m^3 and its Young's modulus is 2.06×10^{11} N/m^2. For the SEMs of the two VAWTs, a fixed constraint is applied at the bottom of the mast, which can simulate connections of the VAWTs with the ground. The method of Block Lanczos is adopted to obtain natural frequencies and mode shapes of the two VAWTs with modal analysis in ANSYS. The results are shown in Section 5.4.1.

4. Forced Vibration Analysis of the Two VAWTs

4.1. Harmonic Response Analysis

Based on the BEM of the H-type and hybrid VAWTs, their harmonic response analysis can be analyzed by ANSYS. Harmonic response analysis is to apply a series of sinusoidal excitations with different frequencies to a linear system and analyze its steady-state responses under sinusoidal excitations without considering transient responses when the excitations are initially added to the system. Amplitude- and phase-frequency characteristics of the two VAWTs at different nodes can be obtained from their harmonic response analysis based on their BEM, as shown in Section 5.3.2.

4.2. Forced Vibration Analysis

Based on the BEM and SEMs of the H-type and hybrid VAWTs, their forced transient responses can be obtained from ANSYS. Forced transient responses of the two VAWTs consist of two parts. One part is transient responses, which decay as time increases, and the other part is steady-state responses. Forced transient response analysis is called transient dynamic analysis in ANSYS. Three methods can be used for transient dynamic analysis: full method, reduced method, and modal superposition method. Use of complete system matrices to calculate transient response is called the full method, which is used here. This method has no matrix reduction and is the most powerful of the three methods, allowing various nonlinear characteristics to be accounted for. Forces applied to the BEM and SEMs of the two VAWTs are aerodynamic loads obtained from CFD simulation in Section 4.3. To obtain forced transient responses of the two VAWTs, the time-varying aerodynamics loads from CFD simulation are applied to the BEM and SEMs. By choosing the Newmark algorithm in ANSYS and setting structural damping factors $\alpha = 0.001$ and $\beta = 0.0005$, displacement responses of the two VAWTs are obtained.

4.3. Time-Varying Aerodynamic Loads on the Two VAWTs

Time-varying aerodynamic loads on the H-type and hybrid VAWTs are calculated from high-fidelity unsteady Reynolds averaged Navier-Stokes (URANS) CFD simulation [7], which calculates aerodynamic loads on the two VAWTs. Aerodynamic loading is important to the performance of the VAWTs. The URANS model is simulated using the ANSYS®Fluent®15.0 high-performance computing software. For pressure-velocity coupling, the semi-implicit method for pressure-linked equations (SIMPLE) algorithm with a second-order upwind spatial discretization scheme is utilized to solve the URANS equations. A second-order implicit transient formulation is selected to ensure the accuracy of time integration. The inlet turbulence viscosity ratio is set to 10 to give a reasonable estimation of the free stream turbulence. Additionally, for each inner iteration, the residual convergence criterion is fixed at 10^{-5} for better converged results. The Reynolds number based on the diameter of the two VAWTs is approximately 4×10^5. The numerical setup is summarized in Table 2 and more details can be found in Ref. [7]. In the simulation, the angular velocity of the two VAWTs is set to 16.625 rad/s when the wind speed is 5 m/s at the steady operational status and the corresponding TSR is 2.3. Time histories of aerodynamic loads on three outer blades (i.e., blades 1, 2, and 3) of the two VAWTs and inner blades of the hybrid VAWT are shown in Figure 8. Note that the free stream direction is denoted as the X-direction, and the direction perpendicular to the free stream and within the cross-sectional plane of a turbine is denoted as the Y-direction. The Z-direction is defined from X- and Y-directions by the right-hand rule. The aerodynamic loads at any specific time are calculated via integration of both the pressure and viscous forces over blade surfaces. In Figure 8, blue solid lines, black dashed lines, and red dotted lines represent forces on the three outer blades, green two-dot chain lines represent forces on the three inner blades of the hybrid VAWT, and navy-blue solid lines represent resultant forces on all the blades of the two VAWTs. All these forces periodically vary with time. Instantaneous vorticity fields around the H-type VAWT from CFD simulation are shown in Figure 9a and those around the hybrid VAWT are shown in Figure 9b.

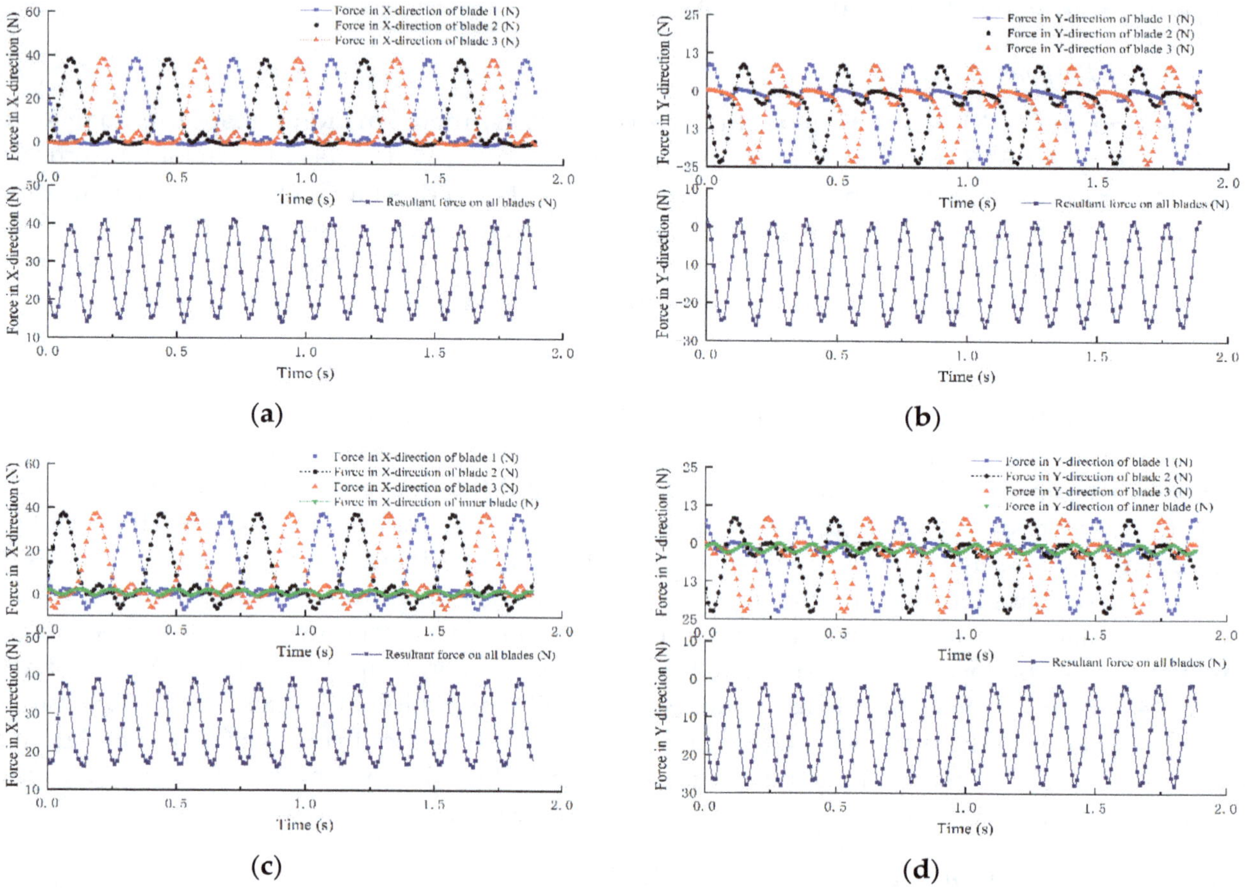

Figure 8. Aerodynamic loads acting on different VAWT blades in X- and Y-directions: (**a**) H-type VAWT in the X-direction; (**b**) H-type VAWT in the Y-direction; (**c**) hybrid VAWT in the X-direction; and (**d**) hybrid VAWT in the Y-direction.

Table 2. Summary of parameters in CFD simulation.

Input Variable	Value/Setting
Turbulence Model	Spalart-Allmaras model
Pressure-velocity coupling	SIMPLE
Spatial discretization scheme	Second-order upwind
Time integration	Second-order implicit
Inlet turbulence viscosity ratio	10
Convergence criterion for residuals	10^{-5}
Reynolds number	4.9×10^5

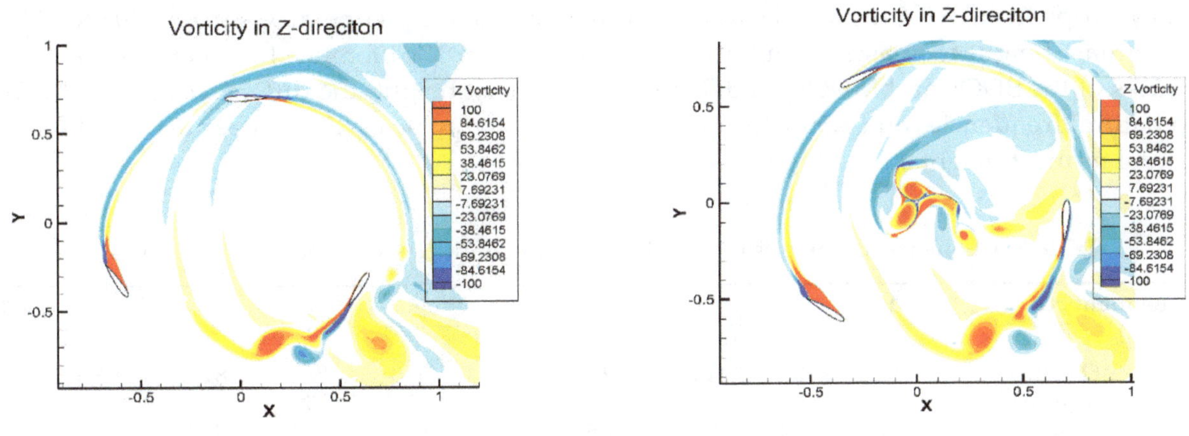

(a) (b)

Figure 9. Instantaneous vorticity fields around the (**a**) H-type and (**b**) hybrid VAWTs.

5. Numerical Results and Discussions

5.1. SDOF Results

Mast parameters of the SDOF model of the H-type and hybrid VAWTs are listed in Table 3. For different mast parameters with the mast height varying from 1 to 4 m and the mast diameter varying from 66.7 to 200 mm, natural frequencies of the SDOF model of the H-type and hybrid VAWTs are obtained from Equation (17), as shown in Table 4, where cross-section dimensions, lengths, and densities of each element of the mast of the two VAWTs are specified. The natural frequency decreases with the mast height for both the H-type and hybrid VAWTs.

Table 3. Mast parameters of the SDOF model of the two VAWTs.

Parameter	Value
Mast height	$L = 4$ m
Outer diameter of the cross-section	$D_0 = 0.2$ m
Mast thickness	$h = 0.005$ m
Inertia of the cross-section	$I_0 = 1.57 \times 10^{-5}$ m^4
Mast mass	$m_0 = 23.6$ Kg/m
Turbine mass for the H-type VAWT	$m_{turbine} = 694.4$ Kg
Turbine mass for the hybrid VAWT	$m_{turbine} = 704.9$ Kg
Young's modulus	$E = 2.06 \times 10^{11}$ N/m^2
Material density of steel	7.85×10^3 kg/m^3

Table 4. Natural frequencies of the SDOF model of the H-type and hybrid VAWTs with different mast parameters.

Case	L (m)	D_0 (mm)	D_1 (mm)	h (mm)	K	I_0 (mm^4)	H-type ω_n (Hz)	Hybrid ω_n (Hz)
Case 1	1	88.9	66.7	5	0.25	1.38×10^6	3.87	3.98
Case 2	2	133	100	5	0.25	4.62×10^6	2.96	2.99
Case 3	4	200	150	5	0.25	1.57×10^7	2.07	2.05

5.2. MDOF Results

Geometric properties of the MDOF model of the H-type and hybrid VAWTs with different mast heights are listed in Table 5. Table 6 shows the first three natural frequencies of the H-type VAWT with different mast heights. Natural frequencies of the MDOF model of the VAWTs decrease with the mast height. More specifically, the first natural frequency drops from 3.73 to 2.07 Hz, the second natural frequency drops from 399.4 to 50.0 Hz, and the third natural frequency drops from 1004 to 150.5 Hz when the mast height increases from 1 to 4 m. There is 3.67% difference between the first natural frequencies of the SDOF and MDOF models for the 1 m-mast height; the differences are 3.12% for the second natural frequencies and 3.50% for the third natural frequencies. Similar results are obtained for the 2 and 4 m-mast heights.

Table 5. Geometric properties of the MDOF model of the H-type and hybrid VAWTs.

	Element No.	1	2	3	4	5	6	7	8	9	10
1 m	D (mm)	40	40	66.7	69.8	72.9	76.0	79.1	82.2	85.3	88.9
	h (mm)	20	20	5	5	5	5	5	5	5	5
	L (mm)	900	100	125	125	125	125	125	125	125	125
	H-type m (kg)	44.9	44.9	604.5	1.3	1.3	1.3	1.3	1.3	1.3	1.3
	Hybrid m (kg)	50.2	50.2	604.5	1.3	1.3	1.3	1.3	1.3	1.3	1.3
2 m	D (mm)	40	40	100	105	109	114	119	123	128	133
	h (mm)	20	20	5	5	5	5	5	5	5	5
	L (mm)	900	100	250	250	250	250	250	250	250	250
	H-type m (kg)	44.9	44.9	604.5	3.9	3.9	3.9	3.9	3.9	3.9	3.9
	Hybrid m (kg)	50.2	50.2	604.5	3.9	3.9	3.9	3.9	3.9	3.9	3.9
4 m	D (mm)	40	40	150	157	164	171	178	185	192	200
	h (mm)	20	20	5	5	5	5	5	5	5	5
	L (mm)	900	100	500	500	500	500	500	500	500	500
	H-type m (kg)	44.9	44.9	604.5	12.0	12.0	12.0	12.0	12.0	12.0	12.0
	Hybrid m (kg)	50.2	50.2	604.5	12.0	12.0	12.0	12.0	12.0	12.0	12.0

Table 6. First three natural frequencies of the SDOF and MDOF models of the H-type VAWT.

Mast Height	Mode	SDOF	MDOF	Difference
1 m	1	3.87	3.73	3.67%
	2		399.4	
	3		1004	
2 m	1	2.96	2.87	3.12%
	2		145.1	
	3		399.2	
4 m	1	2.07	2.00	3.50%
	2		50.0	
	3		150.5	

Table 7 shows the first three natural frequencies of the MDOF model of the hybrid VAWT. Natural frequencies of the MDOF model decrease with the mast height. More specifically, the first natural frequency drops from 3.98 to 2.05 Hz, the second natural frequency drops from 412.4 to 69 Hz, and the third natural frequency drops from 1105 to 166.8 Hz when the mast height increases from 1 to 4 m. There is 3.25% difference between the first natural frequencies of the SDOF and MDOF models for the 1 m-mast height; the differences are 3.53% for the second natural frequencies and 3.02% for the third natural frequencies. Similar results are obtained for the 2 and 4 m-mast heights.

Table 7. First natural frequencies of the SDOF and MDOF models of the hybrid VAWT.

Mast Height	Mode	SDOF	MDOF	Difference
1 m	1	3.98	3.85	3.25%
	2		412.4	
	3		1105	
2 m	1	2.99	2.89	3.53%
	2		170.5	
	3		417.9	
4 m	1	2.05	1.99	3.02%
	2		69.0	
	3		166.8	

5.3. BEM Results

5.3.1. Free Vibration

According to Section 3.3 above, the BEM has 21 nodes and 42 DOFs for the H-type and hybrid VAWTs; this model has more DOFs than the SDOF and MDOF models, which have one DOF for the SDOF model and ten nodes and 20 DOFs for the MDOF model. With numbers of nodes and DOFs increasing, natural frequencies can be more accurately calculated from the BEM. Table 8 shows comparison between the first three natural frequencies of the MDOF model and BEM of the H-type VAWT with different mast heights. For the 1 m-mast height, there is 3.54% difference between the first natural frequencies of the SDOF and MDOF models; the differences are 3.22% for the second natural frequencies and 0.49% for the third natural frequencies. Similar results are obtained for the 2 and 4 m-mast heights. Figure 10a–c show the first, second, and third mode shapes of the BEM of the 4 m-mast H-type VAWT, respectively.

Table 8. First three natural frequencies of the MDOF model and BEM of the H-type VAWT with different mast heights.

Mast Height	Mode	MDOF	BEM	Difference
1 m	1	3.73	3.60	3.54%
	2	399.4	386.9	3.22%
	3	1004	999.4	0.49%
2 m	1	2.87	2.78	3.09%
	2	145.1	141.0	2.88%
	3	399.2	398.7	0.13%
4 m	1	2.00	1.94	3.09%
	2	50.0	48.6	2.88%
	3	150.5	150.3	0.13%

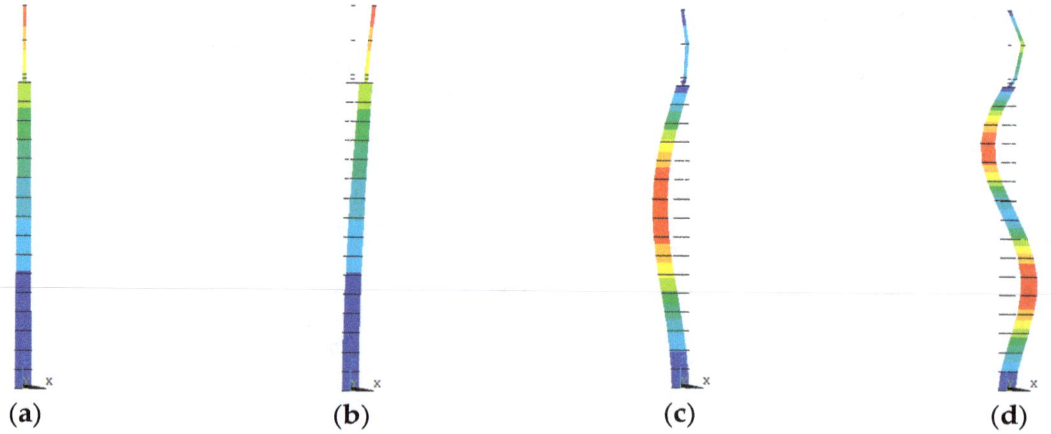

Figure 10. (**a**) BEM and (**b–d**) the first three mode shapes of the 4 m-mast H-type VAWT in the X-Y plane: (**b**) the first mode shape with the natural frequency of 1.941 Hz, (**c**) the second mode shape with the natural frequency of 48.556 Hz, and (**d**) the third mode shape with the natural frequency of 150.54 Hz.

Table 9 shows comparison between the first three natural frequencies of the MDOF model and BEM of the hybrid VAWT with different mast heights. For the 1 m-mast height, there is 3.16% difference between the first natural frequencies of the SDOF and MDOF; the differences are 0.22% for the second natural frequencies and 1.92% for the third natural frequencies. Similar results are obtained for the 2 and 4 m-mast heights. The first three mode shapes of the hybrid VAWT are similar to those of the H-type VAWT shown in Figure 10.

Table 9. First three natural frequencies of the MDOF model and BEM of the hybrid VAWT with different mast heights.

Mast Height	Mode	MDOF	BEM	Difference
1 m	1	3.85	3.73	3.16%
	2	412.4	411.5	0.22%
	3	1105	1084	1.92%
2 m	1	2.89	2.80	3.11%
	2	170.5	164.1	3.92%
	3	417.9	417.4	0.12%
4 m	1	1.99	1.93	3.11%
	2	69.0	66.4	3.92%
	3	166.8	167.0	0.12%

5.3.2. Forced Vibration Results

Amplitude-frequency responses of the H-type and hybrid VAWTs can be obtained from harmonic response analysis in ANSYS. Figure 11 shows amplitude-frequency responses of the 1 m-mast H-type and hybrid VAWTs in X- and Y-directions. From Figure 11a,b, one can find that the first three resonance frequencies of the H-type VAWT are 3.6, 386.9, and 999.4 Hz, respectively, in both X- and Y-directions. The first three resonance frequencies of the hybrid VAWT in both X- and Y-directions are 3.73, 411.5, and 1084 Hz, respectively. The first resonance frequency of the H-type VAWT is 3.6% lower than that of the hybrid VAWT for the 1 m-mast height, since inner blades of the hybrid VAWT provide more overall stiffness effect than overall mass effect compared with the H-type VAWT. The first resonance frequency is the main one, and one can conclude that for the 1 m-mast height, the hybrid VAWT has better performance than the hybrid VAWT in terms of increasing its resonance frequencies. Since the mast is made of steel and inner blades are made of aluminum alloy, increase in the stiffness and natural frequencies of the hybrid VAWT is limited.

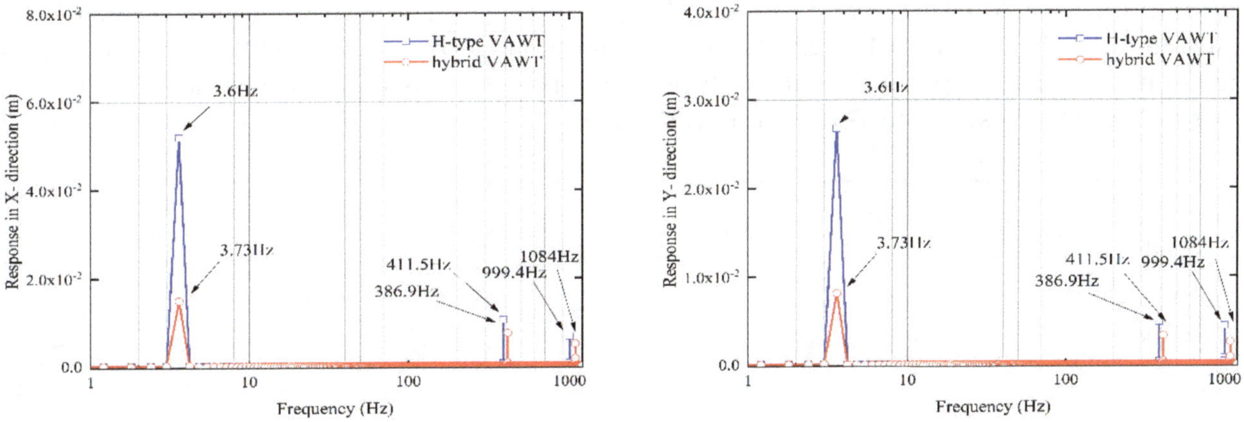

(a) (b)

Figure 11. (**a**) Amplitude-frequency responses of the 1 m-mast H-type and hybrid VAWTs in (**a**) X and (**b**) Y directions.

Figure 12 shows amplitude-frequency responses of the 4 m-mast H-type and hybrid VAWTs in X- and Y-directions. From Figure 12a,b, one can find that the first three resonance frequencies of the H-type VAWT are 1.9, 48.9, and 150.4 Hz, respectively, in both X- and Y-directions. The first three resonance frequencies of the hybrid VAWT in both X- and Y-directions are 1.9, 66.6, and 167.3 Hz, respectively. The first resonance frequency of the H-type VAWT is the same as that of the hybrid VAWT, and the second and third resonance frequencies of the H-type VAWT are smaller than those of the hybrid VAWT for the 4 m-mast height.

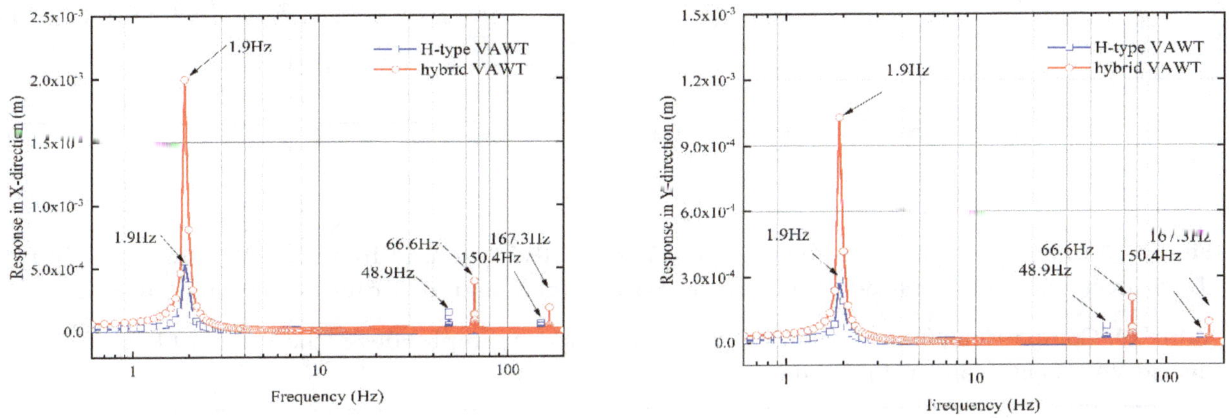

(a) (b)

Figure 12. (**a**) Amplitude-frequency responses of the 4 m-mast H-type and hybrid VAWTs in (**a**) X and (**b**) Y directions.

Figure 13 shows forced transient responses of the 1 m-mast H-type and hybrid VAWTs in X- and Y-directions under aerodynamic loads shown in Figure 9, where Figure 13a,b shows global forced transient responses of the two VAWTs in the X- and Y-directions, and Figure 13c,d shows forced transient responses of the two VAWTs in the X- and Y-directions at the steady state, in which points with the maximum absolute response amplitudes are connected together to better show the steady-state response amplitudes. The maximum steady-state response amplitudes of the two VAWTs are shown in Table 10. As can be seen in Table 10, the maximum steady-state response amplitudes of the H-type and hybrid VAWTs in the X-direction are 1.17×10^{-6} m and 9.22×10^{-7} m, respectively. The maximum steady-state response amplitude of the hybrid VAWT is 26.9% smaller than that of the H-type VAWT in the X-direction. The maximum steady-state response amplitudes of the H-type and hybrid VAWTs in the Y-direction are 1.30×10^{-6} m and 1.06×10^{-6} m, respectively. The maximum steady-state response amplitude of the hybrid VAWT is 22.6% smaller than that of the H-type VAWT in the Y-direction. Steady-state response amplitudes of the H-type and hybrid VAWTs in the X-direction are 11.1% and 14.9% smaller than those in the Y-direction, respectively.

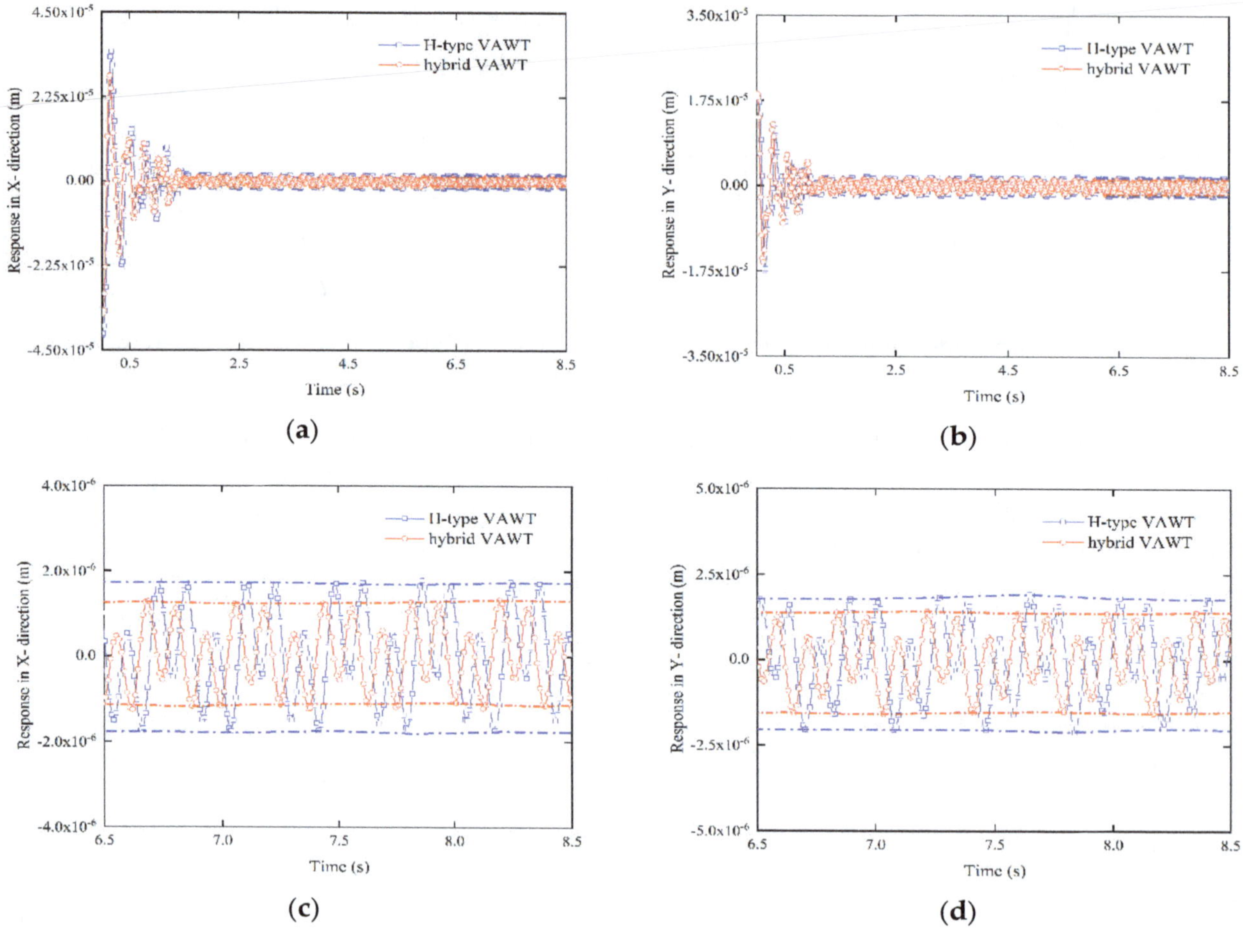

Figure 13. Forced transient responses of the 1 m-mast H-type and hybrid VAWTs in (**a**) X- and (**b**) Y-directions, and their steady-state responses in the (**c**) X- and (**d**) Y-directions from 6.5 to 8.5 s.

Table 10. Comparison of the maximum steady-state response amplitudes of the 1 m-mast H-type and hybrid VAWTs in X- and Y-directions.

Direction	H-type VAWT (m)	Hybrid VAWT (m)	Difference
X-direction	1.17×10^{-6}	9.22×10^{-7}	26.9%
Y-direction	1.30×10^{-6}	1.06×10^{-6}	22.6%
Difference	11.1%	14.9%	

Figure 14 shows forced transient responses of the 4 m-mast H-type and hybrid VAWTs in X- and Y-directions, where Figure 14a,b shows global forced transient responses of the two VAWTs in the X- and Y-directions, and Figure 14c,d shows forced transient responses of the two VAWTs in the X- and Y-directions at the steady state, in which points with the maximum absolute response amplitudes are connected together to better show the maximum steady-state response amplitudes. The maximum steady-state response amplitudes of the two VAWTs are shown in Table 11. As can be seen in Table 11, the maximum steady-state response amplitudes of the H-type and hybrid VAWTs in the X-direction are 9.55×10^{-7} m and 1.19×10^{-6} m, respectively. The maximum steady-state response amplitude of the hybrid VAWT in the X-direction is 24.6% larger than that of the H-type VAWT in the X-direction.

The maximum steady-state response amplitudes of the H-type and hybrid VAWTs in the Y-direction are 1.01×10^{-6} m and 1.33×10^{-6} m, respectively. The maximum steady-state response amplitude of the hybrid VAWT is 31.6% larger than that of the H-type VAWT in the Y-direction. Steady-state response amplitudes of the H-type and hybrid VAWTs in the X-direction are 5.76% and 11.8% smaller than those in the Y-direction, respectively.

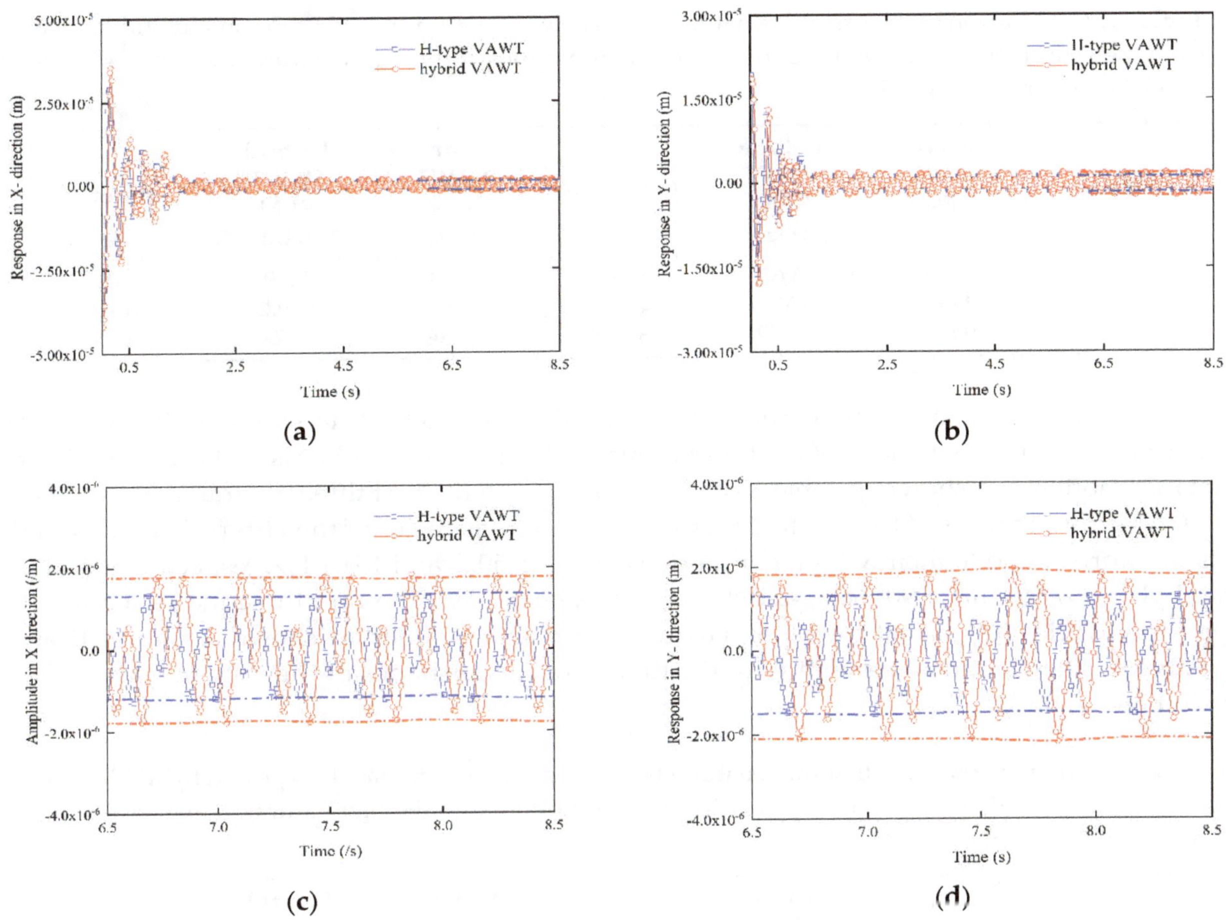

Figure 14. Forced transient responses of the 4 m-mast H-type and hybrid VAWTs in (**a**) X- and (**b**) Y-directions, and their steady-state responses in the (**c**) X- and (**d**) Y-directions from 6.5 to 8.5 s.

Table 11. Comparison of the maximum steady-state response amplitudes of the 4 m-mast H-type and hybrid VAWTs in X- and Y-directions.

Direction	H-type VAWT (m)	Hybrid VAWT (m)	Difference
X-direction	9.55×10^{-7}	1.19×10^{-6}	24.6%
Y-direction	1.01×10^{-6}	1.33×10^{-6}	31.6%
Difference	5.76%	11.8%	

5.4. SEM Results

5.4.1. Free Vibration Results

SEM models of the 1 m-mast H-type and hybrid VAWTs are built to simulate the two VAWTs. The first three natural frequencies of the two VAWTs from their SEMs that correspond to their X-Y planar motions are shown in Table 12 and they are compared with the first three natural frequencies of the two VAWTs from their BEM there. The first three natural frequencies of the H-type VAWT from its SEM that correspond to its X-Y planar motion are 3.65, 402.5, and 1032 Hz, respectively, and those of the hybrid VAWT are 3.79, 429.2, and 1121 Hz, respectively. Since the first natural frequency of the H-type VAWT is lower than that of the H-type VAWT, the hybrid VAWT has better performance than the H-type VAWT. Differences between the first three natural frequencies of the H-type VAWT from its BEM and SEM are 1.39, 4.03, and 3.26%, respectively. Differences between the first three natural frequencies of the hybrid VAWT from its BEM and SEM are 1.61, 4.30, and 3.41%, respectively.

Table 12. Comparison of the first three natural frequencies of the 1 m-mast H-type and hybrid VAWTs from their SEMs that correspond to their X-Y planar motions and the first natural frequencies of the two VAWTs from their BEM.

Mode	H-type VAWT BEM (Hz)	H-type VAWT SEM (Hz)	Difference	Hybrid VAWT BEM (Hz)	Hybrid VAWT SEM (Hz)	Difference
1	3.60	3.65	1.39%	3.73	3.79	1.61%
2	386.9	402.5	4.03%	411.5	429.2	4.30%
3	999.4	1032	3.26%	1084	1121	3.41%

SEMs of the 4 m-mast H-type and hybrid VAWTs are also built to simulate the two VAWTs. The first three natural frequencies of the two 4 m-mast VAWTs from their SEMs that correspond to their X-Y planar motions are shown in Table 13 and compared with the first three natural frequencies of the two VAWTs from their BEM there. The first three natural frequencies of the H-type VAWT from their SEMs that correspond to their X-Y planar motions are 1.92, 50.2, and 154.1 Hz, respectively, and those of the hybrid VAWT are 1.90, 66.9, and 170.4 Hz, respectively. The first three mode shapes of the H-type and hybrid VAWTs that correspond to their X-Y planar motions are shown in Figures 15 and 16, respectively. They are in agreement with the first three mode shapes of the H-type VAWT from its BEM in Figure 10b–d.

Table 13. Comparison of the first three natural frequencies of the 4 m-mast H-type and hybrid VAWTs from their SEMs that correspond to their X-Y planar motions and the first natural frequencies of the two VAWTs from their BEM.

Mode	H-type VAWT BEM (Hz)	H-type VAWT SEM (Hz)	Difference	Hybrid VAWT BEM (Hz)	Hybrid VAWT SEM (Hz)	Difference
1	1.94	1.92	1.04%	1.93	1.90	1.58%
2	48.6	50.2	3.31%	64.4	66.9	3.97%
3	150.3	154.1	2.53%	167.0	170.4	2.02%

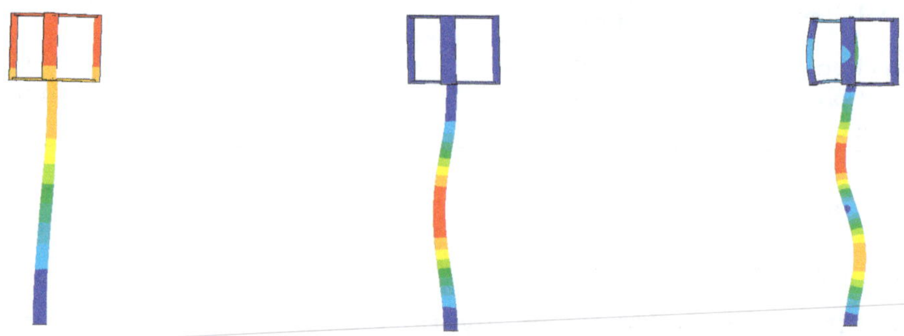

(a) (b) (c)

Figure 15. First three mode shapes of the H-type VAWT from its SEM that correspond to its X-Y planar motion: (**a**) the first mode shape with the natural frequency 1.92 Hz; (**b**) the second mode shape with the natural frequency 50.2 Hz; and (**c**) the third mode shape with the natural frequency 154.1 Hz.

(a) (b) (c)

Figure 16. First three mode shapes of the hybrid VAWT from its SEM that correspond to its X-Y planar motion: (**a**) the first mode shape with the natural frequency 1.90 Hz; (**b**) the second mode shape with the natural frequency 66.9 Hz; and (**c**) the third mode shape with the natural frequency 170.4 Hz.

5.4.2. Forced vibration results

Figure 17 shows forced transient responses of the 1 m-mast H-type and hybrid VAWTs in X- and Y-directions, where Figure 17a,b shows global forced transient responses of the two VAWTs in the X- and Y-directions, and Figure 17c,d shows forced transient responses of the two VAWTs in the X- and Y-directions at the steady state, in which points with the maximum absolute response amplitudes are connected together to better show the steady-state response amplitudes. The maximum steady-state response amplitudes of the two VAWTs are shown in Table 14. As can be seen in Table 14, the maximum steady-state response amplitudes of the H-type and hybrid VAWTs in the X-direction are 1.15×10^{-6} m and 9.29×10^{-7} m, respectively. The maximum steady-state response amplitude of the hybrid VAWT is 23.8% smaller than that of the H-type VAWT in the X-direction. The maximum steady-state response amplitudes of the H-type and hybrid VAWTs in the Y-direction are 1.35×10^{-6} m and 1.12×10^{-6} m, respectively. The maximum steady-state response amplitude of the hybrid VAWT is 20.5% smaller than that of the H-type VAWT in the Y-direction. Steady-state response amplitudes of the H-type and hybrid VAWTs in the X-direction are 17.4% and 20.1% smaller than those in the Y-direction, respectively.

Figure 18 shows forced transient responses of the 4 m-mast H-type and hybrid VAWTs in X- and Y-directions, where Figure 17a,b shows global forced transient responses of the two VAWTs in the X- and Y-directions, and Figure 18c,d shows forced transient responses of the two VAWTs in the X- and Y-directions at the steady state, where points with the maximum absolute response amplitudes are connected together to better show the steady-state response amplitudes. The maximum steady-state response amplitudes of the two VAWTs are shown in Table 15. As can be seen in Table 15, the maximum steady-state response amplitudes of the H-type and hybrid VAWTs in the X-direction are 1.01×10^{-6} m and 1.24×10^{-6} m, respectively. The maximum steady-state response amplitude of the hybrid VAWT is 22.8% larger than that of the H-type VAWT in the X-direction. The maximum steady-state response amplitudes of the H-type VAWT and hybrid VAWT in the Y-direction are 1.15×10^{-6} m and 1.43×10^{-6} m, respectively. The maximum steady-state response amplitude of the hybrid VAWT is 24.3% larger than that of the H-type VAWT in the Y-direction. Steady-state response amplitudes of the H-type and hybrid VAWTs in the X-direction are 13.8% and 15.3% smaller than those in the Y-direction, respectively.

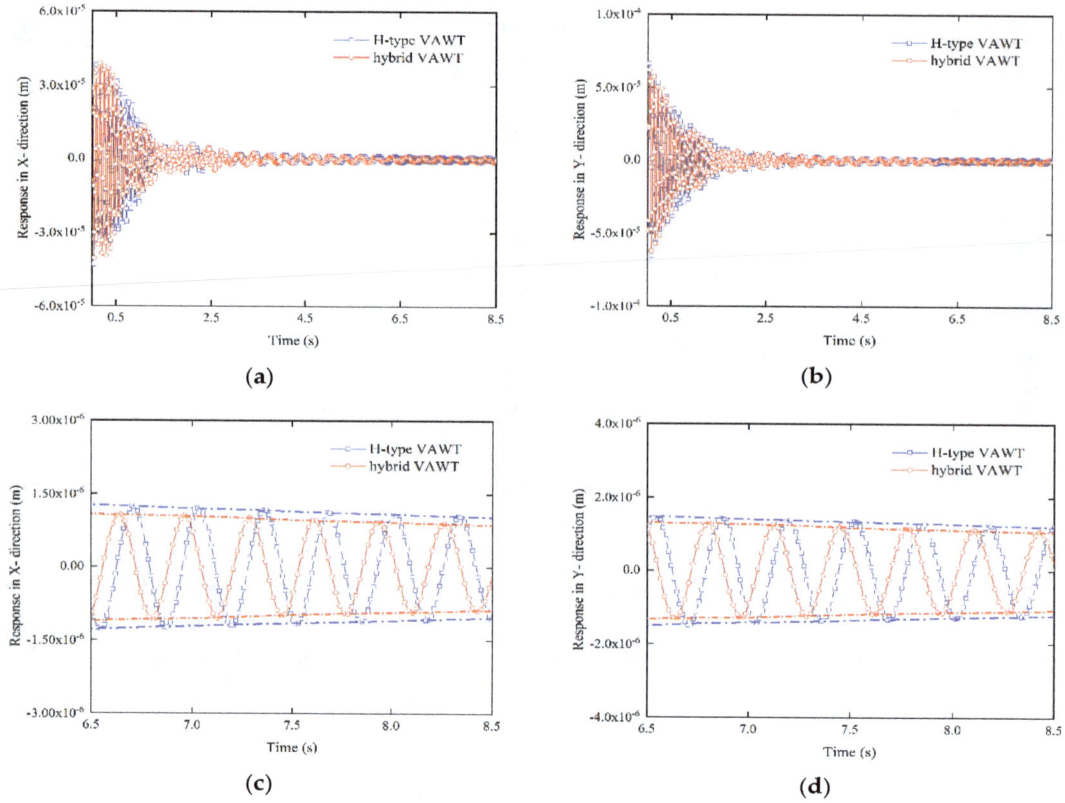

Figure 17. Forced transient responses of the 1 m-mast H-type and hybrid VAWTs in (**a**) X- and (**b**) Y-directions, and their steady-state responses in the (**c**) X- and (d) Y-directions from 6.5 to 8.5 s.

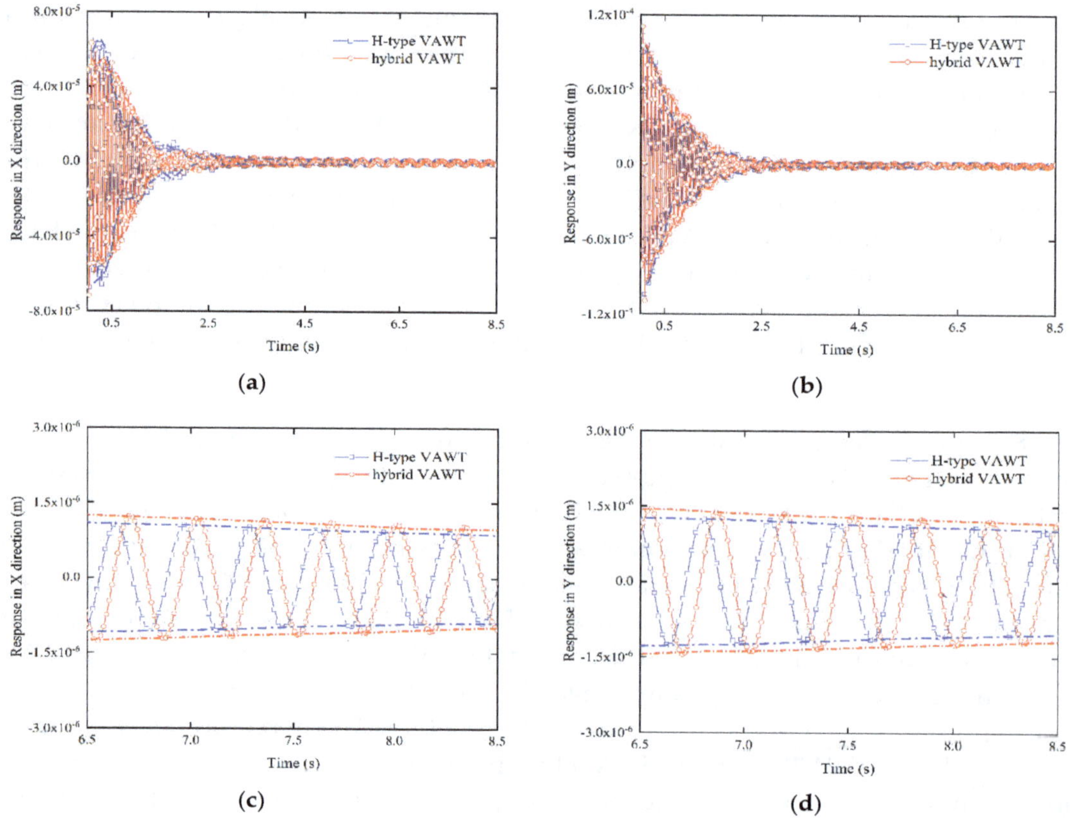

Figure 18. Forced transient responses of the 4 m-mast H-type and hybrid VAWTs in (**a**) X- and (**b**) Y-directions, and their steady-state responses in the (**c**) X- and (**d**) Y-directions from 6.5 to 8.5 s.

Table 14. Comparison of the maximum steady-state response amplitudes of the 1 m-mast H-type and hybrid VAWTs in X- and Y-directions from their SEMs.

Mode	H-type VAWT (m)	Hybrid VAWT (m)	Difference
X-direction	1.15×10^{-6}	9.29×10^{-7}	23.8%
Y-direction	1.35×10^{-6}	1.12×10^{-6}	20.5%
Difference	17.4%	20.1%	

Table 15. Comparison of the maximum steady-state response amplitudes of the 4 m-mast H-type and hybrid VAWTs in X- and Y-directions from their SEMs.

Mode	H-type VAWT (m)	Hybrid VAWT (m)	Difference
X-direction	1.01×10^{-6}	1.24×10^{-6}	22.8%
Y-direction	1.15×10^{-6}	1.43×10^{-6}	24.3%
Difference	13.8%	15.3%	

5.5. *Aerodynamic Excitation Frequency Analysis*

Since aerodynamic loading on the two VAWTs is a periodic excitation, resonance can occur when an excitation frequency is close to some natural frequencies of the two VAWTs. Therefore, excitation frequencies of the aerodynamic loading are determined in this section. Resultant aerodynamic loads acting on all blades of the H-type and hybrid VAWTs in the X-direction and their dominant excitation frequencies are shown in Figure 19, where the fast Fourier transform (FFT) is taken over the aerodynamic loads to find the dominant excitation frequencies. Based on the FFT analysis, dominant excitation frequencies of the resultant aerodynamic loads are 7.9 Hz, 15.9 Hz, and 23.8 Hz. Note that the higher frequencies are about twice or three times of the fundamental frequency, which confirms that the resultant aerodynamic loads have good periodicity. To ensure that aerodynamic excitation does not cause resonance, natural frequencies of the VAWTs should not be close to these excitation frequencies in their structural design. Similarly, resultant aerodynamic loads acting on all blades of the H-type and hybrid VAWTs in the Y-direction and their dominant excitation frequencies are shown in Figure 20, which also confirms that they have good periodicity. Dominant excitation frequencies of the resultant aerodynamic loads on all blades in the Y-direction are the same as those in the X-direction. Note that aerodynamic loads in X- and Y-directions are applied on turbine shafts of the two VAWTs by all their blades. Moments produced by aerodynamic loads in X- and Y-directions along the Z-direction on the two VAWTs are shown in Figure 21. As expected, the aerodynamic moments along the Z-direction have the same dominant excitation frequencies as those of the aerodynamic loads in X- and Y-directions and are hence periodic.

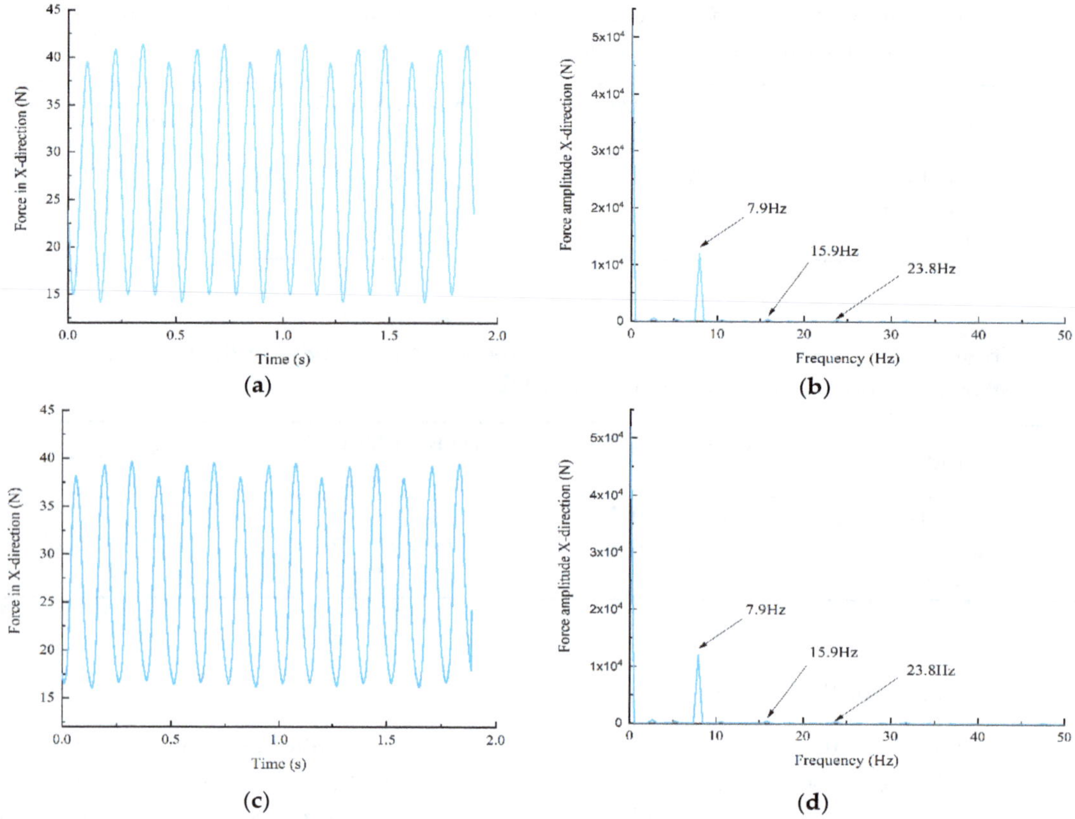

Figure 19. Aerodynamic loads acting on blades of the (**a**) H-type and (**c**) hybrid VAWTs in the X-direction and their dominant excitation frequencies for the (**b**) H-type and (**d**) hybrid VAWTs.

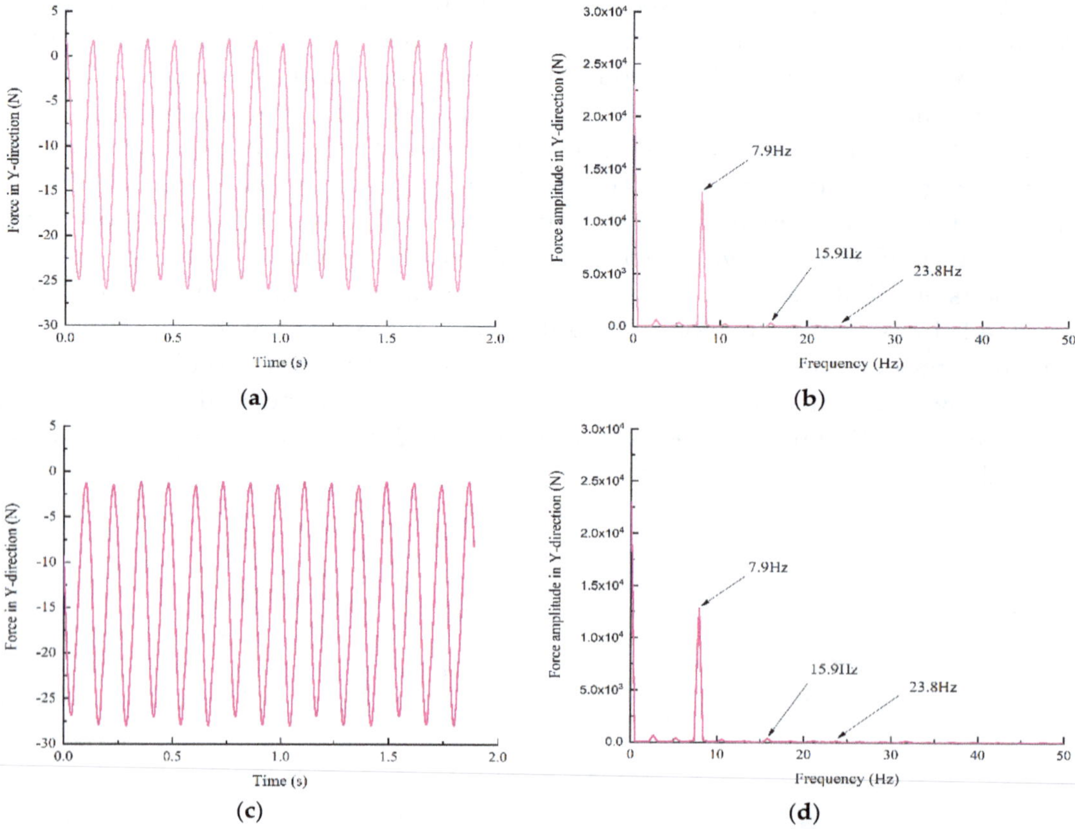

Figure 20. Aerodynamic loads acting on blades of the (**a**) H-type and (**c**) hybrid VAWTs in the Y-direction and their dominant excitation frequencies for the (**b**) H-type and (**d**) hybrid VAWTs.

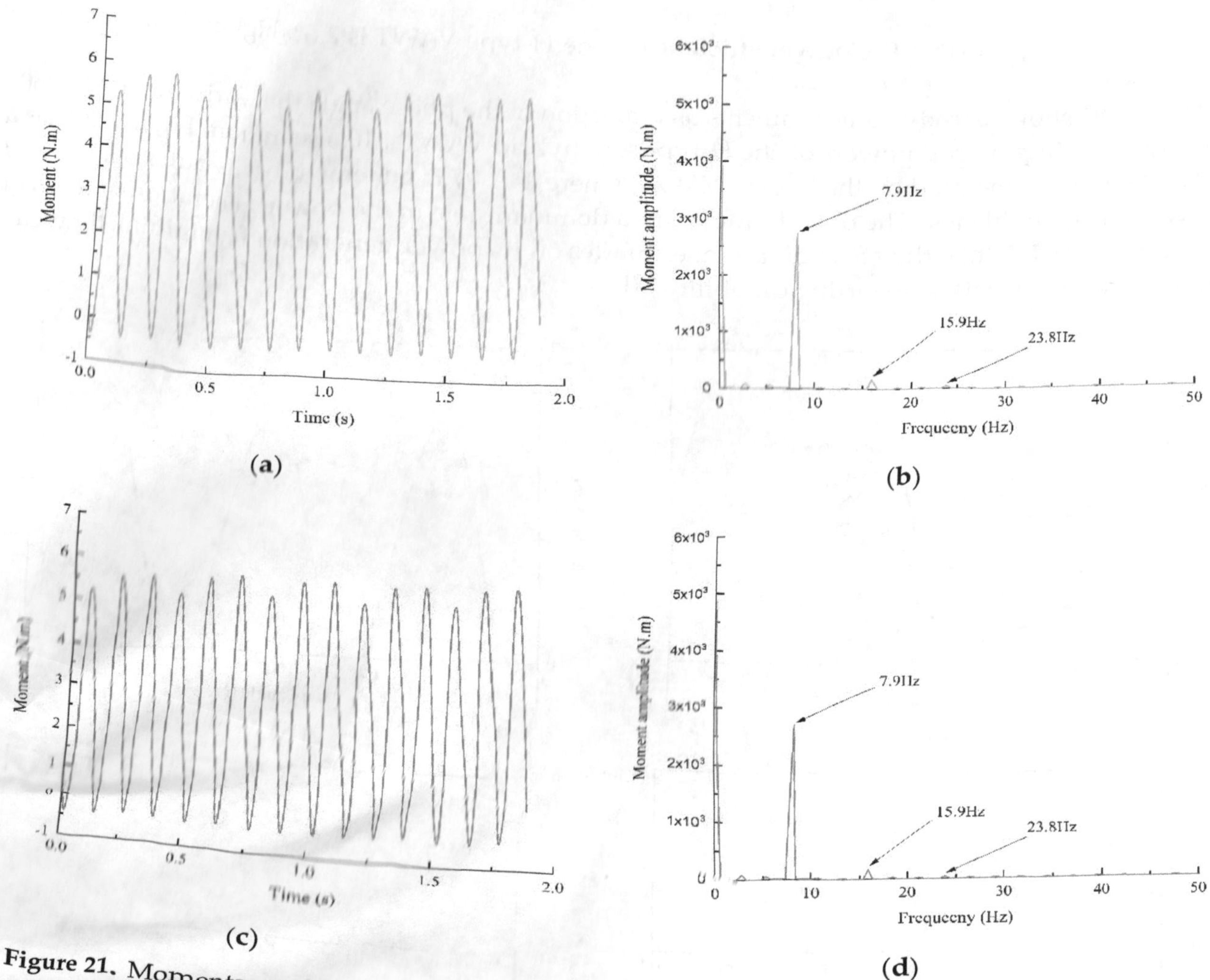

Figure 21. Moments produced by aerodynamic loads in X- and Y-directions along the Z-direction on the (**a**) H-type and (**c**) hybrid VAWTs and their dominant excitation frequencies for the (**b**) H-type and (**d**) hybrid VAWTs.

5.6. Power Efficiency Analysis

According to Ref. [12], the power efficiency or coefficient of the two VAWTs is defined as

$$C_P = \frac{P_t}{P_w}, \tag{38}$$

where P_t is the power output of the VAWTs, and P_w is the power available in incoming wind flow. They are defined as

$$P_t = \omega \frac{1}{nT} \int_0^{nT} M(t) dt, \tag{39}$$

$$P_w = \frac{1}{2} \rho U_\infty^3 2RH, \tag{40}$$

where ω is the angular velocity of the VAWTs, M is the time-dependent moment from all blades of the VAWTs, T is their rotation period, n is the number of periods, ρ is the stream density, U_∞ is the free stream velocity, R is the radius of the VAWTs, and H is the height of their blades. In this work, $\omega = 16.625 \text{rad/s}$, $R = 0.7$ m, $\rho = 1.225$ kg/m^3, $H = 1$ m, $U_\infty = 5$ m/s, and the TSR is 2.3; ...27.1 W, and $P_t = 46.7$ W for the H-type VAWT and $P_t = 43.9$ W for the hybrid VAWT. ...the power coefficient of the H-type VAWT is $C_P = 43.6\%$ and that of the hybrid VAWT is $C_P = 41\%$. The power efficiency of the H-type VAWT is 2.6% higher than that of the hybrid VAWT.

Figure 22 shows aerodynamic moments as a function of the phase angle that varies from 0 to 360° and corresponding output powers of the H-type and hybrid VAWTs. It is seen from Figure 22 that a total of 43.9 W is generated by the hybrid VAWT, where 44.1 W is generated by its outer blades and −0.2 W by its inner blades. The outer blades play a dominant role in the power generation process of the hybrid VAWT. While the effect of the inner blades on its power generation is minimal, they can significantly enhance its self-starting capability [7].

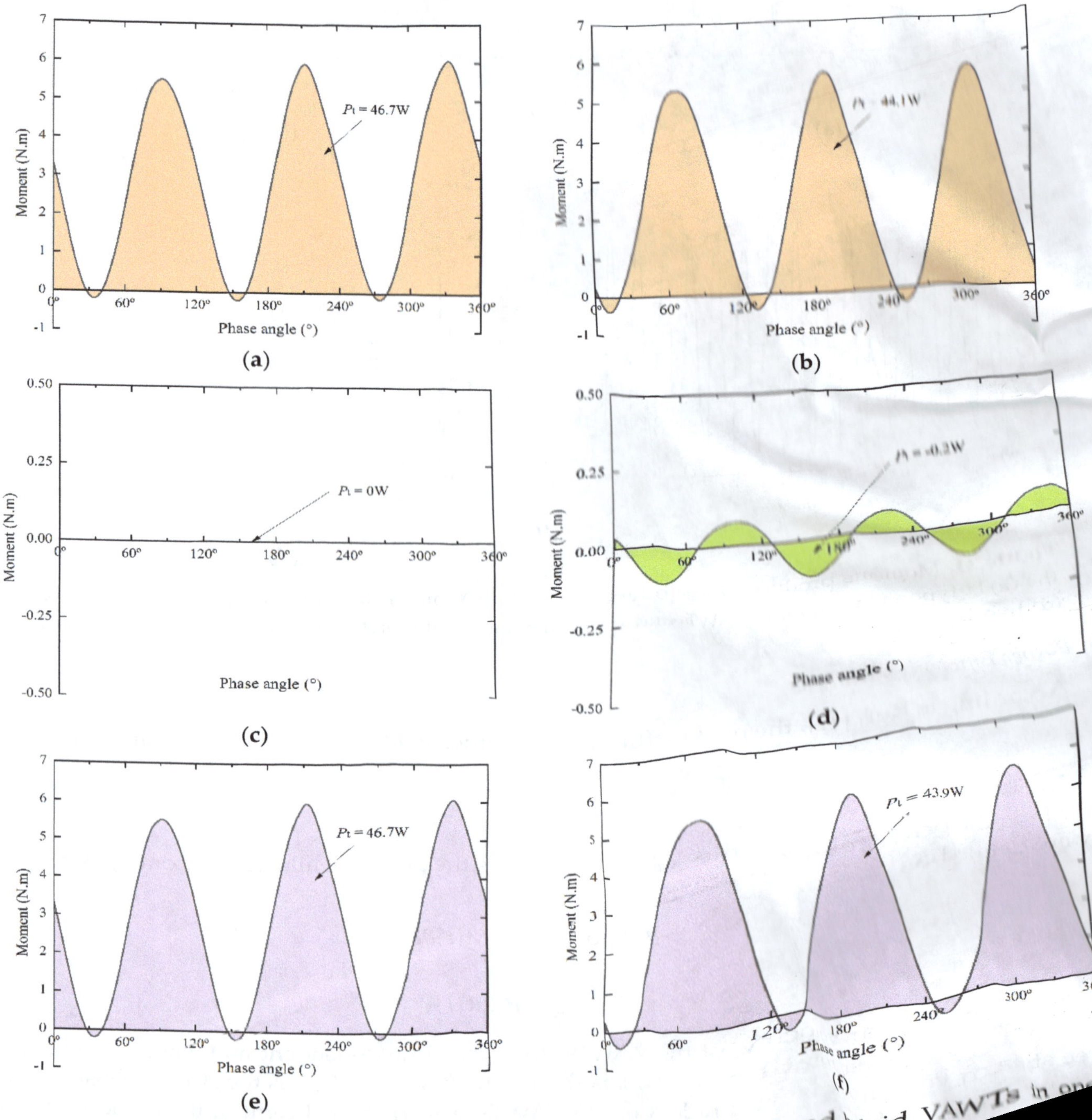

Figure 22. Aerodynamic moments and output powers of the H-type and hybrid VAWTs in one rotation period: (**a**,**b**) output powers from outer blades of the H-type and hybrid VAWTs; (**c**,**d**) output powers from inner blades of the H-type and hybrid VAWTs; powers from both outer and inner blades of the H-type and hy

6. Conclusions

This work studies free and forced vibrations of the H-type and hybrid VAWTs. Based on the above analysis results, some conclusions can be obtained as follows:

(1) Differences between natural frequencies of the SDOF model of the H-type and hybrid VAWTs and the first natural frequencies of their MDOF model range from 3.12% to 3.67% and from 3.02% to 3.53%, respectively, when the mast height varies from 1 to 4 m, which means that the SDOF model is a useful model to analyze the first natural frequencies of the two VAWTs.

(2) Differences between the first three natural frequencies of the MDOF model of the H-type and hybrid VAWTs and those of their BEM range from 0.13% to 3.54% and from 0.12% to 3.92%, respectively, when the mast height varies from 1 to 4 m, which means that the MDOF model is a useful and reasonably simplified model to analyze free vibrations of the two VAWTs.

(3) Differences between the first three natural frequencies of the BEM of the H-type and hybrid VAWTs and corresponding natural frequencies of their SEMs range from 1.39% to 4.03% and from 1.61% to 4.30%, respectively, when the mast height varies from 1 and 4 m, which means that the BEM is a useful and reasonably simplified model to analyze free vibrations of the two VAWTs.

(4) Based on their SEMs, for the 1 m-mast height, the first natural frequency of the H-type VAWT is 3.69% smaller than that of the hybrid VAWT. For the 4m-mast height, the first natural frequency of the H-type VAWT is 1.05% larger than that of the hybrid VAWT. For the 1 m-mast height, steady-state forced response amplitudes of the H-type VAWT are 23.8% and 20.5% larger than those of the hybrid VAWT in X- and Y-directions, respectively. For the 4 m-mast height, steady-state forced response amplitudes of the H-type VAWT are 22.8% and 24.3% smaller than those of the hybrid VAWT in X- and Y-directions, respectively. For the 1 m-mast H-type and hybrid VAWTs, the maximum steady-state response amplitudes in the X-direction are 17.4% and 20.1% smaller than those the Y-direction, respectively. For the 4m-mast H-type and hybrid VAWTs, the maximum steady-state response amplitudes in the X-direction are 13.8% and 15.3% smaller than those in the Y-direction, respectively.

(5) Aerodynamic forces and moments on blades of the two VAWTs are periodic. Outer blades of the hybrid VAWT play a dominant role in its power generation process. While inner blades of the hybrid VAWT have a minimal effect on its power generation, they can significantly enhance its self-starting capability.

Author Contributions: Conceptualization, W.Z. and M.Y.; methodology, W.Z. and X.Z.; software, M.T.; validation, X.Z. and K.L.; formal analysis, X.Z.; investigation, M.T.; resources, G.L.; data curation, K.L. and G.L.; writing—original draft preparation, M.T.; writing—review and editing, W.Z. and M.Y.; visualization, K.L.; supervision, W.Z.; project administration, W.Z.; and funding acquisition, W.Z. and M.Y. All authors have read and agreed to the published version of the manuscript.

References

1. American Wind Energy Association (AWEA). *Wind Powers America Annual Report*; American Wind Energy Association (AWEA): Washington, DC, USA, 2019.
2. Wilburn, D.R. *Wind Energy in the United States and Materials Required for the Land-Based Wind Turbine Industry from 2010 through 2030*; US Department of the Interior: US Geological Survey: Washington, DC, USA, 2011.
3. Bianchini, A.; Balduzzi, F.; Di Rosa, D.; Ferrara, G. On the use of Gurney Flaps for the aerodynamic performance augmentation of Darrieus wind turbines. *Energy Convers. Manag.* **2019**, *184*, 402–415.
4. Han, H.; Liu, L.; Cao, D. Dynamic modeling for rotating composite Timoshenko beam and analysis on its bending-torsion coupled vibration. *Appl. Math. Modell.* **2020**, *78*, 773–791.
5. Castelli, M.R.; Giulia, S.; Ernesto, B. Numerical analysis of the influence of airfoil asymmetry on VAWT performance. *World Acad. Sci. Eng. Technol.* **2012**, *61*, 312–321.
6. Bianchini, A.; Balduzzi, F.; Ferrara, G.; Ferrari, L. Virtual incidence effect on rotating airfoils in Darrieus wind turbines. *Energy Convers. Manag.* **2016**, *111*, 329–338.

7. Liu, K.; Yu, M.; Zhu, W. Enhancing wind energy harvesting performance of vertical axis wind turbines with a new hybrid design: A fluid-Structure interaction study. *Renew. Energy* **2019**, *140*, 912–927.
8. Zanforlin, S.; Nishino, T. Fluid dynamic mechanisms of enhanced power generation by closely spaced vertical axis wind turbines. *Renew. Energy* **2016**, *99*, 1213–1226.
9. Abdalrahman, G.; Melek, W.; Lien, F.-S. Pitch angle control for a small-scale Darrieus vertical axis wind turbine with straight blades (H-Type VAWT). *Renew. Energy* **2017**, *114*, 1353–1362.
10. Chen, W.-H.; Chen, C.-Y.; Huang, C.-Y.; Hwang, C.-J. Power output analysis and optimization of two straight-bladed vertical-axis wind turbines. *Appl. Energy* **2017**, *185*, 223–232.
11. Joo, S.; Choi, H.; Lee, J. Aerodynamic characteristics of two-bladed H-Darrieus at various solidities and rotating speeds. *Energy* **2015**, *90*, 439–451.
12. Bhargav, M.; Kishore, V.R.; Laxman, V. Influence of fluctuating wind conditions on vertical axis wind turbine using a three dimensional CFD model. *J. Wind Eng. Ind. Aerodyn.* **2016**, *158*, 98–108.
13. Dessoky, A.; Lutz, T.; Bangga, G.; Krämer, E. Computational studies on Darrieus VAWT noise mechanisms employing a high order DDES model. *Renew. Energy* **2019**, *143*, 404–425. [CrossRef]
14. Lositaño, I.C.M.; Danao, L.A.M. Steady wind performance of a 5 kW three-Bladed H-Rotor Darrieus Vertical Axis Wind Turbine (VAWT) with cambered tubercle leading edge (TLE) blades. *Energy* **2019**, *175*, 278–291. [CrossRef]
15. Molina, A.C.; De Troyer, T.; Massai, T.; Vergaerde, A.; Runacres, M.C.; Bartoli, G. Effect of turbulence on the performance of VAWTs: An experimental study in two different wind tunnels. *J. Wind Eng. Ind. Aerodyn.* **2019**, *193*, 103969. [CrossRef]
16. Vergaerde, A.; De Troyer, T.; Molina, A.C.; Standaert, L.; Runacres, M. Design, manufacturing and validation of a vertical-axis wind turbine setup for wind tunnel tests. *J. Wind Eng. Ind. Aerodyn.* **2019**, *193*, 103949. [CrossRef]
17. Peng, H.; Lam, H.; Liu, H. Power performance assessment of H-Rotor vertical axis wind turbines with different aspect ratios in turbulent flows via experiments. *Energy* **2019**, *173*, 121–132. [CrossRef]
18. Benner, B.M.; Carlson, D.W.; Seyed-Aghazadeh, B.; Modarres-Sadeghi, Y. Vortex-Induced Vibration of symmetric airfoils used in Vertical-Axis Wind Turbines. *J. Fluids Struct.* **2019**, *91*, 102577. [CrossRef]
19. Buchner, A.; Lohry, M.W.; Martinelli, L.; Soria, J.; Smits, A.J. Dynamic stall in vertical axis wind turbines: Comparing experiments and computations. *J. Wind Eng. Ind. Aerodyn.* **2015**, *146*, 163–171. [CrossRef]
20. Guo, Y.; Li, X.; Sun, L.; Gao, Y.; Gao, Z.; Chen, L. Aerodynamic analysis of a step adjustment method for blade pitch of a VAWT. *J. Wind Eng. Ind. Aerodyn.* **2019**, *188*, 90–101. [CrossRef]
21. Chen, C.-C.; Kuo, C.-H. Effects of pitch angle and blade camber on flow characteristics and performance of small-size Darrieus VAWT. *J. Vis.* **2013**, *16*, 65–74. [CrossRef]
22. Sengupta, A.; Biswas, A.; Gupta, R. Comparison of low wind speed aerodynamics of unsymmetrical blade H-Darrieus rotors-blade camber and curvature signatures for performance improvement. *Renew. Energy* **2019**, *139*, 1412–1427. [CrossRef]
23. Yao, Y.; Tang, Z.; Wang, X. Design based on a parametric analysis of a drag driven VAWT with a tower cowling. *J. Wind Eng. Ind. Aerodyn.* **2013**, *116*, 32–39. [CrossRef]
24. Maeda, T.; Kamada, Y.; Shimizu, K.; Ogasawara, T.; Nakai, A.; Kasuya, T. Effect of rotor aspect ratio and solidity on a straight-Bladed vertical axis wind turbine in three-dimensional analysis by the panel method. *Energy* **2017**, *121*, 1–9.
25. Huang, D.; Zhou, S.; Han, Q.; Litak, G. Response analysis of the nonlinear vibration energy harvester with an uncertain parameter. *Proc. Inst. Mech. Eng. Part K J. Multi-Body Dyn.* **2020**, *234*, 393–407. [CrossRef]
26. Han, H.; Cao, D.; Liu, L. A new approach for steady-State dynamic response of axially functionally graded and non-uniformed beams. *Compos. Struct.* **2019**, *226*, 111270. [CrossRef]
27. Huan, R.; Pu, D.; Wang, X.; Wei, X. Effects of phase delay on synchronization in a nonlinear micromechanical oscillator. *Appl. Phys. Lett.* **2019**, *114*, 233501. [CrossRef]
28. Verkinderen, E.; Imam, B. A simplified dynamic model for mast design of H-Darrieus vertical axis wind turbines (VAWTs). *Eng. Struct.* **2015**, *100*, 564–576. [CrossRef]
29. Wang, Y.; Lu, W.; Dai, K.; Yuan, M.; Chen, S.-E. Dynamic study of a rooftop vertical axis wind turbine tower based on an automated vibration data processing algorithm. *Energies* **2018**, *11*, 3135. [CrossRef]
30. Rahman, M.; Ong, Z.C.; Chong, W.T.; Julai, S.; Khoo, S.Y. Performance enhancement of wind turbine systems with vibration control: A review. *Renew. Sustain. Energy Rev.* **2015**, *51*, 43–54. [CrossRef]

31. Mabrouk, I.B.; El Hami, A.; Walha, L.; Zghal, B.; Haddar, M. Dynamic vibrations in wind energy systems: Application to vertical axis wind turbine. *Mech. Syst. Signal Process.* **2017**, *85*, 396–414. [CrossRef]
32. Yan, Y.; Wiercigroch, M. Dynamics of rotary drilling with non-uniformly distributed blades. *Int. J. Mech. Sci.* **2019**, *160*, 270–281. [CrossRef]
33. Meirovitch, L. *Principles and Techniques of Vibrations*; Prentice Hall: Upper Saddle River, NJ, USA, 1997; Volume 1.
34. Chopra, A.K. *Dynamics of Structures, Theory and Applications to Earthquake Engineering*; Pearson-Prentice Hall: Upper Saddle River, NJ, USA, 2012.
35. Xie, Z.; Liu, H. Experimental research on the interface lubrication regimes transition of water lubricated bearing. *Mech. Syst. Signal Process.* **2020**, *136*, 106522. [CrossRef]
36. Chen, W.; Gunter, E.J. *Introduction to Dynamics of Rotor-Bearing Systems*; Trafford Publishing: Bloomington, IN, USA, 2007.

5

Video-Tachometer Methodology for Wind Turbine Rotor Speed Measurement

Francesco Natili, Francesco Castellani *, Davide Astolfi and Matteo Becchetti

Department of Engineering, University of Perugia, Via G. Duranti 93, 06125 Perugia, Italy; francesco.natili@studenti.unipg.it (F.N.); davide.astolfi@studenti.unipg.it (D.A.); matteobecchetti@alice.it (M.B.)

* Correspondence: francesco.castellani@unipg.it

Abstract: The measurement of the rotational speed of rotating machinery is typically performed based on mechanical adherence; for example, in encoders. Nevertheless, it can be of interest in various types of applications to develop contactless vision-based methodologies to measure the speed of rotating machinery. In particular, contactless rotor speed measurement methods have several potential applications for wind turbine technology, in the context of non-intrusive condition monitoring approaches. The present study is devoted exactly to this problem: a ground level video-tachometer measurement technique and an image analysis algorithm for wind turbine rotor speed estimation are proposed. The methodology is based on the comparison between a reference frame and each frame of the video through the covariance matrix: a covariance time series is thus obtained, from which the rotational speed is estimated by passing to the frequency domain through the spectrogram. This procedure guarantees the robustness of the rotational speed estimation, despite the intrinsic non-stationarity of the system and the possible signal disturbances. The method is tested and discussed based on two experimental environments with different characteristics: the former is a small wind turbine model (with a 0.45 m rotor diameter) in the wind tunnel facility of the University of Perugia, whose critical aspect is the high rotational speed (up to the order of 1500 RPM). The latter test case is a wind turbine with a 44 m rotor diameter which is part of an industrial wind farm: in this case, the critical point regards the fact that measurements are acquired in uncontrolled conditions. It is shown that the method is robust enough to overcome the critical aspects of both test cases and to provide reliable rotational speed estimates.

Keywords: video tachometer; image processing; vision based measurement; wind turbines; non-stationary machines; correlation matrix.

---

1. Introduction

The condition monitoring of non-stationary rotating machines [1,2] is a complex task in general, and it is very important due to the wide industrial applications, in particular as regards renewable energy systems [3]. For wind energy technology, the maintenance costs are crucially important and have to be taken into account in the economic balance of industrial plants [4–6]. A wind turbine failure does not reflect only a lower availability of the power plant or a loss of production, but due to the increasing size of industrial wind turbines [7,8], even organizing and carrying out repair activity at high levels or offshore [9–11] can be extremely expensive. Furthermore, it should be noted that most modern wind turbines transform the slow rotation of the shaft into the fast rotation of a generator through a gearbox: this implies the necessity of monitoring gears and bearings [12,13] and recognizing faulty signatures with respect to gear mesh frequencies in non-stationary operation.

For non-stationary rotating machines in general and for wind turbines in particular, it is therefore crucial to reliably and precisely measure the rotational speed in order to detect future faults with appropriate advance notice. Continuous condition monitoring systems are typically based on the use of encoders or resolvers to measure the rotational speed based on the mechanical contact with the rotating element. Instead, it might be of practical interest to formulate methodologies that are not based on intrusion or contact with the machine itself.

This kind of approach can be particularly useful for spot analysis in industrial systems, as for example in wind farms. In fact, an important issue in industrial sectors characterized by the use of complex systems regards data openness and site accessibility: this often happens, for example, in the wind energy industry, where wind turbine owners and constructors are typically different subjects. As, for example, the operation of wind farms and the access to wind turbines is in general regulated by agreements between owners and technological providers, the possibility of analyzing certain aspects of a wind turbine's dynamical behavior could likely result in the capability of monitoring it reliably, without intruding and possibly without requiring too much information (as, for example, the gearbox geometry and therefore the gear meshing frequencies).

Based on these considerations, previous work by the authors of this study has been devoted to an innovative approach to wind turbine drive-train condition monitoring [14]: the general concept of that study was to collect appropriate vibration measurements at wind turbines in order to circumvent the drawbacks associated with this kind of data source (for example, the proprietary nature of data recorded by industrial turbine condition monitoring systems). The proposal put forward in [14] was to measure vibrations non-intrusively at the tower of the wind turbines at human height. The rationale for this choice is the avoidance of the interruption of the normal operation of the machine and the exclusion of every possible safety issue associated with human intervention in an industrial wind farm. This idea introduces novel drawbacks, because in general the transfer function from the target drive-train to the tower is not known: nevertheless, in [14], the method has been shown to be effective for the diagnosis, with months of advance notice, of bearing faults in an industrial wind farm. Further development of the analysis proposed in [14] shows the necessity of measuring the rotor speed non-intrusively, with precision and adequate sampling time. The present study deals exactly with this issue.

Therefore, the objective of this study is to formulate and test a contactless methodology for the measurement of wind turbine rotor speed from video acquisitions at ground level. There are several possible applications of this kind of approach in the wind energy industry and research: the above discussion indicates that it could complete the drive-train condition monitoring technique for industrial wind farms, as proposed in [14]. Such methods could also be applied in the context of condition monitoring for refurbished wind turbines, which are typically not (and presumably will not be) equipped with condition monitoring systems. Furthermore, interesting applications can be conceived in the context of small wind energy systems.

Few studies in the literature have been devoted to this kind of issue in the context of wind energy, despite the rising trend of vision-based measurement in engineering (as stated in [15]). To the best of the authors' knowledge, a similar methodology is proposed in [16], in which the approach is tested on a permanent magnet synchronous motor; in that study, it is indicated that a valuable application would be non-stationary rotating machines, such as wind turbines. It should be noted that the present study is more general compared to that in [16] in terms of image analysis and signal processing. In [16], a portion of the target image is selected by the users for rotational speed reconstruction, while in the present study, the entire frame captured by the camera is employed; therefore, the method of this study is more easily automatable. Furthermore, in [16] the rotational speed is estimated from the analysis of the average of the Fourier transform, while in this study, the spectrogram has been employed because it has been considered more robust with respect to disturbances in the signal. A recent study dealing with the application of video-tachometer methods to wind turbines is presented in [17]: the approach is based on computer vision, deep learning and video mining. First, a Single Shot Multibox Detector (SSD) is employed to detect the hub of the wind turbine in the image; subsequently, binary classification algorithms are used to discriminate between blade and non-blade areas. This process is repeated for all frames in a specified time window, and it can be used to estimate the angular speed: the method is validated on images and a video (with 30 frames per second) captured at operating wind farms.

An important aspect of the present study is that two different experimental environments are considered and each is characterized by different critical points; thus, the robustness of the measurement and processing methodology is discussed. The considered experimental test cases are as follows:

- The measurement of the rotor speed of a small horizontal-axis wind turbine model (with a 0.45 m rotor diameter) in the wind tunnel facility of the University of Perugia. In this case, the critical points regard the high rotor speed reached by the prototype (hundreds of RPM) and the rapid variability of the rotor speed for this kind of machine [18].
- The measurement of the rotor speed of an industrial wind turbine (with a 44 m rotor diameter) in a wind farm. In this case, the rotor speed is in the order of a few dozens of revolutions per minute, and the critical points are related to the post-processing, because disturbances can be induced by sunlight and varying backgrounds.

The structure of the manuscript is therefore as follows: in Section 2, the method, including the equipment and algorithms, is introduced. The test cases are described in Section 3. In Section 4, results are presented and discussed. Conclusions are drawn and further directions are indicated in Section 5.

2. Methods and Facilities

In this section, a description of the mathematical background for the proposed video-tachometer application is presented. Subsequently, the experimental setup of the laboratory and field tests is shown.

Video-Tachometer Working Principle

The main goal of the proposed contactless video-based tachometer method was to perform an estimation of wind turbine rotor speeds. The general idea was to collect a reference frame in which one of the wind turbine blades was clearly visible and then to record a series of video acquisitions while the rotor was moving. During the post-processing of the data, the reference image was compared with each frame of the video using a correlation algorithm, obtaining a vector with the calculated correlation values in which the maximum value of the correlation occurred each time one of the blades assumed the same position as that in the reference frame. In this way, knowing the camera frame rate, it was possible to construct a virtual tachometer signal showing peaks with a frequency proportional to the rotor's rotational signal, where the proportionality factor depended on the number of blades.

Mathematically, the comparison between the frames was performed by evaluating the Pearson product–moment correlation coefficient. Images were acquired as 8 bit black and white images, with an $m \times n$ dimensional matrix. Subsequently, the matrixes were reshaped in order to obtain a vector, whose dimension was $n \cdot m$. Labeling as $\hat{X}$ and $\hat{Y}$, respectively, the matrix of the reference image and of a sample frame of the video, both having dimensions $m \times n$ equal to the height and width of the selected resolution of the images, Equation (1) can be written as follows:

$$\hat{X} = \begin{bmatrix} x_{1,1} & \cdots & \cdots & x_{1,n} \\ & \cdot & \cdot & \\ & \cdot & \cdot & \\ & \cdot & \cdot & \\ x_{m,1} & \cdots & \cdots & x_{m,n} \end{bmatrix} \qquad \hat{Y} = \begin{bmatrix} y_{1,1} & \cdots & \cdots & y_{1,n} \\ & \cdot & \cdot & \\ & \cdot & \cdot & \\ & \cdot & \cdot & \\ y_{m,1} & \cdots & \cdots & y_{m,n} \end{bmatrix} \tag{1}$$

As the frames were acquired in black and white with an 8 bit depth, the elements of the matrice assumed values between 0 and 255.

A further step was to reshape the matrices into a single dimension vector, as shown in Equations (2) and (3):

$$X = \begin{bmatrix} x_1 = x_{1,1} & x_2 = x_{1,2} & \dots & x_{mxn} = x_{m,n} \end{bmatrix} \tag{2}$$

$$Y = \begin{bmatrix} y_1 = y_{1,1} & y_2 = y_{1,2} & \dots & y_{mxn} = y_{m,n} \end{bmatrix} \tag{3}$$

After the reshaping, the covariance matrix C was calculated according to Equations (4)–(6):

$$C_{1,1} = var(X) = \frac{1}{nm-1}\sum_{i=1}^{nm}(X_i - \bar{X})^2 \tag{4}$$

$$C_{2,2} = var(Y) = \frac{1}{nm-1}\sum_{i=1}^{nm}(Y_i - \bar{Y})^2 \tag{5}$$

$$C_{1,2} = C_{2,1} = cov(X,Y) = \frac{1}{nm-1}\sum_{i=1}^{nm}(X_i - \bar{X})(Y_i - \bar{Y}) \tag{6}$$

where $\bar{X}$ and $\bar{Y}$ are the mean values of X and Y, respectively.

From the covariance matrix, the Pearson correlation coefficient [19] could be calculated as shown in Equation (7):

$$R = \frac{C_{2,1}}{\sqrt{C_{1,1}C_{2,2}}}. \tag{7}$$

In this way, for each frame, a corresponding value of correlation was obtained. The correlation can be considered to be a useful parameter to evaluate the similitude between two objects [20,21], which in this case were the two vectors X and Y, representing the reference and a frame of the video. This property of the correlation was used in the algorithm to determine when the similitude reached its maximum and extract its periodicity, which was expected to be proportional to the rotational value. The correlation coefficient was close to its maximum value of 1 as the blade assumed a position closer to that of the reference image. The minimum value of the correlation, which could be hypothetically equal to zero, in the practical case was found at about 0.7–0.8, as there were some parts of the pictures that remained unchanged in all the frames (for example, the background and part of the hub). Figure 1 shows how the value of the correlation varied during the revolution of the rotor.

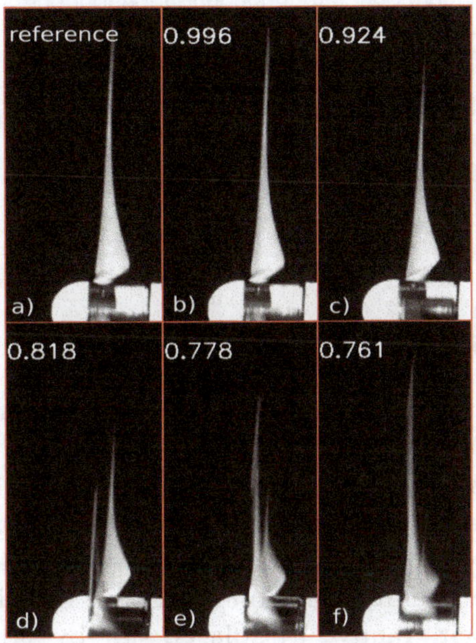

Figure 1. Referenceimage (**a**) and variation of correlation coefficient (**b–f**).

Figure 2 shows an excerpt of a virtual tachometer signal obtained from the correlation of a video acquisition: it can be seen that, from one revolution to another, the maximum value of the correlation, as well as the minimum value, did not assume a constant value but showed some variability. This effect can be caused by many factors, such as vibrations, blade deflection, a slight eccentricity of the rotor or even from a difference between the azimuth position of the blade in the reference image and in the acquired frame. All these occurrences led to a non-constant value of peaks, introducing a subtle challenge to the determination of the effective rotational speed that should be evaluated as the number of peaks in a given temporal interval. As the maximum (or minimum) value variable, two thresholds were defined as extreme limits above or under which the maximum or minimum were considered to be reached.

This approach, although feasible, is not recommended in this application because it would add an element of uncertainty, given the fact that the calculated speed may give inaccurate results depending on the choice of the threshold values. A more robust approach for the extraction of rotational speed from the virtual tachometer can be obtained using a spectrogram, or rather by dividing the signal into multiple time intervals and calculating the spectrum at each section. In this way, the spectrogram shows a peak corresponding to the blade passing frequency, which is three times the rotational frequency. The spectrogram is formally computed through a Short-Time Fourier Transform (STFT), which in its discrete formulation is given in Equation (8):

$$STFT(T,f) = \sum_{t=0}^{k-1} z(t)h(t-T)e^{\frac{-2\pi i t f}{k}}, \tag{8}$$

where $z(t)$ is the input signal, $h(t)$ is a Tukey window function, T and f are the time and frequency indexes, respectively, and k is the number of samples used for the Fourier transform.

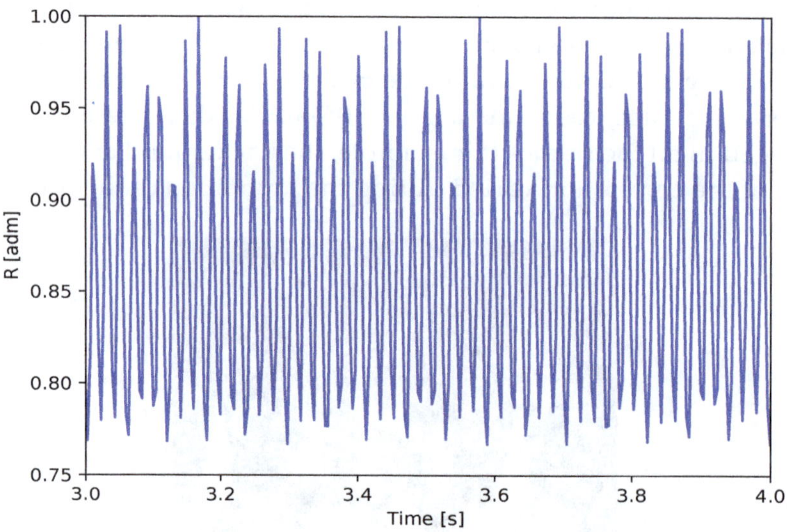

Figure 2. Example of virtual tacho signal.

The spectrogram gives an array whose elements are the square modules of the $STFT(T,f)$ as a result.

If a single FFT that composes the spectrogram is considered, a plot similar to Figure 3 is obtained, where peaks related to different frequencies are present. As expected, the maximum value of each FFT is in correspondence with the blade passing frequency; it is then possible to track all these values to calculate the rotation speed for a machine operating in non-stationary conditions. In Figure 4, the principle of maximum tracking is shown with many FFT splaced side-by-side in the time–frequency domain. The spectrograms in the following sections are 2D representations of Figure 4 with higher resolution in frequency, and the amplitudes are depicted with a color intensity scale. The tracking algorithm may fail if some peaks in FFT are higher than that related to the blade passing frequency: this aspect is discussed in detail in Section 4, after calculating the signal-to-noise ratio (SNR) value and percentage difference.

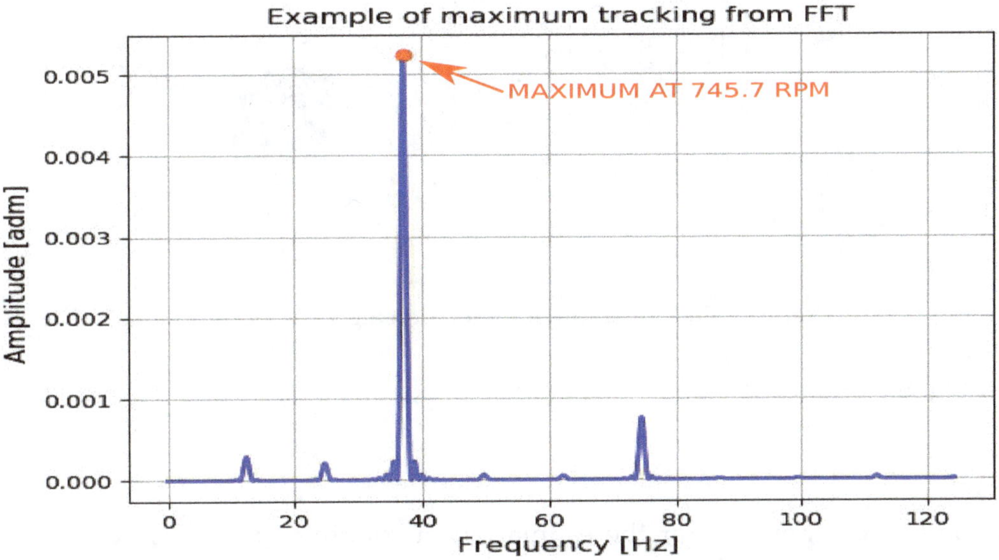

Figure 3. Example of maximum tracking from the Fast Fourier Transform (FFT): the value of 745.7 RPM has been computed by determining the frequency at which the maximum peak appears (blade passing frequency) and dividing by $\frac{no.of blades}{60} = \frac{3}{60} = 0.05$.

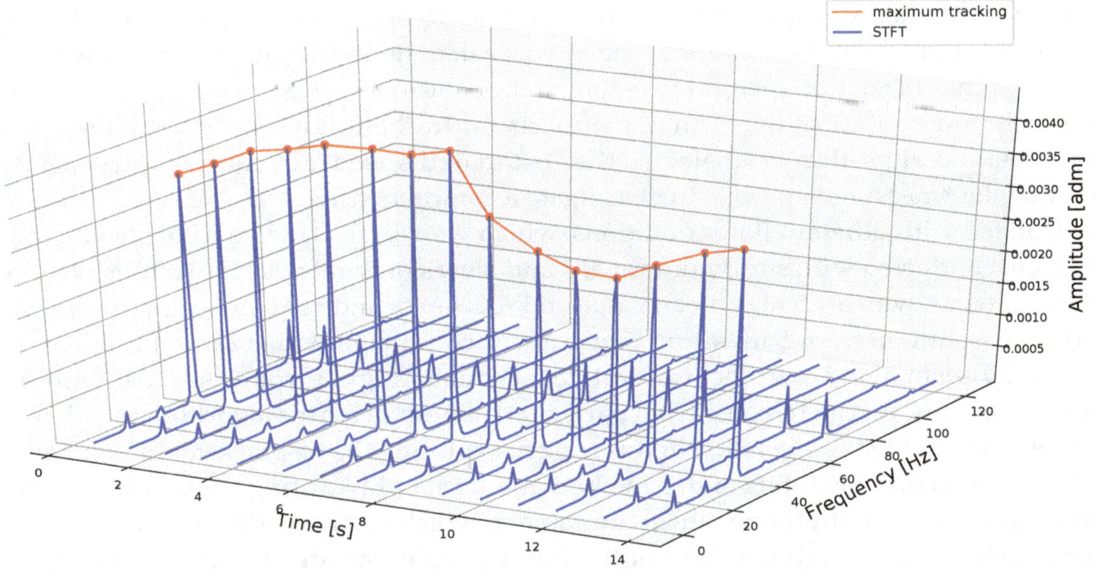

Figure 4. Visualization of maximum tracking with multiple Short-Time Fourier Transforms (STFTs).

Figure 5 shows all of the steps of the correlation algorithm used to estimate the rotational speed of wind turbines. In Section 4, the method is first validated to ensure its reliability through wind tunnel tests and then applied to a full-sized turbine operating in the field.

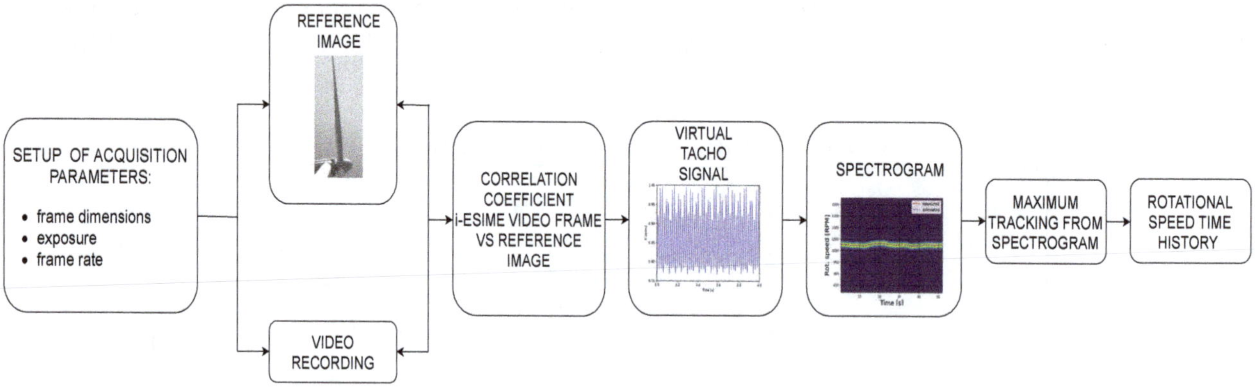

Figure 5. Block diagram of correlation algorithm.

3. Test Cases and Experimental Set-Up

3.1. Wind Tunnel Tests

The correlation-based algorithm for the estimation of rotating speed has been tested in the "R.Balli" wind tunnel facility in the Department of Engineering at the University of Perugia (Italy) . The closed-loop tunnel has an open test chamber with a squared inlet and outlet ducts of 2.2 × 2.2 m and 2.7 × 2.7 m, respectively. Wind speeds can reach up to 45 m/s thanks to a fan driven by a 375 kW electric motor. More details about the wind tunnel facility can be found, for example, in [18].

A small-scale wind turbine with a 0.45 m rotor diameter has been used for this test: its blades are constructed from titanium and the hub, as well as the nacelle, has been internally designed for research purposes and 3D printed. The rotational speed of the turbine can be controlled by varying the electrical resistance applied to the generator; in this case, the electric load was kept constant for all tests at 3 Ω. To validate the video tachometer method, an optical tachometer was installed on the nacelle of the turbine, and a marker was applied to the rotor in order to obtain a transistor–transistor-logic signal from the sensor, from which the rotational speed was calculated. The aim of the test was to evaluate the capability of the correlation method to estimate the rotational speed, with the measurement from the optical tachometer used as the term of comparison. In the following, the rotor speed obtained from the correlation algorithm is labeled as the "estimated speed"; in contrast, the speed from the sensor is called the "measured speed". To investigate a comprehensive series of case studies, many test runs were realized with different flow conditions, which were both steady and unsteady. In particular, three tests were conducted with steady flow at 4, 5 and 6 m/s of wind speed; two tests were conducted with sinusoidal wind, with a 4.5 m/s mean value and ±0.3 m/s and ±0.6 m/s amplitude; one test was conducted with a shut off from 5 m/s; and one test was conducted with a ramp from 3 m/s to 8 m/s.

As a small-scale wind turbine can rotate at much higher rotor speeds with respect to full-sized turbines, this test can be considered particularly stressful for the algorithm as video acquisition has to be captured at a high frame rate. For this reason, exploring the response to different air flows and speeds allowed us to find the limits of this method and its stability during transients. One of the most critical points is the maximum frame rate of the camera, which is strictly dependent on the acquisition parameters, such as the pixel dimensions of the frames and exposure. To achieve an adequate trade off between performance and image quality, we decided to acquire an array of 450 × 200 px with an exposure of 4 ms and a frame rate reaching 250 FPS. Figure 6 shows the arrangement of the equipment in the wind tunnel: the turbine was centered with respect to the inlet section of the flow, and the camera was installed on the left side of the chamber, pointing towards a blade of the rotor. Because of the low exposure time of the acquisition, an external LED light source was used in order to guarantee sufficient illumination.

Figure 6. Experimental setup for wind tunnel tests.

Finally, some considerations need to be addressed regarding the optimal orientation of the camera with respect to the rotor. Even if an extensive sensitivity assessment regarding this aspect exceeds the purposes of the present study, it has to be taken in account that some preliminary tests were executed with different experimental setups, light conditions (laboratory or external environment) and turbine sizes in order to find the best location for the camera. The best results were obtained by placing the camera on one side of the turbine and pointing it tangentially towards a blade. The reasons for considering this arrangement as the most suitable are the fact that if the number of pixels undergoing a noticeable variation during a rotor revolution increases, the difference between the maximum and minimum values of the correlation increases; thus, the effect of noise or harmonics, which is discussed in detail in Sections 4 and 5, decreases. Therefore, the general recommendation is to limit the amount of image occupied by non-rotating elements as much as possible in the background, tower or nacelle.

3.2. Field Measurement Campaign

Field measurement campaigns were performed in September 2019 at a wind farm sited in Italy comprising two 44 m rotor diameter turbines with 750 kW of power each. This test case can be considered peculiar, and thus worthy to be studied, because of its location and its long operation lifetime. Being situated at a mountain ridge, the wind turbines are often affected by icing and are in general subjected to a harsh environment consisting of strong gusts with non-negligible vertical components. According to this, and taking into consideration the fact that the wind farm has been operating since 1999, it is important to constantly survey these wind turbines with innovative, unintrusive methods without interfering with their energy production activity. In addition, the tested wind turbine model has a particular drive-train configuration with a variable speed gearbox with two alternating and fixed rotor speeds of 27 and 16 rotor RPM; this is therefore a useful property to test the video-tachometer algorithm.

An advantage of performing this kind of measurements with a full-scale wind turbine is the low rotational speed of the rotor with respect to that of the wind tunnel tests. This can be considered a facilitating element because the required frame rate from the camera decreases consistently to 25 RPM, allowing the exposure to be increased and for sufficiently bright pictures to be acquiredeven with weak solar light.

Critical issues to be noted in this context include the uncontrollable environmental variables that may affect the calculation of the correlation factor, such as the luminosity changes or the yaw movements of the wind turbine nacelle. As the correlation algorithm is based on the comparison of video frames with respect to a reference image, if some alteration in lighting or in nacelle position appears during the acquisition, the computed correlation would be compromised. For this reason, for the field test, the algorithm was upgraded in order to refresh the reference image as the mean value of correlation was subjected to drift.

Figure 7 shows how the equipment was installed at the base tower of the wind turbine to perform the measurement of the rotational speed. For this test, it was not possible to obtain a measured signal for an instantaneous rotational speed with the same sampling time as the estimates; in this sense, this application was particularly important because it compensated for the lack of information. From the wind turbine data sheets, it was possible to determine that the reference speed during the measurement campaign was 27 revolutions per minute, with possible oscillations due to wind gusts.

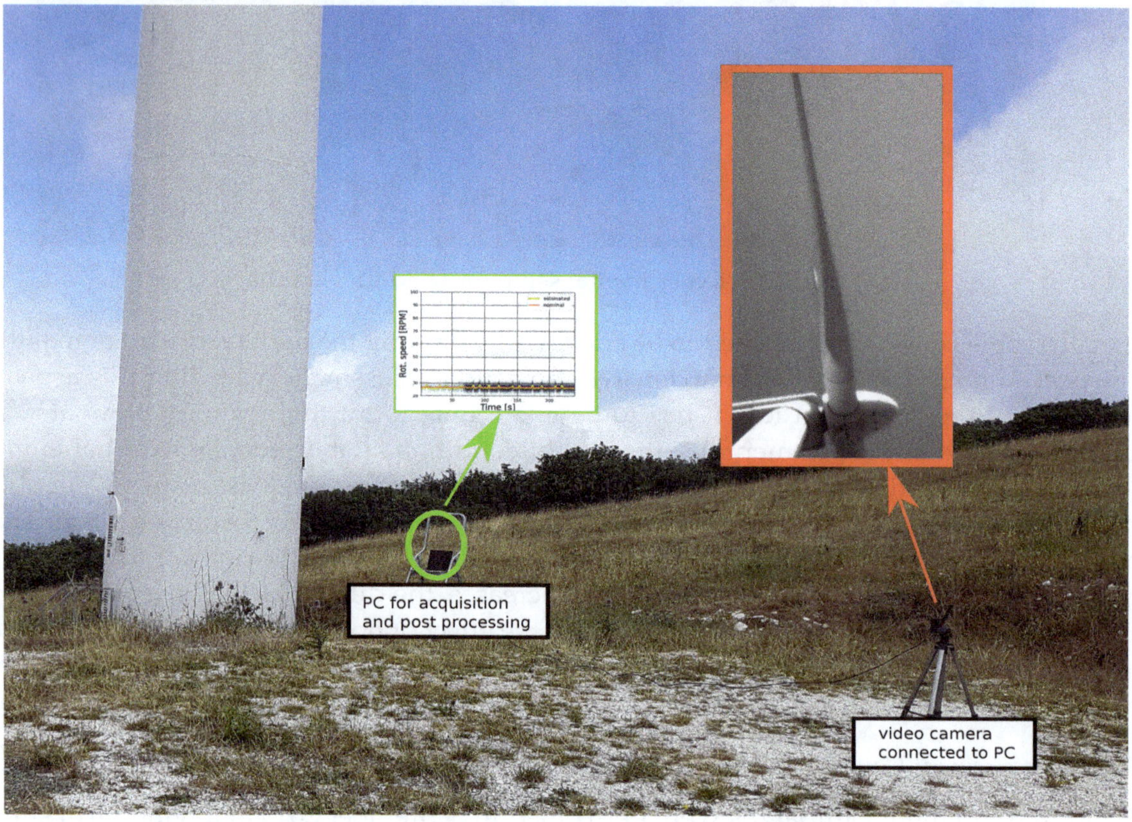

Figure 7. Instruments layout for in-site measurements.

4. Results

The estimation of rotational speed for the wind tunnel tests was realized by processing the spectrograms of the virtual tachometer signals acquired from six different runs; the corresponding time–frequency plots are shown in Figure 8a–f. The accuracy of results from STFT usually depends on the choice of some parameters, such as the number of samples, which reflect the frequency and time resolution. For the following analysis, the spectrograms were realized using a Tukey window, with a 0.25 shape parameter, on 256 sample-length segments with an overlap of 200 samples. The effective resolution of the spectrogram was 0.977 Hz. By adding zero padding to the sample segment, it was possible to finely interpolate the signal in the frequency domain. At least 744 points of zero padding should be added to the 256 sample in order to reach a reliable measurement (i.e., an *RMSE* lower than 1.4 RPM). Increasing the length of the zero padding leads to a better estimation (lower *RMSE*) at the cost of higher computational effort. In Figures 8 and 9, 9744 zero padding points have been used; an interval length greater than 256 samples leads to an insufficient time resolution.

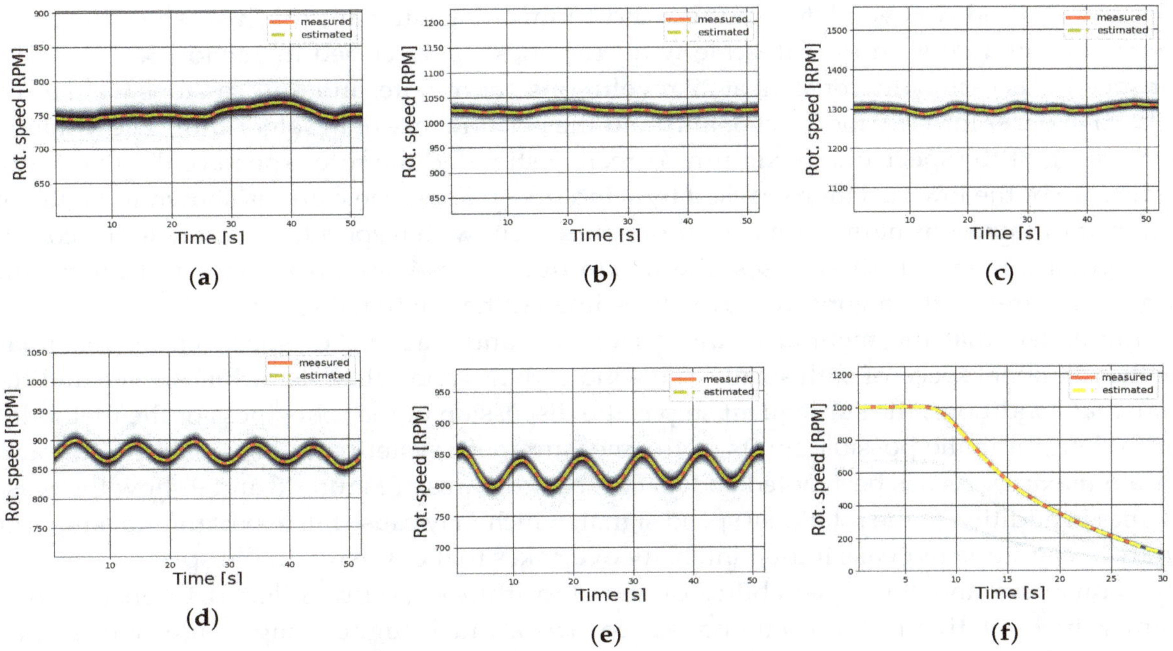

Figure 8. Estimated rotational speed from wind tunnel tests. (**a**) 4 m/s; (**b**) 5 m/s; (**c**) 6 m/s; (**d**) sinusoidal wind with a 4.5 m/s mean value and 0.3 m/s amplitude; (**e**) as the previous run, with 0.6 m/s of amplitude; (**f**) transitory from 5 m/s.

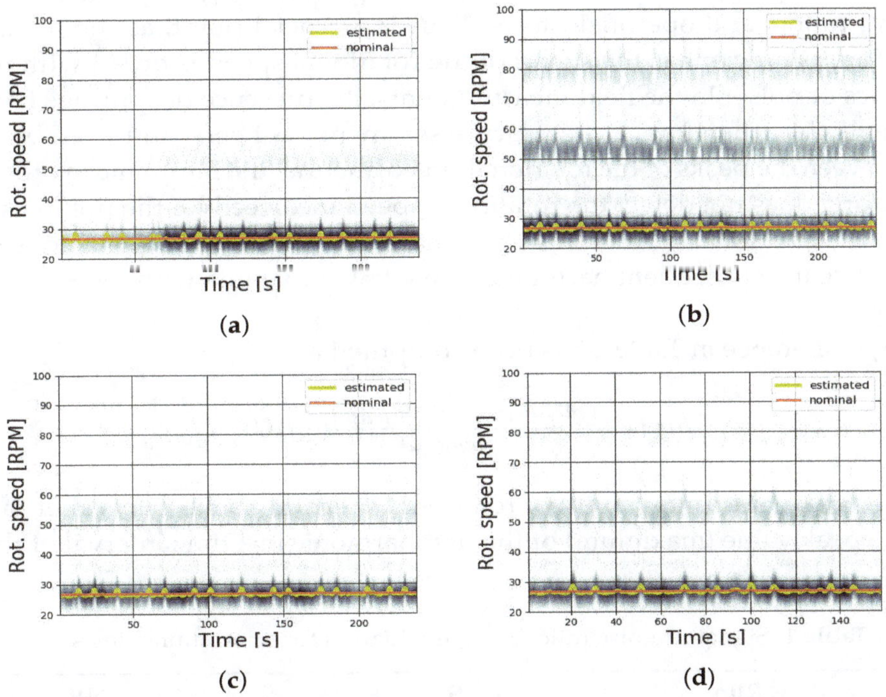

Figure 9. Spectrograms from four sample field measurements. The wind turbine has a constant nominal speed of 27 revolutions per minute.

The estimated rotational speed curves (shown by a dotted yellow line) were obtained by tracking the instantaneous maximum value of the spectrogram and applying the scale factor to calculate the shaft speed from the blade passing speed. The presence of harmonics is visible in most of the acquisitions, in particular in Figure 8f, even if their low intensity did not affect the algorithm, as the maximum value was always related to the rotational frequency.

Once the method was validated in the wind tunnel, the same approach was used in the field measurement campaign with the full-scale wind turbines, as described in Section 3.1. For this test, the rotor speed was nominally constant at 27 revolutions per minute, and four sessions were recorded (Figure 9a–d) in order to verify the functionality and the performance of the algorithm. The results show that the tracking of the spectrum maximum correctly estimated the rotor speed of the wind turbine, even in the case of the low variations caused by wind gusts. For some runs (as shown in Figure 9b,d), it appears that the spurious harmonics were more prominent with respect to the cases in the controlled laboratory environment. In these cases, the maximum values were also reached at the rotational frequency, and therefore the algorithm correctly estimated the rotational speed.

Having stated that the method of the correlation and tracking of spectrogram maxima can estimate the rotational speed of both small-scale wind turbines and full-scale turbines, even in different environmental conditions, the subsequent step is the discussion of the robustness of the methodology in order to determine the possible limits of the measurement system and its capability to track the spectrogram maxima. As has been noted above, the spectrograms of Figures 8 and 9 show the presence of harmonics in addition to the rotational speed signal, which may cause the maximum tracking process to fail in the recognition of peaks if their intensity overtakes that of the rotational speed.

The robustness and the repeatability of the algorithm were investigated to ensure that the maximum values relative to the rotational speed had a much higher magnitude with respect to the background noise introduced by the harmonics. For this reason, in Tables 1 and 2, for each test run, the mean value of the signal-to-noise ratio (SNR) is reported. The signal S is intended to be the maximum value of the spectrogram, relative to the tracked rotational speed, and the noise N is the mean value of $STFT(T,f)$ (Equation (8)). This provides a metric to assess the predominance of the required signal with respect to noise. The results shown in Tables 1 and 2 confirm that the signal related to the rotational speed is at least one order of magnitude larger than the mean value of the $STFT$ both in wind tunnel and open-field tests. This is a crucial point, as the maximum tracking algorithm is therefore as efficient as the signal of the rotational speed is broad with respect to other undesired harmonics or noise. In the field measurements, the presence of spurious harmonics can be considered non-negligible: therefore, a necessary assessment has been conducted by calculating the mean difference between the peaks of the rotational speed and the first harmonic in order to ensure that the maximum tracking does not detect the rotational speed incorrectly if the first harmonic assumes values higher than the rotational speed. This check has been conducted only in consideration of the first harmonic, because the subsequent harmonics have lower maxima with respect to the first for all the conducted tests.

The percentage difference in Table 2 has been computed as

$$diff_{\%} = \frac{peak_{rot} - peak_{harmI}}{peak_{rot}} \times 100 \tag{9}$$

where $peak_{rot}$ is the peak value (maximum) of rotational speed in a time interval of the spectrogram, and $peak_{harmI}$ is the peak value (maximum) of the first harmonic in a time interval of the spectrogram.

Table 1. Signal-to-noise ratio (SNR) evaluation for wind tunnel tests.

Run	S	N	SNR
4 m/s	5.220×10^{-3}	5.742×10^{-5}	9.089×10^{1}
5 m/s	4.013×10^{-3}	3.831×10^{-5}	1.047×10^{2}
6 m/s	2.860×10^{-3}	2.935×10^{-5}	9.742×10^{1}
Sinusoidal 0.3 m/s amplitude	2.783×10^{-3}	3.098×10^{-5}	8.984×10^{1}
Sinusoidal 0.6 m/s amplitude	2.326×10^{-3}	2.772×10^{-5}	8.389×10^{1}
Shut down	1.892×10^{-3}	2.768×10^{-5}	6.209×10^{1}

Table 2. SNR evaluation for in-field tests.

Run	S	N	S/N	% Difference (Mean)
1	2.269×10^{-2}	2.234×10^{-4}	9.987×10^{1}	9.277×10^{1}
2	1.343×10^{-2}	2.172×10^{-4}	6.189×10^{1}	5.072×10^{1}
3	1.201×10^{-2}	1.364×10^{-4}	8.785×10^{1}	8.124×10^{1}
4	4.236×10^{-3}	5.176×10^{-5}	8.184×10^{1}	7.522×10^{1}

A stress test for the measurement equipment was performed in the wind tunnel facility, subjecting the test case wind turbine model to a sharp wind ramp from 0 m/s to 8 m/s and causing the rotational speed to reach up to 1900 RPM in about 50 s from the beginning of the run. As can be seen in Figure 10, the algorithm was able to follow the measured speed up to order of 1500 RPM with low error; beyond this limit, the estimated rotational signal began to be unstable as the maximum value of the spectrogram passed from the rotational speed to its first lower harmonic. This means that the operation of the maximum search does not guarantee that the highest value belongs to the required harmonic. As soon as the wind decreased and the speed slowed to under the level of 1500 RPM, the algorithm again correctly estimated the rotational speed. The impossibility of measuring rotor speeds above the order of 1500 RPM is due to the frame rate of the acquisition rather than the data analysis method.

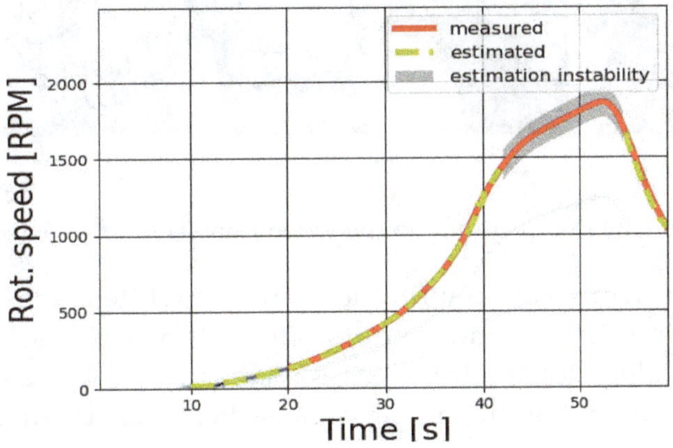

Figure 10. Wind ramp test at 0–8 m/s: the rotational speed quickly reaches 1900 RPM

Finally, an analysis has been conducted to determine the sensitivity of the presented algorithm to external factors concerning the noise that may affect the acquisition of images, especially in conditions of low illumination; i.e., when a low exposition time is required or when some meteorological variability appears in the open field.

To include random noise in this study, a Gaussian white noise signal was digitally added to the frames in one of the wind tunnel tests. A Gaussian distribution $p(x)$ can be described by the following Equation (10):

$$p(x) = \frac{1}{\sqrt{2\pi\sigma^2}} e^{-\frac{(x-\mu)^2}{2\sigma^2}} \tag{10}$$

where the mean value μ is equal to zero for the considered white noise and σ^2 is the variance. As a first test, one matrix with the same dimensions as the frames was generated with random values and then added to the reference image and to all the frames of the video acquisition. Three test values for the variance of the white noise were considered: $\sigma^2 = 5$, $\sigma^2 = 50$, $\sigma^2 = 500$ (Figure 11). This was done in order to determine the highest value admissible by the correlation algorithm. The effect of external noise was computed by using the root mean squared error (*RMSE*) between the time history of the measured and estimated rotational speed. In this situation, even in the case of the heaviest noise with σ^2 = 500, no noticeable deviations in the $RMSE$ were found: its values were found to be 0.256 in all cases.

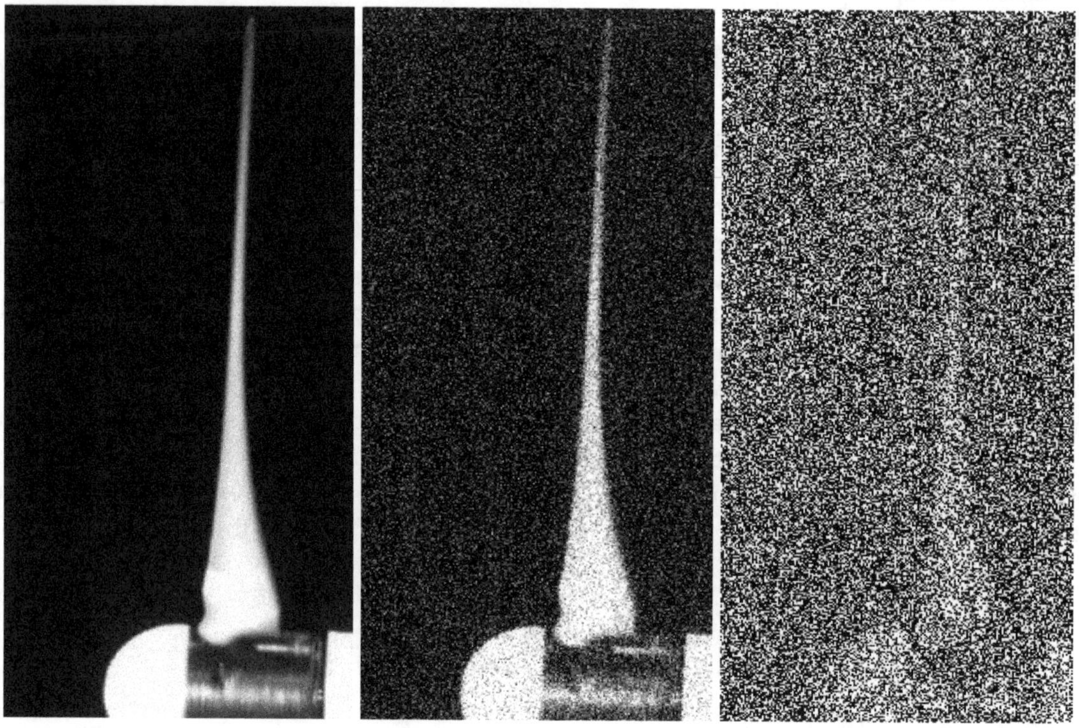

(a) (b) (c)

Figure 11. Different levels of white noise added to the video frames: (**a**) σ^2 = 5; (**b**) υ^2 = 50; (**c**) σ^2 = 500.

A second test was subsequently set up with the matrix of white noise, which was recalculated for each frame of the video; in this case, the random effect of the noise acted both in space and in time. The calculated $RMSE$ valeus for the three cases were $RMSE_5 = 0.259, RMSE_{50} = 0.255$ and $RMSE_{500} = 17.136$. We can observe that the impact of noise becomes non-negligible when a random, time-varying white noise is added whose σ^2 becomes greater than 500. This can be considered to be a qualitative evaluation of the acceptable noise threshold for the algorithm to obtain reliable results.

5. Conclusions

In this study, a method to estimate the rotational speed of a wind turbine through contactless video acquisition has been presented. A straightforward motivation of this work is its possible application in the context of non-intrusive wind turbine condition monitoring techniques, but the potentiality of the proposed method is more general.

In the present study, a correlation-based algorithm has been proposed that employs a reference image of wind turbine blades and a series of frames extracted from a continuous recording while the machine is operating. Knowing the frame rate of the camera, the correlation coefficient with respect to the reference image is computed for each frame. As a result, a virtual tachometer signal is obtained that presents a periodicity corresponding to the blade passing frequency. Finally, the rotor speed time history is extracted by post-processing the virtual tachometer signal with STFT and tracking the maximum values.

The proposed method has been tested in a wind tunnel facility with a small-scale wind turbine model and then applied in a full-scale, industrial-size wind farm. The validation of the algorithm was set up in wind tunnel tests, because the wind turbine was equipped with an optical tachometer sensor, and therefore a comparison between the estimated rotational speed and the measured speed was possible. In the full-scale tests, no reference rotor speed signals were available, even if a reference nominal rotor speed was known from the wind turbine data sheet.

The response of the algorithm has been shown to be robust and repeatable even in consideration of the different conditions of the laboratory environment and the wind farm. On one hand (wind tunnel), there were conditions of controlled and constant lighting, and the critical points regarded the extremely high rotor speed reached by the test case wind turbine device (up to 1500 RPM) and its possible relevant variability in time in the case of transient wind or gusts. On the other hand (industrial wind farm tests), the nominal rotor speed was 27 RPM, and therefore there were no critical aspects as regards the management of the camera frame rate; in this case, the difficulties were constituted by the environmental variability in luminosity and by the dynamics of the rotor (such as yaw movements). In both cases, the speed estimation was considered to be reliable: the behavior of the signal-to-noise ratio shows that the effect of noise did not disturb the algorithm of maximum tracking as much as it compromised the quality of the rotor speed estimation. In fact, the harmonic required to estimate the rotational speed (the signal) was shown to be predominant, up to a factor of 10^2, with respect to the mean value of the module in each STFT (the noise).

Further developments of the present study will involve the use of this methodology in combination with base tower vibration measurements (in a manner similar to [14]) for wind turbine drive-train condition monitoring applications in industrial wind farms. The approach proposed in this work could be particularly valid in the context of fault diagnosis for wind turbines which have not been equipped at their installation with condition monitoring systems because of economic considerations. These kinds of machines are at present typically reaching the end of their expected lifetime, and for this reason, they are typically operating out of warranties and full service agreements with the constructor company; therefore, condition monitoring for these wind turbines is mostly an overlooked topic, and the methodology proposed in this work might constitute a basis for setting up periodic inspections of these turbines.

Author Contributions: Conceptualization, F.N., F.C., D.A. and M.B.; methodology, F.N. and F.C.; software, F.N., F.C. and M.B.; validation, F.N., F.C., D.A. and M.B.; formal analysis, F.N. and F.C.; investigation, F.N., F.C., D.A. and M.B.; resources, F.C.; data curation, F.N. and F.C.; writing–original draft preparation, F.N. and D.A.; writing–review and editing, F.N, F.C., D.A. and M.B.; visualization, F.N., F.C., D.A. and M.B.; supervision, F.C.; project administration, F.C.; funding acquisition, F.C. All authors have read and agreed to the published version of the manuscript.

Acknowledgments: The authors acknowledge Fondazione "Cassa di Risparmio di Perugia" for the funded research project WIND4EV (Wind Turbine Technology Evolution for Lifecycle Optimization) and the Basic Research Plan 2019 of the Department of Engineering of the University of Perugia for a financed project entitled "Operational characterization of mechanical components through the analysis of the correlation between the surface and production characteristics with vibratory phenomena in couplings subject to wear".

References

1. Vishwakarma, M.; Purohit, R.; Harshlata, V.; Rajput, P. Vibration analysis & condition monitoring for rotating machines: A review. *Mater. Today Proc.* **2017**, *4*, 2659–2664.
2. Sghaier, E.; Bourdon, A.; Remond, D.; Dion, J.L.; Peyret, N. Non-stationary operating conditions of rotating machines: Assumptions and their consequences. In Proceedings of the International Conference on Condition Monitoring of Machinery in Non-Stationary Operation, Santander, Spain, 20–22 June 2018; pp. 11–19.
3. Maheswari, R.U.; Umamaheswari, R. Trends in non-stationary signal processing techniques applied to vibration analysis of wind turbine drive train—A contemporary survey. *Mech. Syst. Signal Process.* **2017**, *85*, 296–311. [CrossRef]
4. Dao, C.; Kazemtabrizi, B.; Crabtree, C. Wind turbine reliability data review and impacts on levelised cost of energy. *Wind Energy* **2019**, *22*, 1848–1871. [CrossRef]
5. Bhardwaj, U.; Teixeira, A.; Soares, C.G. Reliability prediction of an offshore wind turbine gearbox. *Renew. Energy* **2019**, *141*, 693–706. [CrossRef]
6. Dao, C.D.; Kazemtabrizi, B.; Crabtree, C.J. Modelling the Effects of Reliability and Maintenance on Levelised Cost of Wind Energy. In Proceedings of the Turbo Expo: Power for Land, Sea, and Air, Phoenix, AZ, USA, 17–21 June 2019; American Society of Mechanical Engineers: New York, NY, USA, 2019; Volume 58721, p. V009T48A001.

7. Barthelmie, R.J.; Shepherd, T.J.; Aird, J.A.; Pryor, S.C. Power and Wind Shear Implications of Large Wind Turbine Scenarios in the US Central Plains. *Energies* **2020**, *13*, 4269. [CrossRef]
8. Cao, Q.; Xiao, L.; Cheng, Z.; Liu, M.; Wen, B. Operational and extreme responses of a new concept of 10 MW semi-submersible wind turbine in intermediate water depth: An experimental study. *Ocean Eng.* **2020**, *217*, 108003. [CrossRef]
9. Carroll, J.; McDonald, A.; Dinwoodie, I.; McMillan, D.; Revie, M.; Lazakis, I. Availability, operation and maintenance costs of offshore wind turbines with different drive train configurations. *Wind Energy* **2017**, *20*, 361–378. [CrossRef]
10. Kang, J.; Wang, Z.; Soares, C.G. Condition-based maintenance for offshore wind turbines based on support vector machine. *Energies* **2020**, *13*, 3518. [CrossRef]
11. Yeter, B.; Garbatov, Y.; Soares, C.G. Risk-based maintenance planning of offshore wind turbine farms. *Reliab. Eng. Syst. Saf.* **2020**, 107062. [CrossRef]
12. Vamsi, I.; Sabareesh, G.; Penumakala, P. Comparison of condition monitoring techniques in assessing fault severity for a wind turbine gearbox under non-stationary loading. *Mech. Syst. Signal Process.* **2019**, *124*, 1–20. [CrossRef]
13. Yin, A.; Yan, Y.; Zhang, Z.; Li, C.; Sánchez, R.V. Fault Diagnosis of Wind Turbine Gearbox Based on the Optimized LSTM Neural Network with Cosine Loss. *Sensors* **2020**, *20*, 2339. [CrossRef] [PubMed]
14. Castellani, F.; Garibaldi, L.; Daga, A.P.; Astolfi, D.; Natili, F. Diagnosis of Faulty Wind Turbine Bearings Using Tower Vibration Measurements. *Energies* **2020**, *13*, 1474. [CrossRef]
15. Shirmohammadi, S.; Ferrero, A. Camera as the instrument: The rising trend of vision based measurement. *IEEE Instrum. Meas. Mag.* **2014**, *17*, 41–47. [CrossRef]
16. Ferreira, F.J.; Duarte, A.F.; Lopes, F.J. Experimental Evaluation of a Novel Webcam-Based Tachometer for In-Situ Rotational Speed Measurement. In Proceedings of the 2020 IEEE International Conference on Industrial Technology (ICIT), Buenos Aires, Argentina, 26–28 February 2020; pp. 917–924.
17. Bahaghighat, M.; Xin, Q.; Motamedi, S.A.; Zanjireh, M.M.; Vacavant, A. Estimation of Wind Turbine Angular Velocity Remotely Found on Video Mining and Convolutional Neural Network. *Appl. Sci.* **2020**, *10*, 3544. [CrossRef]
18. Castellani, F.; Astolfi, D.; Becchetti, M.; Berno, F.; Cianetti, F.; Cetrini, A. Experimental and numerical vibrational analysis of a horizontal-axis micro-wind turbine. *Energies* **2018**, *11*, 456. [CrossRef]
19. Edelmann, D.; Móri, T.; Székely, G. On relationships between the Pearson and the distance correlation coefficients. *Stat. Probab. Lett.* **2021**, *169*, [CrossRef]
20. Asuero, A.G.; Sayago, A.; González, A.G. The Correlation Coefficient: An Overview. *Crit. Rev. Anal. Chem.* **2006**, *36*, 41–59, [CrossRef]
21. De Maesschalck, R.; Jouan-Rimbaud, D.; Massart, D. The Mahalanobis distance. *Chemom. Intell. Lab. Syst.* **2000**, *50*, 1–18, [CrossRef]

6

Surrogate Models for Wind Turbine Electrical Power and Fatigue Loads in Wind Farm

Georgios Gasparis, Wai Hou Lio and Fanzhong Meng *

Department of Wind Energy, Technical University of Denmark, 4000 Roskilde, Denmark; giorgos.sp18@gmail.com (G.G.); wali@dtu.dk (W.H.L.)

* Correspondence: famen@dtu.dk

Abstract: Fatigue damage of turbine components is typically computed by running a rain-flow counting algorithm on the load signals of the components. This process is not linear and time consuming, thus, it is non-trivial for an application of wind farm control design and optimisation. To compensate this limitation, this paper will develop and compare different types of surrogate models that can predict the short term damage equivalent loads and electrical power of wind turbines, with respect to various wind conditions and down regulation set-points, in a wind farm. More specifically, Linear Regression, Artificial Neural Network and Gaussian Process Regression are the types of the developed surrogate models in this work. The results showed that Gaussian Process Regression outperforms the other types of surrogate models and can effectively estimate the aforementioned target variables.

Keywords: surrogate model; fatigue load; wind turbine; wind farm

1. Introduction

In a wind farm, wind turbines are typically positioned closely together to ensure several economic benefits such as reductions in the deployment, operation and maintenance cost. However, in a row of wind turbines, the upstream turbines extract the largest amount of energy from the wind, create velocity deficits and add more turbulence to the flow for the downstream turbines. This region of flow is known as wake. The reduction in wind speeds and added turbulence in a wake will cause the downstream turbine to produce less power and accelerate the structural damage. Thus, a number of wind farm control strategies has been proposed in the literature.

A substantial amount of studies investigated the possibility to increase the aggregated wind farm power output by manipulating the wake behaviour (see [1]). These methods are mostly relevant for wind farms in low wind speeds and can be classified into two distinct classes: induction control and wake steering control. In induction control (e.g., [2–4]), the upstream turbines are down-regulated to reduce the thrust coefficient, aiming to reduce the velocity deficits and create a faster recovery of the wake. In wake steering (e.g., [5,6]), the upstream turbine rotor is misaligned with the ambient wind direction, with the purpose to steer the wake away from the downstream turbine. In contrast to methods increasing the power output, some works (e.g., [7–9]) focused on developing methods to optimise the load situation among turbines in a wind farm, which are mostly relevant for wind farms in above-rated wind conditions. Several works investigated the fatigue loads impact of turbines operating in down-regulations (e.g., [10–12]).

One of the key concerns in wind farm control is the turbine fatigue loads. Calculation of turbine loads, including turbine wake, typically requires Computational Fluid Dynamics (CFD) and aero-servo-elastic turbine simulations that are computationally expensive. Thus, development of a simplified and control-oriented surrogate model is motivated, which offers fast prediction of electrical power and fatigue loads in a wind farm for given control set-point of turbines.

One of the early models was developed by Frandsen [13], that maps the wake effects to an equivalent ambient turbulence. This equivalent ambient turbulence should cause the same level of fatigue load accumulation as the actual wake-induced turbulence and deficit. The Frandsen [13] approach is widely popular because of its simplicity; however, some studies (e.g., [14]) pointed out that the methods provided a conservative load result. Later, a more advanced engineering model [15], the Dynamic Wake Meandering (DWM) model, was developed. The model considers the wake velocity deficit, meandering dynamics and added wake turbulence. This model is accurate at predicting the wake-induced loads and has been validated in a number of studies (e.g., [16]). However, aero-servo-elastic turbine simulations with the DWM model are still not sufficiently fast enough for control-oriented purposes. In recent years, some simpler surrogate load models consider the various atmospheric conditions such as wind speeds, turbulence intensities and wake properties. For example, a model developed by Dimitrov [17] maps the wake properties into turbine loads, while a study by Mendez Reyes et al. [18] developed a load surrogate model that describes the turbine fatigue loads in terms of changes in yaw and wind inflow conditions. A recent study by Galinos et al. [19] considered different down-regulation power set-points in their load surrogate models.

Of the many existing works, training load surrogate models requires a substantial amount of turbine data. However, the required amount of data set of turbine loads might not always be available. Therefore, the aim of this work is to define and demonstrate a simple procedure that maps the outputs of the short-term damage equivalent loads and electrical power of a wind turbine in a wind farm, in computationally efficient surrogate models approximations. More importantly, the surrogate model developed in this work requires a small size of data set with fewer than a thousand samples. Specifically, we investigate three regression methods and identify which methods offer the best and reliable performance based on a small data set. The presented methods are Linear Regression (LR), Artificial Neural Network (ANN) and Gaussian Process Regression (GPR). In addition, this work further investigates the link between the sample size and the accuracy of surrogate models. The proposed surrogate model produces accurate and fast turbine power and load predictions in a wind farm, which is particularly useful for wind industry with a limited amount of measurement data to develop wind farm control strategy.

The remainder of this paper is organised as follows. In Section 2, we present the procedure of collecting the training data. Section 3 discusses three regression methods that are used in this study. In Section 4, evaluation of model preference and discussions are presented. Finally, it is followed by conclusions in Section 5.

2. Definition of the Surrogate Load Modeling Procedure

To acquire the input data set of this work, DTU 10MW reference turbine [20] and a high fidelity aero-elastic simulation code HAWC2 [21] were used. HAWC2 simulates wind turbines with large and flexible blades in time domain. It handles non-linear structure dynamic with arbitrary large rotation. The aerodynamic part of the code is based on the Blade Element Momentum (BEM) theory [22], but extended from the classic approach to handle dynamic inflow, dynamic stall, skew inflow, shear effects on the induction and effects from large deflections. The DWM model, which is a medium-fidelity model providing detailed information of the non-stationary inflow wind field to each individual wind turbine within a wind farm, is implemented in HAWC2 for performing wind farm simulation. Three different wind classes, namely IA, IB and IC representing different turbulence intensity reference values (I_{ref}) of 0.16, 0.14 and 0.12, were used in the simulation. The mean wind speed range ($\bar{U}$) was from 4 m/s to 24 m/s in 2 m/s increments, and six random turbulent seeds were used per wind speed. The simulation time was 600 seconds using time step of 0.01 s. This simulation time of 600 s is sufficiently long enough for the wake of the upwind turbine to hit the downwind turbine in the second row. The down-regulation method based on minimizing the

Thrust Coefficient (C_T) [23] was used. The down-regulation level (δ_d) is defined as the ratio between the power demanded to generate in derating operation (P_derated) and rated power (P_rated):

$$\delta_\text{d} = \frac{P_\text{derated}}{P_\text{rated}} \tag{1}$$

For the upstream turbine (WT-1), the down-regulation level in percentage were set as 60%, 70%, 80%, 90%, 100%. For the downstream wind turbine (WT-2) the down regulation levels were 70%, 80%, 90%, 100% for each down regulation levels of WT-1. The wind turbines in both rows were combined with all three wind classes. The Wohler exponent values (m) of 3, 4 and 10 were chosen for the bearings (yaw and shaft), the tower and the blades respectively. The space distance between the first and the second row of the wind turbines was chosen four rotor diameters (4D). This spacing represented a compromise between compactness, which minimises capital cost, and the need for adequate separations to reduce energy loss through the wind shadowing from upstream wind turbines. The ambient wind direction was set to zero degrees, which means yaw misalignment was not considered.

The results from the simulations were post-processed. The short term damage equivalent load $S_\text{eq,ST}$ for each time-series was computed using the rain-flow counting method [24]. Subsequently, the average short term Damage Equivalent Load (DEL) was computed as follows

$$\bar{S}_\text{eq,ST} = \left(\sum_i p_i \cdot S^m_\text{eq,ST}\right)^{\frac{1}{m}} \tag{2}$$

where $p_i = \frac{1}{6}$ represents the identical weight for six turbulent seeds, m is the Wohler exponent value mentioned above. For each wind speed the short-term DELs for each turbulent seed $S_\text{eq,ST}$ were combined as shown in (2) into the average short-term DEL $\bar{S}_\text{eq,ST}$ corresponded to each wind speed.

The input data set $\mathcal{D}_\text{in} := \{\mathcal{U}, I_\text{ref}, \delta_\text{d,up}, \delta_\text{d,down}\}$ consists of four features. The features are: wind speed ($\bar{U}$), turbulence intensity reference (I_ref), down regulation level of the upstream and downstream wind turbine ($\delta_\text{d,up}$ and $\delta_\text{d,down}$). The wind speed and turbulence intensity reference features consist of values that were previously described. The other two features consist of values of down regulation levels. In cases when only the loads and power of one single turbine are of interest, the power set-point for the upstream turbine is set to be 0. For example, if there was only an upstream wind turbine down regulated at 80%, then the power set-point of upstream wind turbine had to be set to 0 and the power set-point of downstream wind turbine to 0.8. In this way an extra feature addition that shows if the wind turbine is upstream or downstream, was avoided.

In Table 1 the different input values are presented. First, we have turbulent wind cases with 11 different wind speeds and three turbulent classes. For one single turbine case, the upstream turbine is operated at 60%, 70%, 80%, 90% and 100% of the rated power, that provides five variations. In a case of two turbines, five variations are for the upstream turbine and four variations are for the downstream turbine. Therefore, the total number of DEL sets that were calculated from post processing is 825 for each load channels. For the electrical power, the number of post processed results is the same since the six turbulent seeds for each wind speed were combined and the mean value was taken. The input data was a matrix of 825 × 4 elements. A general schematic diagram that represents the components of the training process is given in Figure 1.

Table 1. Input variables for surrogate models.

Parameters	Values	Number of Samples
Wind Speed $\bar{U}$ [m/s]	4, 6, 8, 10, 12, 14, 16, 18, 20, 22, 24	11
Turbulence intensity reference I_{ref} [-]	0.12, 0.14, 0.16	$\times 3$
Upstream down-regulation level $\delta_{\text{d,up}}$ [-]	0.00, 0.60, 0.70, 0.80, 0.90, 1.00	$\times(5 + 5 \times 4)$
Downstream down-regulation level $\delta_{\text{d,down}}$ [-]	0.70, 0.80, 0.90, 1.00	
Total number of samples		825

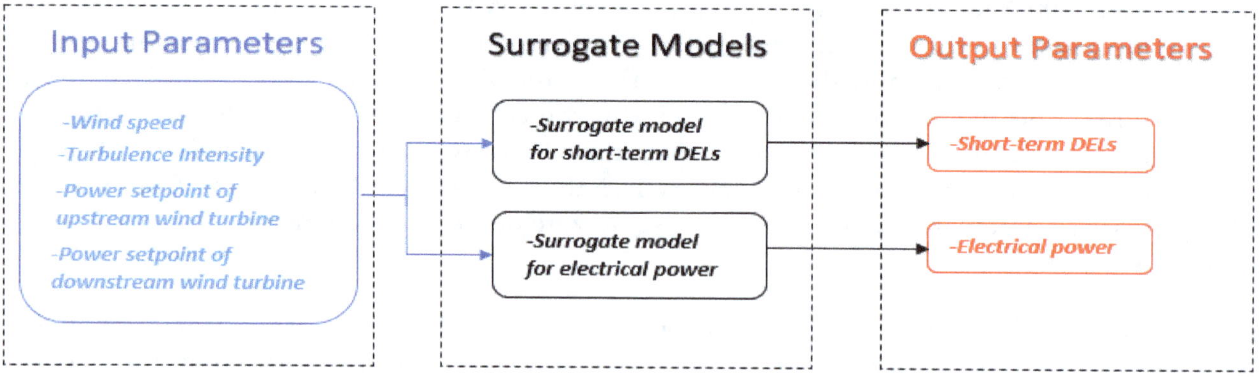

Figure 1. General schematic diagram of surrogate models.

3. Methodology

3.1. Linear Regression

In this supervised learning task, a training set is given, comprised of N observations $\mathcal{D}_{\text{ob}} := \{X, y\}$, where $X = [x_1, x_2, \ldots, x_N]^T \in \mathbb{R}^{N \times p}$ represents the input data set and $y = [y_1, y_2, \cdots, y_N]^T \in \mathbb{R}^N$ is denoted as the output data. The input vector $x_i \in \mathbb{R}^p$ has p elements of input. The goal is to predict new y_i from input data x_i. In Linear Regression, the predicted output $\hat{y}_i$ is formulated as follows.

$$\hat{y}_i(x, w) = w_0 + w_1 x_{i,1} + \cdots + w_p x_{i,p} \tag{3}$$

where $w = [w_0, w_1, \cdots, w_p]^T \in \mathbb{R}^p$ is a vector of tunable coefficients and $x_i = [x_{i,1}, \ldots, x_{i,p}]$ are the input features. Learning then consists of selecting the parameters w based on the training data X and y.

In this work, the input data X is the variables listed in Table 1. With respect to the output data, for the short-term DELs surrogate model, y is the short-term DELs for each load channels. For the electrical power surrogate model, the output data y is the electrical power. A standardization was made in the data set. A 10-fold cross-validation was used and the data was divided in 90% training and 10% test set for each fold.

To predict the target variables $\hat{y}$, `sklearn.linear.model.LinearRegression` package in high-level programming language Python was used [25]. This model uses Ordinary Least Squares method by fitting a linear model with coefficients w to minimise the residual sum of squares between the observed targets in the data set, and the targets predicted by the linear approximation. To find the tunable coefficients w in (3), the optimisation problem is formulated as follows.

$$\hat{w} = \min_{w} ||y - Xw||^2 \tag{4}$$

The estimation of the coefficients for Ordinary Least Squares relies on the independence of the features. For each fold, the model was trained and the least squares optimisation (4) was computed using the singular value decomposition of X. To avoid developing an over-fitting model, a general technique for controlling model complexity known as regularization was considered. The type of regularization that was considered is commonly referred to as L_2 regularization, and the resulting model is commonly called Ridge regression [25]. For the purpose of this study, the regularization constant was tweaked in an internal cross validation. With this internal cross validation the best regularization term is chosen for each outer fold.

3.2. *Artificial Neural Networks*

Artificial Neural Networks (ANNs) consist of a net of various connected information processing units, so-called artificial neurons. For the purpose of this work, a feed-forward network was created. First, the input vector $x = [x_1, x_2 \ldots, x_p]$ is presented to the input layer in a way that neuron i in the input layer is given an activation equal to x_i. The activation is then propagated to one or several hidden layers and eventually to the output layer consisting of neurons corresponding to the components of the output vector. An example of an ANN can be seen in Figure 2.

A more general form of Equation (3) can be written as:

$$\hat{y} = \tilde{x}^T w, \quad \tilde{x} = [1 \;\; x_1 \;\; x_2 \;\; \ldots \;\; x_p]^T \tag{5}$$

The various-weights of the model can be defined as: $W^{(1)} = [w_1^{(1)} \; w_2^{(1)} \; \ldots \; w_H^{(1)}]$ and $W^{(2)} = [w_1^{(2)} \; w_2^{(2)} \; \ldots \; w_D^{(2)}]$, where D is the number of output neurons. By having a nonlinear activation $h^{(1)}$ and a number of H hidden neurons the activation of the kth output neuron can be written as:

$$\hat{y}_k(x, w) = h^{(2)} \left(\sum_{j=1}^{H} W_{k,j}^{(2)} h^{(1)} \left(\tilde{x}^T w_j^{(1)} \right) \right) \tag{6}$$

where $h^{(1)}$ is the activation function for the hidden layer, $h^{(2)}$ is the final transfer function for the output neuron.

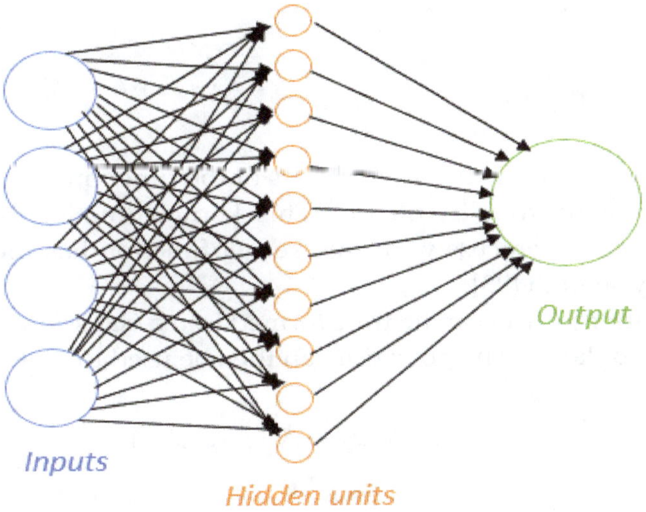

Figure 2. Artificial Neural Network schematic diagram with four inputs, ten hidden units and one output.

For the purpose of this study, the open source machine learning library based on the Torch library `PyTorch` [26] was used. The Hyperbolic Tangent function was selected as the activation function and a linear function as the final transfer function for the output layer. A standardization was made in the data set. For each fold, the validation data is divided in 90% training data and 10% test data. Since the data set has a limited number of samples, this division of the data set may play a vital role to the behaviour of the model. After dividing the data, the model is trained using back-propagation and then the performance of the model is tested in each fold. The Stochastic Gradient Descent was selected as the optimizer with an early stop when the learning curve was converged. The learning rate was set to be 0.005. Two replicates of the model for each fold were created in order to train the model with the minimum loss and different starting point of backward pass. For selecting the optimal network architecture an inner loop was added to the 10-fold cross validation. Inside the loop, networks with one or two hidden layers and one to twenty neurons in each layer were trained and tested. The model with ten neurons and one hidden layer was selected as the model with the best performance and a weight decay (L2 equivalent) regularisation was implemented in order to reduce the error and prevent over-fitting [26].

3.3. Gaussian Process Regression

A Gaussian process is a random process in which any point $x \in \mathbb{R}^d$ is assigned a random variable $f(x)$ and the joint distribution of a finite number of these variables $p(f(x_1), \ldots, f(x_N))$ is itself Gaussian:

$$p(f|X) = \mathcal{N}(f|\mu, K) \tag{7}$$

where $f = (f(x_1), \ldots, f(x_N))$ is a vector of function values of random variable x, $\mu = (m(x_1), \ldots, m(x_N))$ is a vector of mean values of random variable x and $K = K_{ij} = \kappa(x_i, x_j)$ is a positive definite kernel function or covariance function [27,28]. Given these points, a Gaussian process is a distribution over functions whose shape is defined by a positive definite kernel function or covariance function K. In general terms, a kernel or covariance function specifies the statistical relationship between two points x_i, x_j in the input space; that is, how distinctly a change in the value of the Gaussian process at x_i correlates with a change in the Gaussian process at x_j. If points x_i and x_j are considered to be similar by the kernel or covariance, the function values at these points, $f(x_i)$ and $f(x_j)$, can be expected to be similar too.

In this work, since the short-term DELs for all the load channels and electrical power output are expected to be smooth, the squared exponential kernel, also known as Gaussian kernel or Radial Basis Function (RBF) kernel was used:

$$K = K_{ij} = \kappa(x_i, x_j) = \sigma^2 e^{-\frac{1}{2l^2}(x_i - x_j)^T (x_i - x_j)} \tag{8}$$

where the length parameter l controls the smoothness of the function and σ determines the average distance of the function away from its mean. This kernel function is infinitely differentiable, which means that the Gaussian process with this kernel function has mean square derivatives of all orders, and is thus very smooth [28].

A Gaussian process prior $p(f|X)$ can be transformed into a Gaussian process posterior $p(f|X, y)$ by including some observed data y. The posterior can then be used to make predictions f_* given new input data X_*:

$$p(f_*|X_*, X, y) = \mathcal{N}(f_*|\mu_*, \Sigma_*) \tag{9}$$

$$\mu_* = K_*^T K_y^{-1} y \tag{10}$$

$$\Sigma_* = K_{**} - K_*^T K_y^{-1} K_* \tag{11}$$

where $K_y = \kappa(X, X) + \sigma_n^2 I$, $K_* = \kappa(X, X_*)$, $K_{**} = \kappa(X_*, X_*)$ and σ_n^2 is the noise term in the diagonal of K_y. This term is set to zero if training targets are noise-free and to a value greater than zero if observations are noisy.

The implementation of the Gaussian Process Regression was made with the help of Python package named `sklearn.gaussian_process.GaussianProcessRegressor` [25] that is developed based on [28]. In this case, a standardization was not made in the data set. Again, 10-fold cross-validation was used and the data was divided in 90% training and 10% test set for each fold. The noise level was taken relatively small $\sigma_n^2 = 10^{-10}$. Furthermore, the default optimiser was used and the number of restarts of the optimiser for finding the kernel's parameters which maximise the log-marginal likelihood, was set to two [25].

In order to test the robustness of Gaussian Process Regression model, the performance of the model was evaluated using different number of input data sets by excluding some sample points randomly. The different input data sets consisted 742, 660, 577, 495 and 412 sample points, the results are discussed in Section 4.3.

4. Results and Discussions

4.1. Model Comparison

A quantitative comparison is carried out by computing the Normalized-Root-Mean-Square Error NRMSE:

$$\text{NRMSE} = \frac{\sqrt{\frac{\sum_{i=1}^{N_k}(\hat{y}_i - y_i)^2}{N_k}}}{\mathbb{E}(y)} \quad (12)$$

where N_k is the number of observations, $\hat{y}_i$ is the prediction variable, y_i is the observed variable and $\mathbb{E}(y)$ is the expected value of the observation variables y.

In Figure 3, the NRMSEs for all predicted variables for all models are presented. The load channels that were used are: blade root flapwise (BRflap), blade root edgewise (BRedge), blade root torsional (BRtor), tower bottom fore-aft (TBFA), tower bottom side-side (TBSS), tower bottom torsional (TBtor), yaw bearing tilt (YBtilt), yaw bearing roll (YBroll), yaw bearing yaw (YByaw), main bearing tilt (MBtilt), main bearing yaw (MByaw) and main bearing torsional (MBtor) moments. In addition to the loads, the prediction of electrical power (Power) is also of interest in this work.

	LR NRMSE [%]	ANN NRMSE [%]	GPR NRMSE [%]
Power	24.33	54.66	0.07
BRflap	7.92	16.85	1.41
BRedge	2.41	5.14	0.3
BRTor	9.21	5.42	1.27
TBFA	13.58	18.99	2.11
TBSS	18.25	32.27	8.3
TBtor	7.51	15.75	0.48
YBtilt	8.04	13.85	0.41
YBroll	18.99	14.74	2.25
YByaw	7.66	13.95	0.41
MBtilt	7.88	12.73	0.61
MByaw	7.83	12.34	0.35
MBtor	28.41	22.7	1.16

Figure 3. Normalised root mean squared error (NRMSE) for all predicted variables for Linear Regression (LR), Artificial Neural Network (ANN) and Gaussian Process Regression (GPR).

In Linear Regression (LR), for all the prediction variables, the NRMSE are relatively big. This is as expected as the trend of the prediction variables tends to be non-linear. The only prediction variable that has relatively small error is BRedge. This is due to the fact that BRedge of short term DELs has linear trend and it is mainly affected by gravity. In Artificial Neural Network (ANN), the NRMSE for many of the predicted short term DELs of the load channels and electrical power are higher than the Linear Regression model. More specifically, the ANN has better performance than Linear Regression only in the predicted short-term DELs of BRtor, YBroll and MBtor. Artificial Neural Networks is supposed to have the ability to learn and model non-linear and complex relationships. With this in mind, it would be expected that the ANN would have greater performance than Linear Regression. This is not the case because of the small data set (825 samples). To determine if the ANN model over-fits, the relationship between training and test error was observed. Moreover, the ANN model was tested in a holdout data set. These procedures could not give a valid proof that the model over-fits due to the small training and test sample. When the sample for testing is small, there is a possibility that the model behaves well in this specific small test sample but fails to capture the overall performance. As other studies have shown in many scientific fields [29,30], ANN typically have poor performance in complex problems with a limited number of training cases. In this specific case, the ANN has too many parameters to fit properly and therefore Linear Regression is more robust and outperforms ANN [31]. In Gaussian Process Regression (GPR), the NRMSE for most of the predicted variables is significantly lower in comparison to the other two models. More specifically, the NRMSEs for electrical power and BRedge, TBtor, YBtilt, YByaw, MBtilt, MByaw short-term DELs are below 1%, whereas BRflap, BRtor, TBFA and MBtor are below 2%. The outlier is TBSS, which gives a relatively high NRMSE of 8.3%. These results indicate that the Gaussian Process Regression model has high accuracy and can be trusted to predict with this (relative small) input data set, the electrical power and short DEL of a small wind farm. The scatter plots of the estimated variables against the simulated loads and electrical power are shown in Figures 4–7. In these scatter plots in many cases, the results show that the GPR method gives a relative good prediction of the target variables.

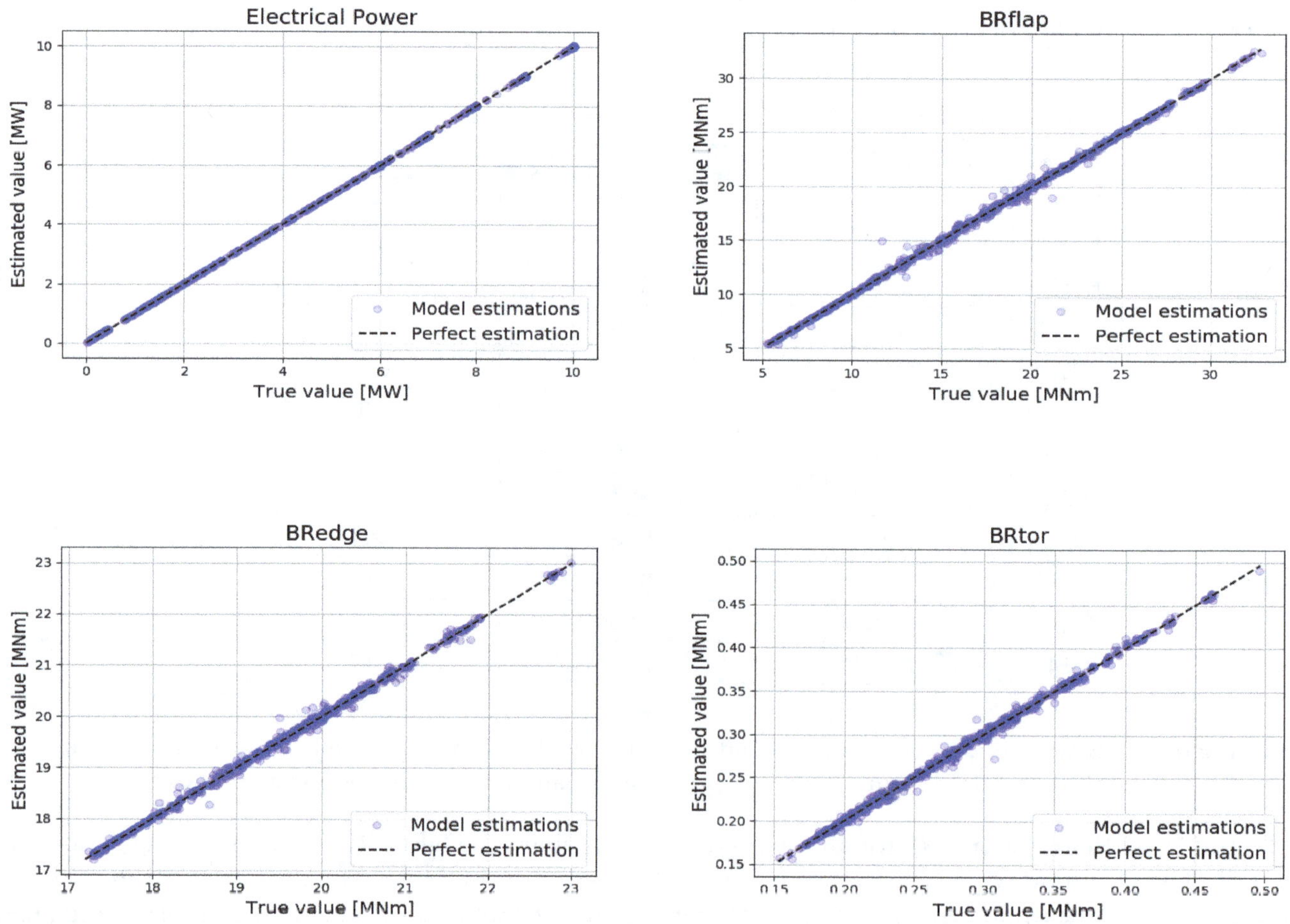

Figure 4. Electrical power, BRflap, BRedge and BRtor estimated and observed values for GPR method.

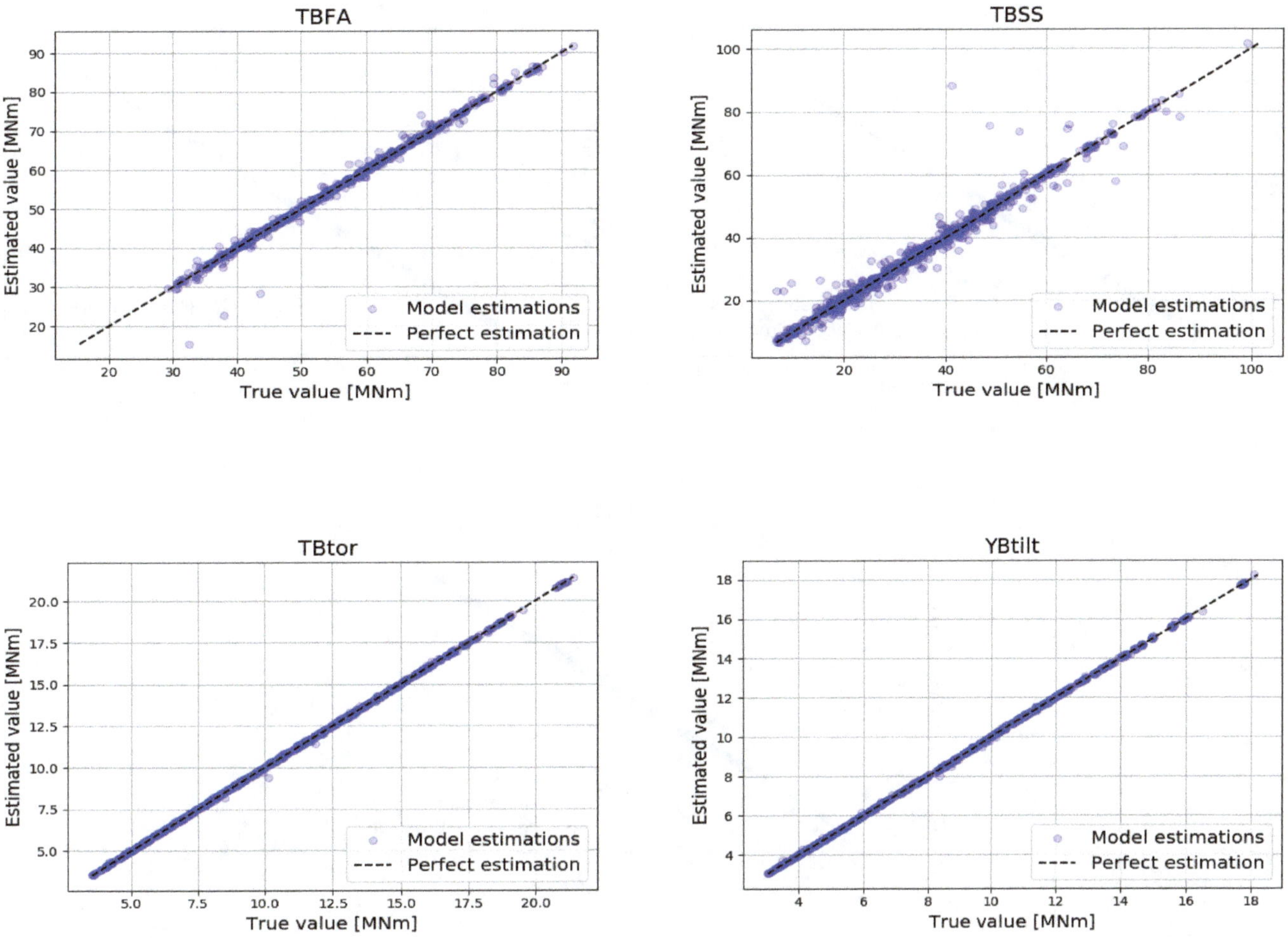

Figure 5. TBFA, TBSS, TBtor and YBtilt estimated and observed values for GPR method.

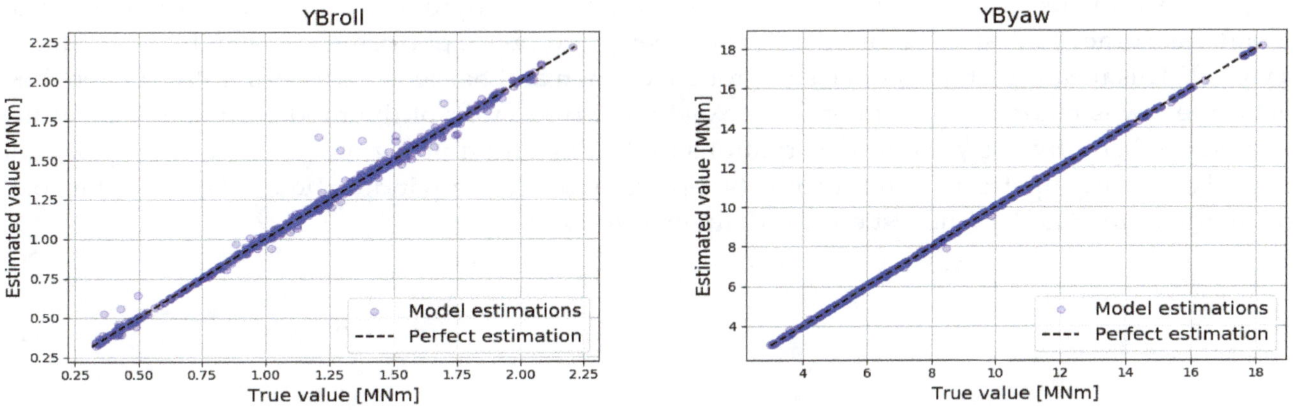

Figure 6. YBroll and YByaw estimated and observed values for GPR method.

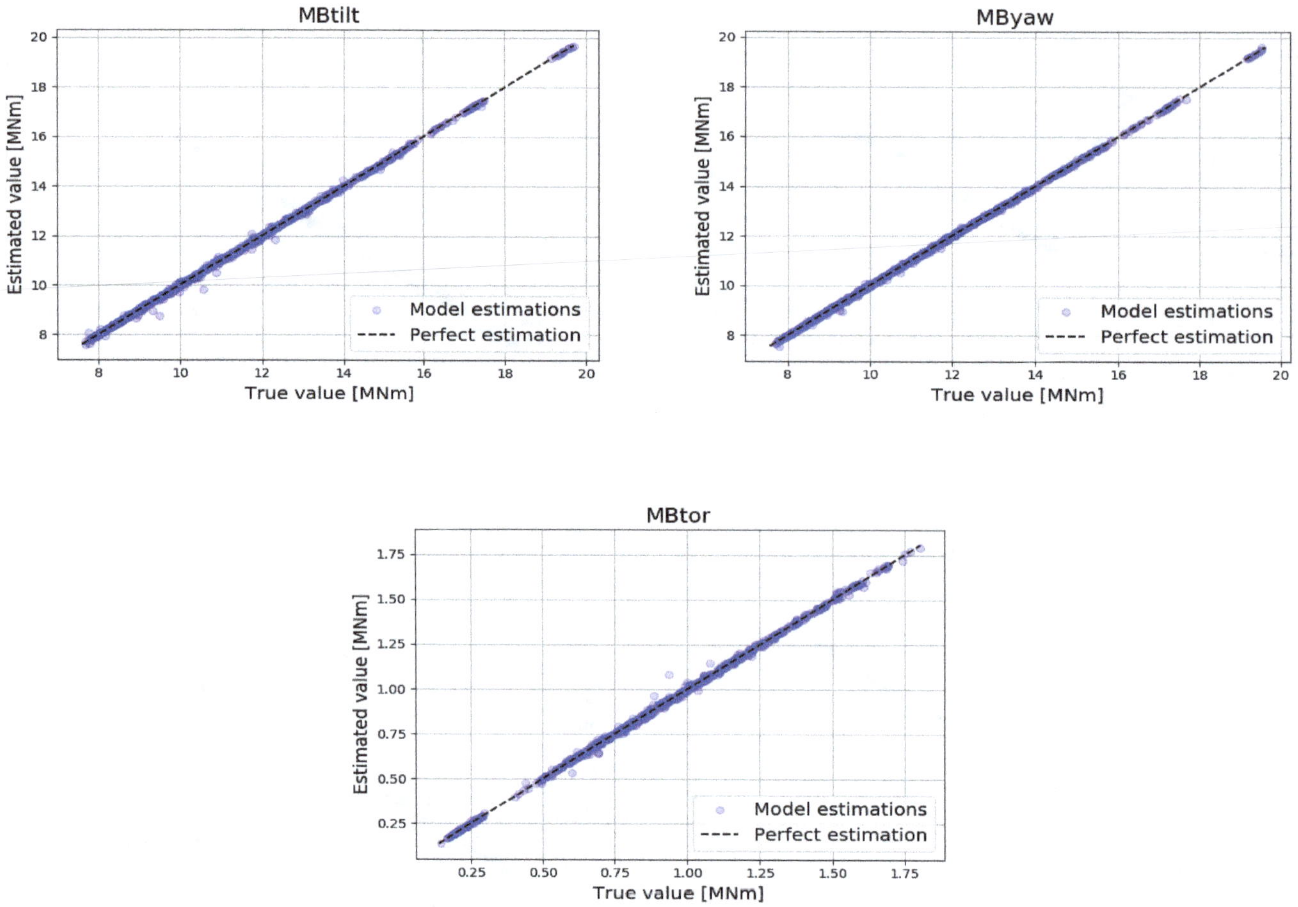

Figure 7. MBtilt, MByaw and MBtor estimated and observed values for GPR method.

4.2. Performance Evaluation of Gaussian Process Regression: Full Data Set

As previously discussed, Gaussian Process Regression has high accuracy in the prediction of short DELs and electrical power. However, due to uncertain nature of data, the model may result in better accuracy but fail to realise the data properly and hence performs poorly when the data set is varied. This means that the model is not robust enough and hence its usage is limited. Given these facts, a test was made in order to observe exactly the behaviour of the model for specific features. More specifically, by using all the observations for testing and training the model, at $I_{\text{ref}} = 0.12$ and normal operating upstream wind turbine (set-point of upstream wind turbine = 1), the test error is given in Figures 8 and 9. The test error is defined as follows.

$$e = \frac{|y - \hat{y}|}{y} \tag{13}$$

where $\hat{y}$ denotes the estimate of the observation variable y.

As it is observed, the highest error (12%) that is found in this model is when the prediction variable is TBSS, the wind speed is 10 m/s and the downstream wind turbine is operating at 100%. The test errors of the rest of prediction variables including the electrical power shown in Figure 9 are small. Similar trends have been also found for the other two turbulence intensities of 0.14 and 0.16, while in this paper we only show the results for turbulence intensity of $I_{\text{ref}} = 0.12$.

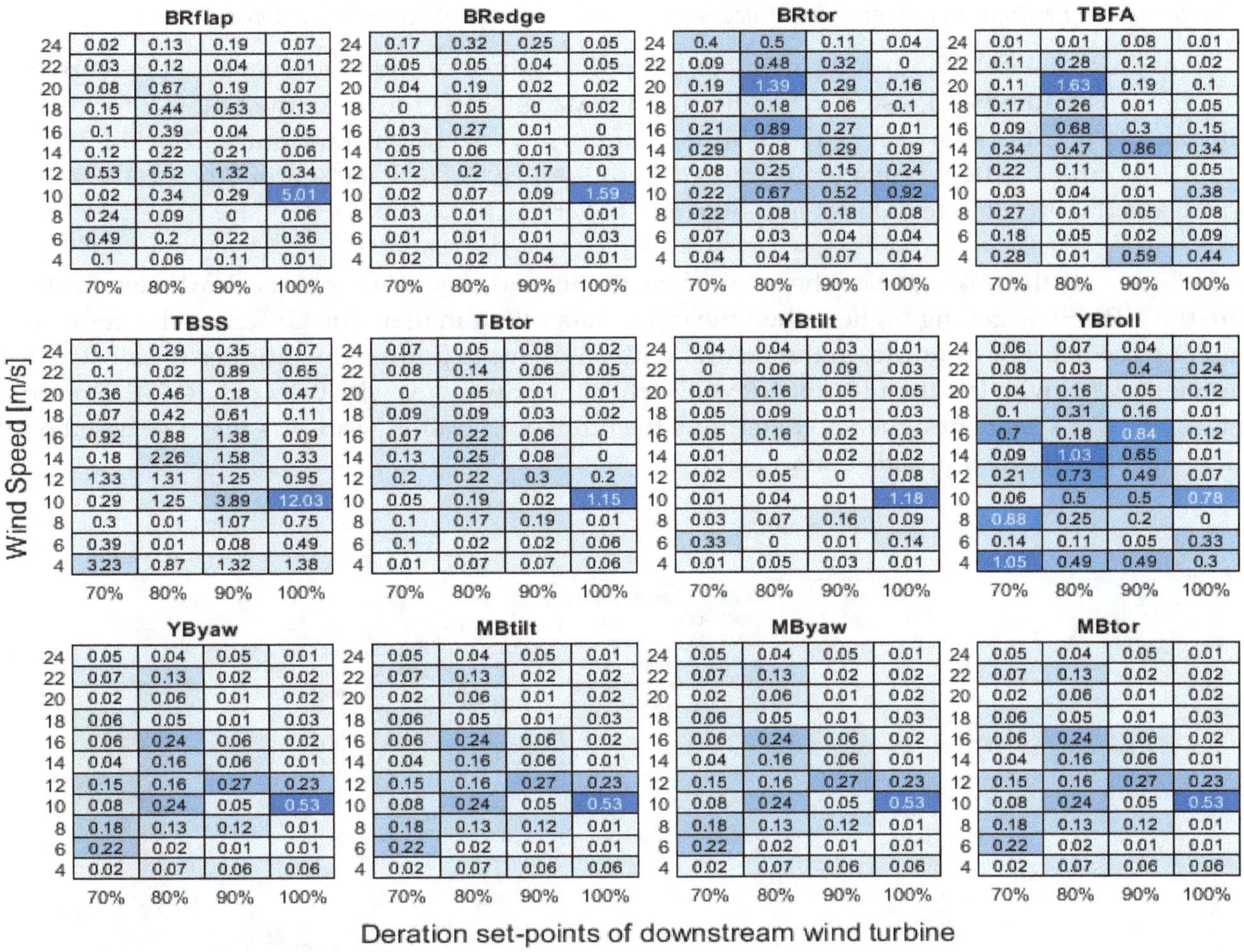

Figure 8. Test error of GPR for all DELs of different deration set-points of downstream wind turbine (x-axis) at $I_{\text{ref}} = 0.12$ with normal operating upstream wind turbine. The model was trained and tested by all input data set.

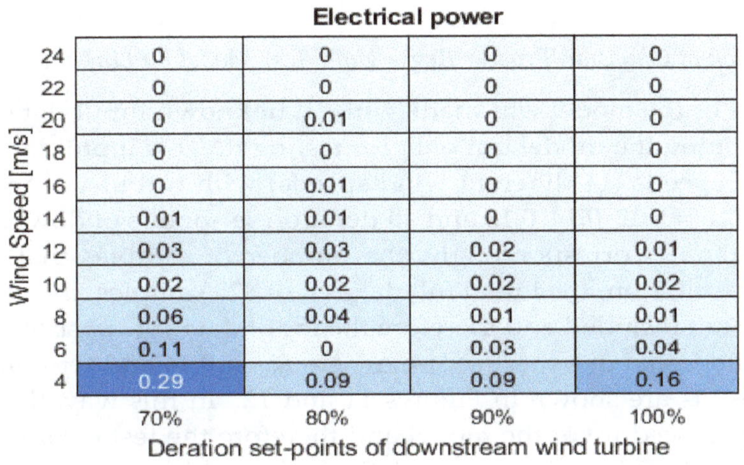

Figure 9. Test error of GPR electrical power of different deration set-points of downstream wind turbines (x-axis) at $I_{\text{ref}} = 0.12$ with normal operating upstream wind turbine. The model was trained and tested by all input data set.

4.3. Performance Evaluation of Gaussian Process Regression: Various Size of Input Data Set

To capture the behaviour and evaluate the performance of the Gaussian Process Regression, the model was trained and tested in smaller input data sets. In Figure 10, the NRMSE for different input sample points can be seen. The training samples were chosen randomly from the full data set of 825 samples. The models were trained with 90%, 80%, 70%, 60% and 50% of 825 samples. Model trained with full size data set is shown as a benchmark.

In Figure 10, the link between the error of the model and size of samples is clearly demonstrated, where the NRMSE is getting higher when the input data set is smaller. The GPR model demonstrated promising results. For example, in the results with 660 sample points, the fatigues for all structural components were with less than 10% of NRMSEs while the power was with 0.2% of NRMSE. Eventually, it is the choice of system designers to determine the number of training samples they use based on the accuracy needed by their applications.

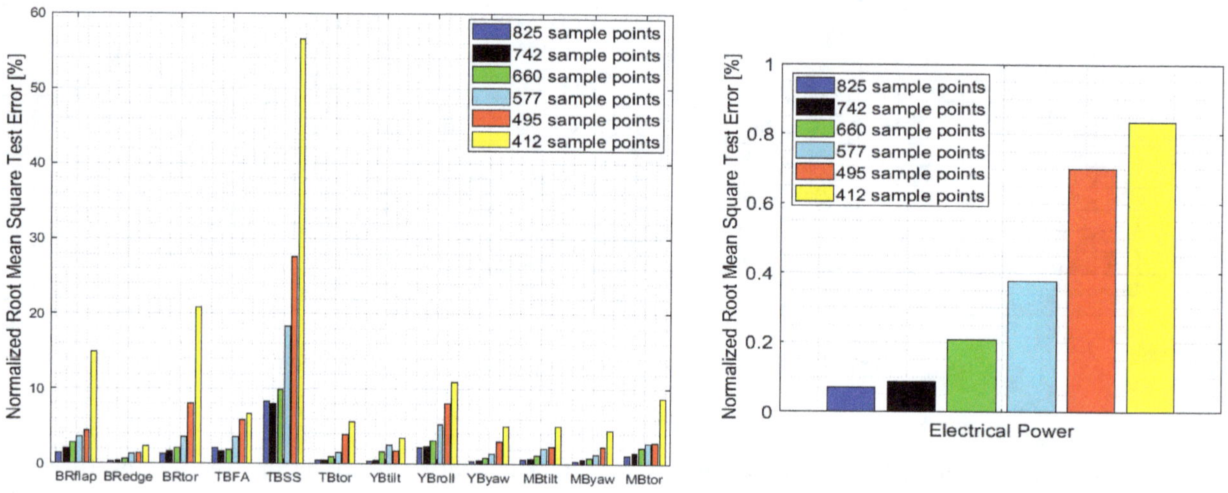

Figure 10. NRMSE of GPR for all DELs (**left**) and electrical power (**right**) for different sample points.

4.4. Performance Evaluation of Gaussian Process Regression: Test Model by Unknown Input Data Set

Finally, the last test of the model was made with an unknown input data set. This input data set was not used for training the model but only for testing it. This input data set did not contain features of all the wind speeds (11 different wind speeds) with turbulence intensities (3 different turbulence intensities) $I_{\text{ref}} = 0.12, 0.14, 0.16$ and all deration set-points of downstream wind turbine (70%, 80%, 90% and 100%). Correspondingly, the number of samples that were excluded was $11 \times 3 \times 4 = 132$. As a result, from the initial full data set of 825 samples, the previously mentioned 132 sample points were not included and therefore the input data set consisted of 693 observations. The model was saved and tested in a specific case, where $I_{\text{ref}} = 0.14$ and set-point of upstream wind turbine at 90%. The results are shown in Figures 11 and 12. In this way, this specific case is not included in the training procedure of the model and therefore the test error of this new unknown data can show how robust the model is. In general, the results are promising, where the NRMSEs for loads and electrical power are relatively small. For some channels, for example, power, the highest test error occurs at the lowest wind speeds and especially at 4 m/s. Nonetheless, for wind farm load re-balancing, load surrogate models are typically employed in the above-rated wind conditions.

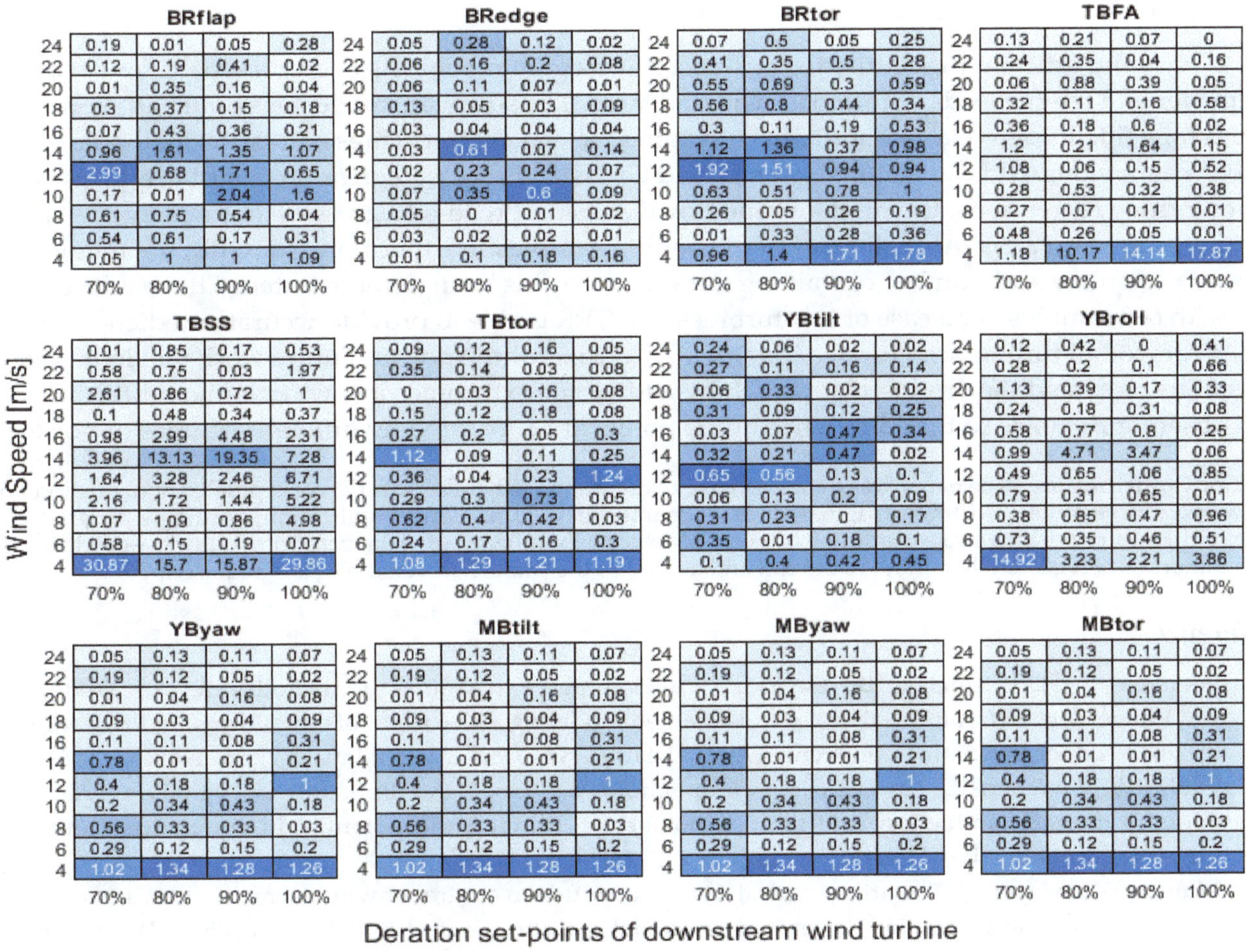

Figure 11. Test error of GPR for all DELs of different deration set-points of downstream wind turbines at $I_{ref} = 0.14$ with operating upstream wind turbine at 90%. The model was trained and tested in an input data set that does not contain input data of $I_{ref} = 0.12, 0.14, 0.16$ and downstream wind turbines at 70%, 80%, 90% and 100%.

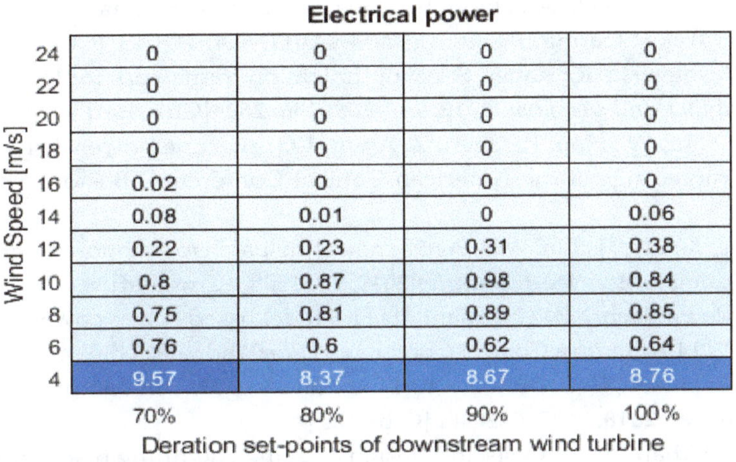

Figure 12. Test error of GPR for electrical power of different deration set-points of downstream wind turbines at $I_{ref} = 0.14$ with operating upstream wind turbine at 90%. The model was trained and tested in an input data set that does not contain input data of $I_{ref} = 0.12, 0.14, 0.16$ and downstream wind turbines at 70%, 80%, 90% and 100%.

5. Conclusions

A fast and efficient procedure to construct a surrogate model has been presented for predicting the power and fatigue loads of turbines in a wind farm. The surrogate model was built based on results from a high fidelity aero-elastic simulation including the wake effect. Three regression methods, namely, Linear Regression, Artificial Neural Network and Gaussian Process Regression, have been investigated. Among three regressions, the model trained by Gaussian Process Regression has achieved superior accuracy based on a given small amount of training samples. Furthermore, for GPR, the relationship between accuracy and number of training data samples has been demonstrated. The result revealed that with 660 samples, for a case of two turbines, the GPR is able to provide accurate predictions within 10% errors for all turbine structural loads and 0.2% errors for the power. Future work will look into the possibility to extend the proposed approach by adding more turbines in a wind farm and furthermore, add observation noise so that different sample points are corrupted by different degrees of noise.

Author Contributions: The paper was a collaborative effort among the authors. G.G.: conceptualization, methodology, software, investigation, writing—original draft preparation; W.H.L.: conceptualization, writing—original draft preparation, supervision, project administration; F.M.: conceptualization, data curation, writing—original draft preparation, supervision. All authors have read and agreed to the published version of the manuscript.

References

1. Boersma, S.; Doekemeijer, B.; Gebraad, P.; Fleming, P.; Annoni, J.; Scholbrock, A.; Frederik, J.; van Wingerden, J.W. A tutorial on control-oriented modeling and control of wind farms. In Proceedings of the 2017 American Control Conference (ACC), Seattle, WA, USA, 24–26 May 2017; pp. 1–18.
2. Frederik, J.; Weber, R.; Cacciola, S.; Campagnolo, F.; Croce, A.; Bottasso, C.; van Wingerden, J.W. Periodic dynamic induction control of wind farms: Proving the potential in simulations and wind tunnel experiments. *Wind Energy Sci.* **2020**, *5*, 245–257. [CrossRef]
3. Munters, W.; Meyers, J. Towards practical dynamic induction control of wind farms: Analysis of optimally controlled wind-farm boundary layers and sinusoidal induction control of first-row turbines. *Wind Energy Sci.* **2018**, *3*, 409–425. [CrossRef]
4. Annoni, J.; Gebraad, P.M.O.; Scholbrock, A.K.; Fleming, P.A.; van Wingerden, J.W. Analysis of axial-induction-based wind plant control using an engineering and a high-order wind plant model. *Wind Energy* **2016**, *19*, 1135–1150. [CrossRef]
5. Raach, S.; Schlipf, D.; Cheng, P.W. Lidar-based wake tracking for closed-loop wind farm control. *J. Phys. Conf. Ser.* **2016**, *753*, 052009. [CrossRef]
6. Fleming, P.; Annoni, J.; Scholbrock, A.; Quon, E.; Dana, S.; Schreck, S.; Raach, S.; Haizmann, F.; Schlipf, D. Full-Scale Field Test of Wake Steering. *J. Phys. Conf. Ser.* **2017**, *854*, 012013. [CrossRef]
7. Soleimanzadeh, M.; Wisniewski, R.; Kanev, S. An optimization framework for load and power distribution in wind farms. *J. Wind Eng. Ind. Aerodyn.* **2012**, *107–108*, 256–262. [CrossRef]
8. Zhang, B.; Soltani, M.; Hu, W.; Hou, P.; Chen, Z. A wind farm active power dispatch strategy for fatigue load reduction. In Proceedings of the American Control Conference, Boston, MA, USA, 6–8 July 2016; pp. 5879–5884.
9. Liu, Y.; Wang, Y.; Wang, X.; Zhu, J.; Lio, W.H. Active power dispatch for supporting grid frequency regulation in wind farms considering fatigue load. *Energies* **2019**, *12*, 1508. [CrossRef]
10. Mirzaei, M.; Soltani, M.; Poulsen, N.K.; Niemann, H.H. Model based active power control of a wind turbine. In Proceedings of the 2014 American Control Conference, Portland, OR, USA, 4–6 June 2014; pp. 5037–5042.
11. Lio, W.H.; Mirzaei, M.; Larsen, G.C. On wind turbine down-regulation control strategies and rotor speed set-point. *J. Phys. Conf. Ser.* **2018**, *1037*, 032040. [CrossRef]
12. Lio, W.H.; Galinos, C.; Urban, A. Analysis and design of gain-scheduling blade-pitch controllers for wind turbine down-regulation. In Proceedings of the 2019 IEEE 15th International Conference on Control and Automation (ICCA), Edinburgh, UK, 16–19 July 2019.
13. Frandsen, S. *Turbulence and Turbulence Generated Fatigue in Wind Turbine Clusters*; Technical Report Risø-R-1188; Risø National Laboratory: Roskilde, Denmark, 2007.
14. Keck, R.E.; de Maré, M.; Churchfield, M.J.; Lee, S.; Larsen, G.; Madsen, H.A. Two improvements to the dynamic wake meandering model: Including the effects of atmospheric shear on wake turbulence and incorporating turbulence build-up in a row of wind turbines. *Wind Energy* **2015**, *18*, 113–132. [CrossRef]

15. Larsen, G.C.; Madsen, H.A.; Thomsen, K.; Larsen, T.J. Wake meandering: A pragmatic approach. *Wind Energy* **2008**, *11*, 377–395. [CrossRef]
16. Larsen, T.J.; Madsen, H.A.; Larsen, G.C.; Hansen, K.S. Validation of the dynamic wake meander model for loads and power production in the Egmond aan Zee wind farm. *Wind Energy* **2013**, *16*, 605–624. [CrossRef]
17. Dimitrov, N. Surrogate models for parameterized representation of wake-induced loads in wind farms. *Wind Energy* **2019**, *22*, 1371–1389. [CrossRef]
18. Mendez Reyes, H.; Kanev, S.; Doekemeijer, B.; van Wingerden, J.W. Validation of a lookup-table approach to modeling turbine fatigue loads in wind farms under active wake control. *Wind Energy Sci.* **2019**, *4*, 549–561. [CrossRef]
19. Galinos, C.; Kazda, J.; Lio, W.H.; Giebel, G. T2FL: An efficient model for wind turbine fatigue damage prediction for the two-turbine case. *Energies* **2020**, *13*, 1306. [CrossRef]
20. Bak, C.; Zahle, F.; Bitsche, R.; Yde, A.; Henriksen, L.C.; Natarajan, A.; Hansen, M.H. *Description of the DTU 10 MW Reference Wind Turbine*; Technical Report-I-0092; DTU Wind Energy: Roskilde, Denmark, 2013.
21. Larsen, T.J.; Hansen, A.M. *How 2 HAWC2, the User's Manual*; Technical Report; DTU Wind Energy: Roskilde, Denmark, 2019.
22. Glauert, H. Airplane Propellers. In *Aerodynamic Theory*; Springer: Berlin/Heidelberg, Germany, 1935; pp. 169–360.
23. Meng, F.; Lio, A.W.H.; Liew, J. The effect of minimum thrust coefficient control strategy on power output and loads of a wind farm. *J. Phys. Conf. Ser.* **2020**, *1452*, 012009. [CrossRef]
24. Downing, S.D.; Socie, D.F. Simple rainflow counting algorithms. *Int. J. Fatigue* **1982**, *4*, 31–40. [CrossRef]
25. Pedregosa, F.; Varoquaux, G.; Gramfort, A.; Michel, V.; Thirion, B.; Grisel, O.; Blondel, M.; Prettenhofer, P.; Weiss, R.; Dubourg, V.; et al. Scikit-learn: Machine Learning in Python. *J. Mach. Learn. Res.* **2011**, *12*, 2825–2830.
26. Paszke, A.; Gross, S.; Massa, F.; Lerer, A.; Bradbury, J.; Chanan, G.; Killeen, T.; Lin, Z.; Gimelshein, N.; Antiga, L.; et al. PyTorch: An Imperative Style, High-Performance Deep Learning Library. In *Advances in Neural Information Processing Systems 32*; Wallach, H., Larochelle, H., Beygelzimer, A., d'Alché-Buc, F., Fox, E., Garnett, R., Eds.; Curran Associates, Inc.: New York, NY, USA, 2019; pp. 8024–8035.
27. Murphy, K.P. *Machine Learning: A Probabilistic Perspective*; MIT Press: Cambridge, MA, USA, 2013.
28. Rasmussen, C.; Williams, C. *Gaussian Processes for Machine Learning*; MIT Press: Cambridge, MA, USA, 2006.
29. Keshari, R.; Ghosh, S.; Chhabra, S.; Vatsa, M.; Singh, R. Unravelling Small Sample Size Problems in the Deep Learning World. In Proceedings of the 2020 IEEE Sixth International Conference on Multimedia Big Data (BigMM), New Delhi, India, 24–26 September 2020; IEEE Computer Society: Los Alamitos, CA, USA, 2020, pp. 134–143.
30. Bataineh, M.; Marler, T. Neural network for regression problems with reduced training sets. *Neural Netw.* **2017**, *95*, 1–9. [CrossRef]
31. Van der Ploeg, T.; Austin, P.; Steyerberg, E. Modern modelling techniques are data hungry: A simulation study for predicting dichotomous endpoints. *BMC Med. Res. Methodol.* **2014**, *14*, 137. [CrossRef] [PubMed]

7

Optimizing the Recognition and Feature Extraction of Wind Turbines through Hybrid Semantic Segmentation Architectures

Miguel-Ángel Manso-Callejo *, Calimanut-Ionut Cira, Ramón Alcarria and José-Juan Arranz-Justel

Departamento de Ingeniería Topográfica y Cartografía, E.T.S.I. en Topografía, Geodesia y Cartografía, Universidad Politécnica de Madrid, 28031 Madrid, Spain; ionut.cira@upm.es (C.-I.C.); ramon.alcarria@upm.es (R.A.); josejuan.arranz@upm.es (J.-J.A.-J.)

* Correspondence: m.manso@upm.es

Abstract: Updating the mapping of wind turbines farms—found in constant expansion—is important to predict energy production or to minimize the risk of these infrastructures during storms. This geoinformation is not usually provided by public mapping agencies, and the alternative sources are usually consortiums or individuals interested in mapping and study. However, they do not offer metadata or genealogy, and their quality is unknown. This article presents a methodology oriented to optimize the recognition and extraction of features (wind turbines) using hybrid architectures of semantic segmentation. The aim is to characterize the quality of these datasets and help to improve and update them automatically at a large-scale. To this end, we intend to evaluate the capacity of hybrid semantic segmentation networks trained to extract features representing wind turbines from high-resolution images and to characterize the positional accuracy and completeness of a dataset whose genealogy and quality are unknown. We built a training dataset composed of 5140 tiles of aerial images and their cartography to train six different neural network architectures. The networks were evaluated on five test areas (covering 520 km^2 of the Spanish territory) to identify the best segmentation architecture (in our case, LinkNet as base architecture and EfficientNet-b3 as the backbone). This hybrid segmentation model allowed us to characterize the completeness—both by commission and by omission—of the available georeferenced wind turbine dataset, as well as its geometric quality.

Keywords: semantic segmentation; quality control; wind turbines; feature extraction

1. Introduction

The representation of geographic elements in vector models is generally produced using various capturing and digitization technologies in which the human factor intervenes, that is, they are not free from errors. The human factor also means that it is not possible to faithfully represent a geographical reality because it is in constant change due to extreme natural actions (earthquakes, tsunamis, fires) or continuous (tectonic movements, climate change, etc.), or due to anthropic activities. This lack of accuracy in cartography and geographic information sources is manifested by omission and commission errors, and by geometric location errors.

Thanks to the advances in parallel processing technologies (Graphics Processing Units or GPUs) and artificial intelligence (AI), methodologies are able to be developed for the efficient large-scale processing of data with artificial neural networks (ANN), and obtain high success rates in predictions in different areas. One such example is Deep Learning (DL) applications in remote sensing, where it is possible to process aerial images through semantic segmentation techniques and assign land cover labels to every pixel in an image [1].

There are several works in the literature related to the use of DL techniques for monitoring the condition of wind turbines [2], or for their identification in images captured from autonomous vehicles (drones) for supporting maintenance tasks [3]. However, we did not find papers describing experiences related to the use of semantic segmentation for the extraction of wind turbines. These types of features can have different dimensions (blade diameters from 30 m to 140 m, hub heights from 40 m to 100 m), depending on their capacity to produce energy. The complexity of these structures derives from difficulties in identifying them in high-resolution aerial images due to the large areas of shadow generated, the orientations of the shadows (depending on the time the aerial image was captured), and the errors generated in the orthorectification process.

Interest in applying computer vision techniques to remotely sensed imagery has increased following the introduction of convolutional neural networks like AlexNet [4], VGGNet [5], or Inception [6]. Supported by these advancements, modern segmentation techniques emerged. Current semantic segmentation implementations use convolutional layers to downsample (encode) the images and transposed convolutions in the upsampling part (decoder) to resize the image to the original dimensions.

The two most popular semantic segmentation architectures are U-Net [7] and LinkNet [8]. They both follow the encoder–decoder structure (presented in Figure 1), which allows the use of backbone networks at each encoding step and introduce skip connections between two corresponding paths as an intent to provide local context and transfer information. State-of-the-art segmentation models focus on improving the predictions and exploring the information over larger receptive fields, on developing new links between parts of the segmentation model to provide more local context or on introducing lateral connections to allow the detection of objects at multiple scales.

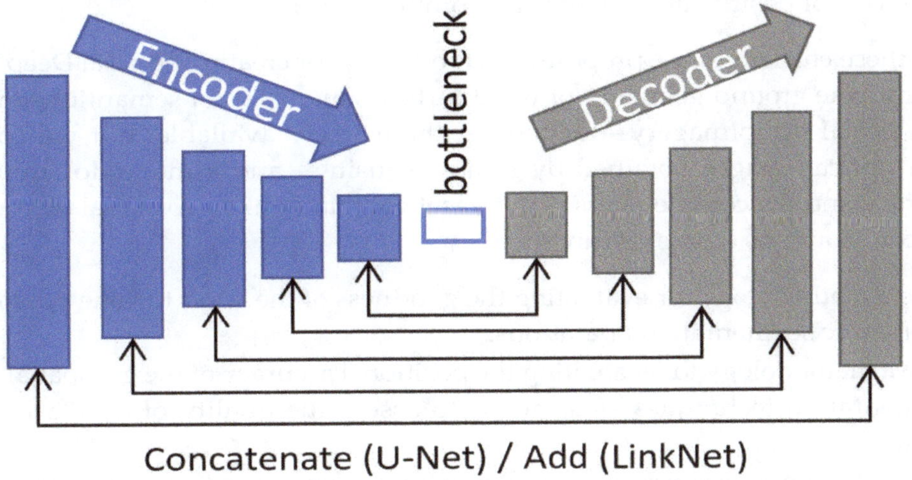

Figure 1. Simplified representation of the Encoder–Decoder Structure present in U-Net and LinkNet. Note: The encoder downsamples the input image using convolutional layers, while the decoder part upsamples the representation from the bottleneck to the original input size using transposed convolutions and enriches it using the skip connections.

The backbone networks also evolved to improve the efficiency of segmentation operation. Moving away from standard CNN architectures like VGGNet [5], new backbone networks specialized in segmentation were introduced. Some of the most popular backbone networks specialized in semantic segmentation are EfficientNet [9] and SEResNeXt [10] (based on ResNetXt [11] but using "Squeeze-and-Excitation" (SE) blocks [10]).

In remote sensing applications, semantic segmentation aims to assign a land cover class to every pixel of an image. This task becomes the operation of extracting geospatial objects from remote sensing images and is challenging due to the high variety of geospatial objects and their complexity (many displaying similar hyperspectral signatures and structural properties) and due to the strong trade-off between downsampling (which enables the abstraction and obtaining of richer information) and objects extraction (requiring a strong local context). Most of the existing methods use supervised learning to extract the geometries using their radiometric, spatial, and photometric features.

Deep learning applied to remote sensing has proven to be useful for aerospatial objects extraction [12] and aircraft recognition [13,14], for automobile detection [15], for building detection and mapping [16,17], for roof segmentation [18,19], and for vegetation and buildings segmentation [20]. Other works in the relevant literature focused on building frameworks for extracting multi-class geospatial objects (including bridges, ships, tennis courts) [21], on extracting pylons of electric transport lines [22,23], on detecting vehicles [24,25], on extracting photovoltaic panels using random forest techniques [26] or through semantic segmentation [27], or road segmentation using adversarial spatial pyramid networks [28]. One of the main drawbacks is related to the small study areas taken into account, featuring favorable scenarios (as also pointed out in [29]).

For this reason, in this article, we will explore the feasibility of training various segmentation models to extract and map wind turbines in optical high-resolution remote sensing orthorectified images. We will use an available dataset (point-type geometry) that georeferences 90% of the wind turbines in Spain, but whose genesis and sources are unknown, and therefore their quality is also unknown. We set ourselves the tasks of exploring the ability to use DL models in extracting wind turbines from aerial imagery and to characterize the quality (positional accuracy and completeness) of the said dataset in order to answer the following research questions:

i. Which semantic segmentation network generates the best results in the task of wind turbines segmentation from high-resolution aerial imagery?
ii. Can a DL model trained to extract wind turbines help in characterizing the quality of an available dataset in terms of positional accuracy and completeness?

To address these challenges, we propose a methodology for creating optimal Deep Learning-based solutions to extract the ground geometry of wind turbines by means of semantic segmentation using high-resolution aerial orthoimagery—for having them freely available, but we could have used high-resolution optical images obtained by remote sensing—and a methodology to evaluate the positional accuracy and the completeness of a dataset containing the geospatial objects.

The main contributions of this research are:

- We propose a methodology for evaluating the goodness of the wind turbine predictions extracted through semantic segmentation operations.
- We propose a methodology for evaluating the positional accuracy of the geospatial object extracted using segmentation techniques as a means to assess the quality of a dataset containing the wind turbines.
- We train five different semantic segmentation networks on a novel dataset composed of 5,140 tiles of high-resolution aerial orthoimagery and their corresponding cartographic mask containing the wind turbine feature.
- We study how changes in a network's structure (depthwise) and learning techniques affect a segmentation model's performance in the wind turbine extraction task.

The use of the best performing network, through the proposed methodology, allows to extract the features that represent wind turbines with few false-negatives and generate a set of data to use to compare and characterize the quality of another dataset of which the quality is unknown.

2. Data and Methodology

No official source with georeferenced wind turbines has been found. Some businesses in the field offer lists of their windmill farms registries, in some cases, even mentioning the coordinates of the centroids of the wind farms (e.g., "The Wind Power", by the Spanish Eolic Business Association, or the Spanish electric transport operator, e.sios). Unfortunately, they are not very useful for detecting the individual wind turbines that form the windmill farms. Nonetheless, we identified the websites of the Association of Renewable Energy Maintenance Companies ("Asociación de Empresas de Mantenimiento de Energías Renovables", AEMER) and the database generated by Antonio Marín [30], and obtained a dataset containing 18,000 georeferenced wind turbines distributed among the entire Spanish mainland and the Balearic Islands (as seen in Figure 2).

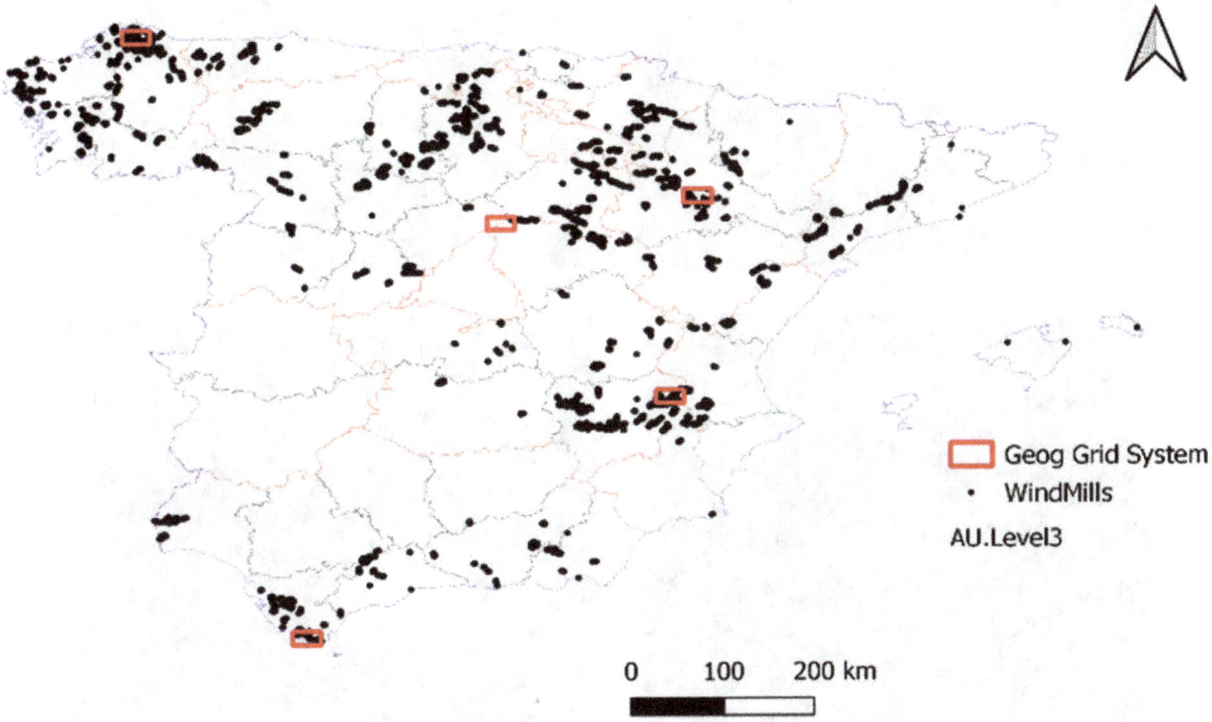

Figure 2. Geographical distribution of wind power facilities in the Spanish mainland and the Balearic Islands. Note: The location of the five areas studied in this work is marked with a red rectangle.

However, we promptly identified various quality problems related to this data repository: location errors consisting of systematic misplacements of the objects, wind turbines that do not exist in the aerial images (they may have been planned and even built, but are inexistent in the latest orthoimage freely obtained from the National Geographic Information Center ("Centro Nacional de Información Geográfica", CNIG), and wind turbines not registered in the database. In Figure 3a–d, we can find examples of georeferencing problems and the challenges generated by the differences in types, sizes, and shadows of wind turbines.

To generate the training data bank, we conducted a visual inspection of the available wind turbine dataset (vector format) superimposed on the orthoimages. We applied a general criterion of representation for the footprint of the wind turbine's base (a circle with a radius of 3 m), regardless of their size and height (since no such information was available at the individual level). The criterion of 3 m of radius for the bases of the wind turbines was adopted, after we analyzed the number of turbines aggregated by power (from 100 KW to 20 MW) and verified that 80% of the installed turbines had powers between 500 KW and 2 MW, typically using the last one, with a base diameter of 5 m. A radius of 3 m was selected for its representation so that the positioning errors of the centers, on an orthophoto of 0.5 m resolution, do not affect and the base is contained in that circle. This way, we generated segmentation masks (raster format) that contain information about the classes of each pixel "Wind Turbine Exists" or "No wind turbine", the pixels defining the base of the wind turbines being easy to identify.

Figure 3. (**a**) Misplaced coordinates resulting in positional errors, (**b**) the lower wind turbine is not georeferenced, (**c**) the turbine on the left has not yet been built and (**d**) an element that is not a turbine, but a pole for measuring atmospheric parameters and/or a lightning rod. Notes: We can also observe that the shadows are different in each image, due to the sun's inclination at the moment the image was captured. Since all the imagery has the same scale, we can also observe the differences in the sizes of the wind turbines.

For the tagging operation, we divided the latest orthoimage tiles of the PNOA [31] (Spanish "Plan Nacional de Ortofotografía Aérea de España") into tiles of 256 × 256 pixels at a scale of scale 1:2132 (~0.6 m pixel resolution) using a web application built for the tagging purpose. This web tool allows the simultaneous visualization of the aerial orthoimagery and its cartographic representation, and provides tools that allow the operators to work in parallel during the labelling operations. This same work logic has been successfully employed in other works related to the binary recognition of continuous geospatial elements (road network) in orthophotographs using deep neural networks ([32,33]). This way, we obtained a dataset of tiles containing 5140 aerial images of the wind turbines and their cartographic representations. This training dataset represents 28% of the wind turbines whose positional error and completeness accurate that we want to analyze. In Figure 4, we can find some illustrative examples of differences in size and textures around the turbines.

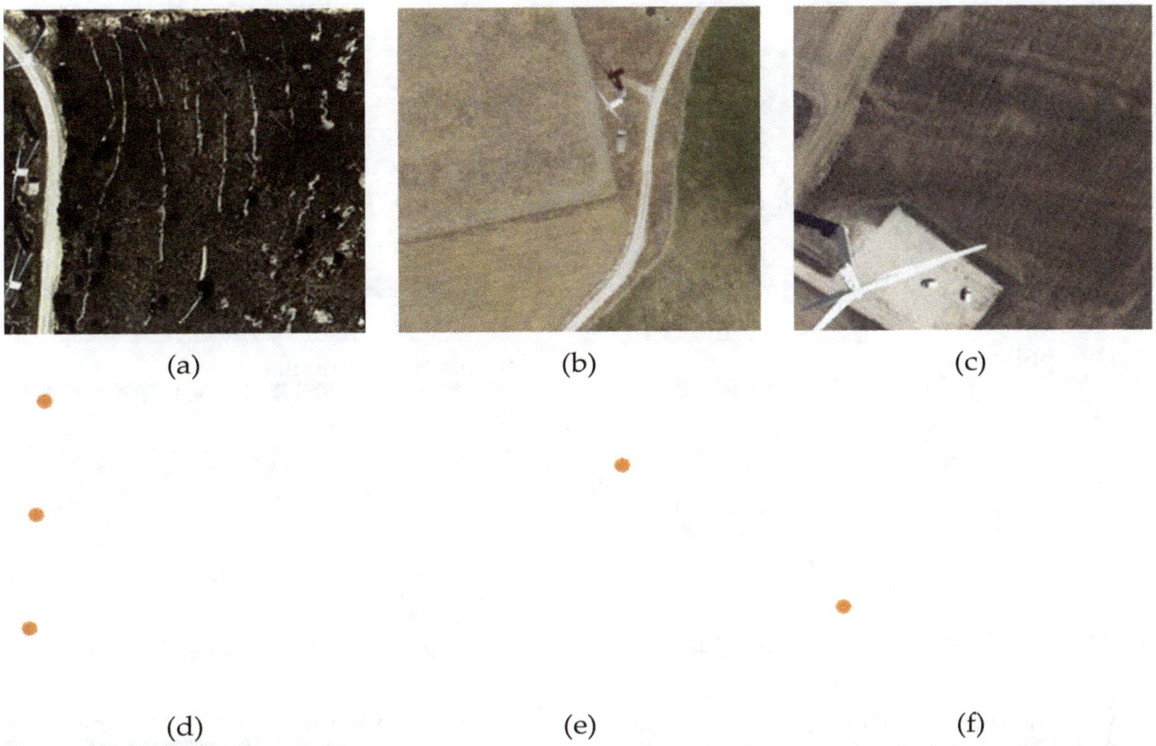

Figure 4. Examples of sets of aerial images divided into tiles of 256 × 256 pixels (**a–c**) and their correspondent cartographic representation (**d–f**) used for training the models. The differences in size of the wind turbines can be observed in the left and central images, as opposed to the one on the right.

The methodology proposed in this paper allows the design of Deep Learning solutions to segment wind turbines in high-resolution aerial orthoimagery through feature extraction.

As specified in the introduction, segmentation models are usually built using a segmentation architecture coupled with a base network (or backbone). In [34], we studied the appropriateness of using hybrid segmentation networks for extracting complex geospatial elements and contrasted their performance in comparison to state of the-art segmentation architectures, obtaining improvements in performance metrics of 2.7–3.5% when compared to the original architectures trained from scratch. The best performing hybrid models were featuring U-Net [7] as segmentation architecture with SEResNeXt50 [10] as a backbone network and LinkNet [8] coupled with EfficientNet [9] as backbone networks and will be used in this study. We also considered the b0, b1, b2 and b3 variations of EfficientNet as backbones to analyze how the depth of the backbone will impact the segmentation results.

2.1. Methodology for Evaluating the Goodness of the Semantic Segmentation Predictions

To evaluate the goodness of the predictions returned by the semantic segmentation models, we selected five areas of Spanish geography with extensions of 28 × 19 km each (National Geographical Grid System at 1:50,000 scale) from the H0008, H0383, H0432, H0767 and H1077 cartographic sheets (represented by red rectangles in Figure 2). The methodology applied for their evaluation consisted of eight steps, which are graphically summarized in Figure 5 and described below.

As a first step, traverse from North to South and West to East the latest orthoimage downloaded from the CNIG in ECW format, and extract the tiles. Given that the latest orthoimages have a spatial resolution of 0.25 m × 0.25 m and the training of the network was performed on tiles with a pixel size of 0.597 m, the resolution of the latest orthophoto was halved (to 0.5 m/ pixel). For this purpose, tiles of 512 × 512 pixels were extracted, these images being collapsed afterwards by choosing one out of every 4 pixels.

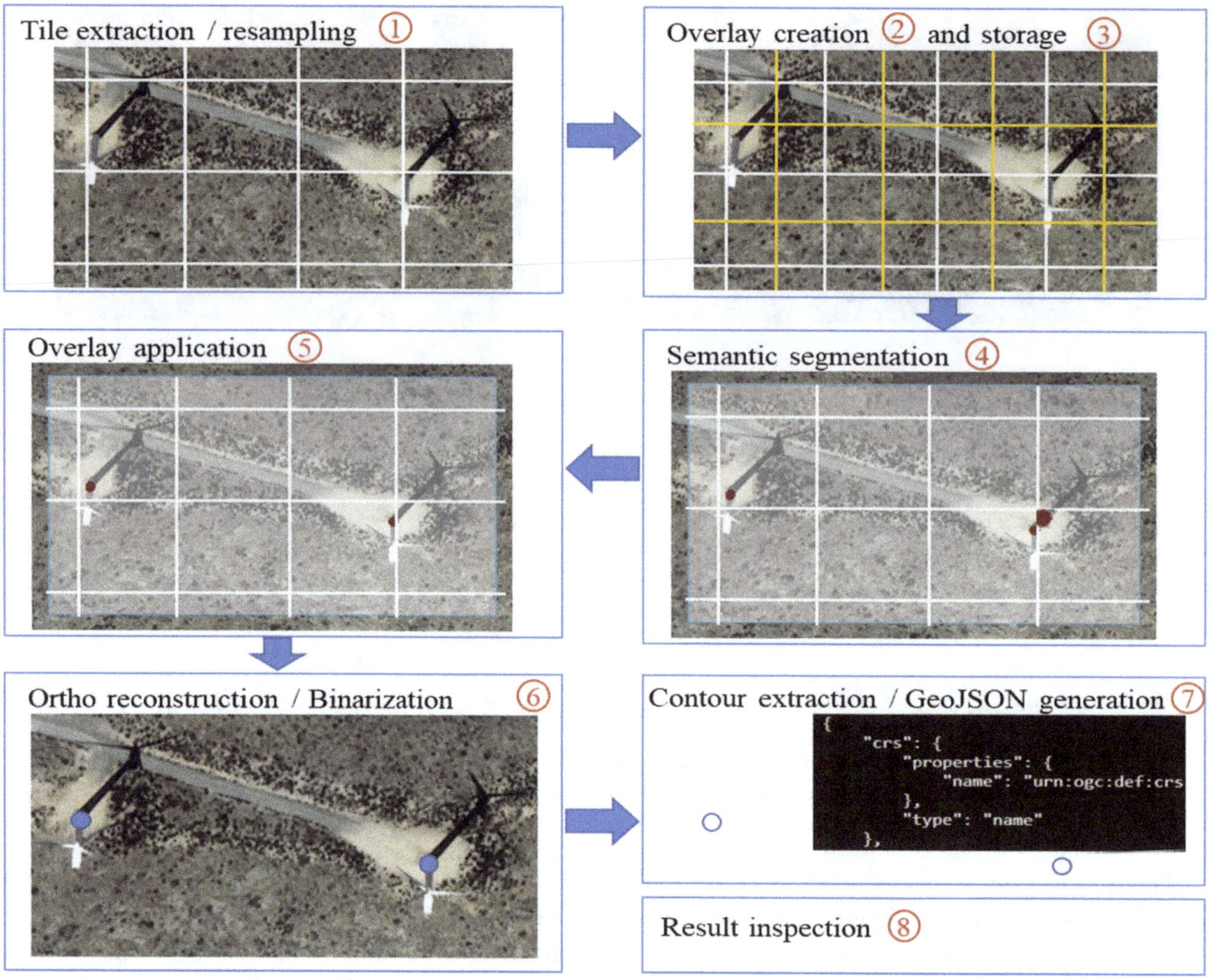

Figure 5. Graphical representation of methodology for evaluating the goodness of the semantic segmentation predictions.

Next, to avoid the common semantic segmentation problem where objects near the boundaries of an image are inaccurately extracted due to the lack of context, we apply a modified version of the method proposed by [29], where the predictions are averaged with overlays. Our proposed method consists of processing for each original 256 × 256-pixel tile, four secondary tiles of the same size, their center being the centers of the four quarters of the original tile (as shown in Figure 6). In this way, we are able to keep the central 64 × 64 pixels and discard the pixels farthest from the center. This strategy also penalizes the computation as proposed by [29], but does not average or mask results.

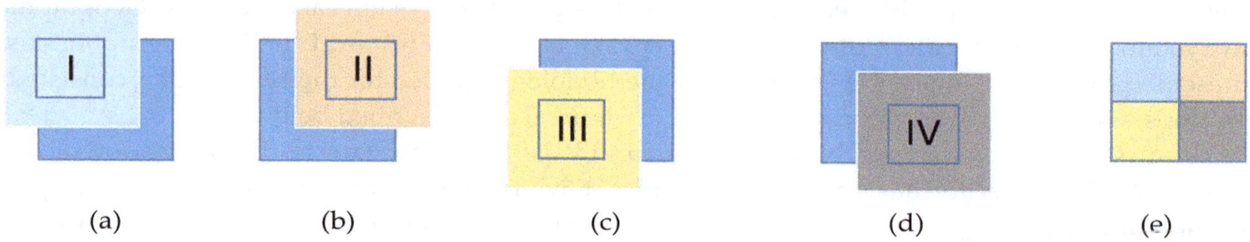

Figure 6. Abstractization of applying the methodology proposed in Section 2.1 to a random tile. We can see how four subimages are centered in each quadrant of the final tile (I, II, III, IV) of size 256 × 256 (**a**–**d**), and how the peripheral part of all of them is discarded to form an image of the same size in which the predictions close to the corners are not considered (**e**).

As a third step, a georeferenced 256 × 256 tiff image is constructed from the four matrices of 64 × 64 cells (featuring numerical values in the range (0,1) and stored on the hard disk (Figure 6).

Next, the algorithm runs through the entire orthoimage in a Z-shape (generating approximately 32,000 prediction images per analyzed region).

In the fourth step, the created overlay is used to reconstruct a prediction image with dimensions similar to the original one, through a merge operation (by generating a mosaic). The prediction image and the original will not have the same dimensions, since a 256 × 256-pixels image cannot be centered at the edges of the original image and a 64-pixel strip is wasted at each edge. On a practical level, this is not a problem since orthoimages provided by CNIG are openly available and extend beyond the boundaries of the sheets, in such a way that two adjacent orthoimages always overlap in a strip.

Next, once the orthoimage is reconstructed, a binarization operation is carried out over the mosaic of predictions, by saturating the probabilities that exceed a parameterizable threshold to 1, and the rest to 0.

As a penultimate step, the binarized image is vectorized to generate the polygons of the bases of the identified wind turbines, in an operation similar to the extraction of the contour lines from a digital elevation model. The centroids of these polygons are calculated and stored as turbine coordinates in a GeoJSON file, to facilitate their visualization and subsequent analysis.

At last, each turbine identified in the five selected regions is analyzed and contrasted with the orthophotography itself to determine errors in the identification, whether they are false-positives or false-negatives, thus evaluating the goodness of the trained hybrid segmentation model.

2.2. Methodology for Evaluating the Positional Accuracy of the Dataset

The data generated by the semantic segmentation operation in areas where wind turbines are presumed to exist was considered as reference data to determine the positional accuracy of the initial dataset. In this operation, the methodology applied consisted of:

We start by iterating to evaluate the latest orthoimage available (centered on the initial known positions of the wind turbines) with the trained semantic segmentation model to obtain a matrix of 256 rows × 256 columns containing the probability values in the range (0,1).

Next, binarize the probability matrix by saturating the high probability values to 1 and the low values to 0 in a way similar to the "nearblack or nearwhite" function.

Then, build a single band geotiff image that shares the geometric characteristics of the input orthoimage (pixel size and origin).

Afterwards, vectorize the binary image to extract the elliptical polygons (ideally circular) representing the wind turbine bases.

Finally, calculate and store the centroids of these polygons as the true coordinates of the wind turbines, adding as feature attributes the original coordinates in a GeoJSON file, to facilitate their subsequent analysis and visualization.

Once these five steps are completed, analyze the pairs of coordinates obtained to eliminate pairs of mismatched coordinates (where several turbines are identified in the same tile and the calculated coordinates correspond to only one of them) and statistically analyze the errors. The most precise data are considered those obtained from the segmentation operation, as they are obtained from orthoimages generated by an official governmental institution and their error is limited since the source of the data, its genealogy and its geometric quality are known.

It can be observed that steps 2–5 of the methodology proposed for evaluating the positional accuracy are the same as those of the methodology proposed in Section 2.1. However, this time, we work with smaller volumes of information (tiles of 256 × 256 pixels), instead of images of large dimensions (54,000 columns × 28,000 rows).

3. Implementation of the Proposed Methodology

The implementation of the methodology started with the creation of the training and test sets. Training neural networks require a large number of training examples to obtain a model capable of delivering good predictions. Creating these training banks is costly, not only due to the volume of the images required, but also because we need to ensure a correct correspondence between the segmentation masks (cartographic representation) of the wind turbines and the aerial images containing them. In Figure 4, we can find some examples of the aerial images and the representation of the bases of the wind turbines used for the training process. The dataset resulting from the tagging operation (5140 sets of tiles and their correspondent geographic representation) was randomly split into the training set (containing 4112 sets) and the test set (containing 1028 sets), following the 80–20% division criteria.

We built the networks using the deep learning "Segmentation Models" library [35] (based on Keras [36], with TensorFlow 1.14 [37] as backend) and trained them on an NVIDIA 2060 GPU. We modified the standard configurations of U-Net and LinkNet by replacing the default backbones both in the encoder and the decoder parts with networks that proved to be better suited for image segmentation tasks. The following hybrid segmentation models were considered for training: LinkNet-EfficientNet-b0, LinkNet-EfficientNet-b1, LinkNet-EfficientNet-b2, LinkNet-EfficientNet-b3 and U-Net-SEResNeXt50.

These models were trained with sets of input tiles and their segmentation mask via stochastic gradient descent to optimize the segmentation function over the training set. This method performs an update of the model´s weights and combined with backpropagation is the standard training algorithm for ANN. As hyperparameters, we considered Adam [38] to optimize the binary cross-entropy Jaccard loss function. This loss function measures the similarity between the prediction (*pr*) and the ground-truth (*gt*) when working with a network ending in a sigmoid function, $L(gt, pr) = -gt * log(pr) - (1 - gt) * log(1 - pr)$.

The segmentation model takes an RGB image of size 256 × 256 × 3 and outputs a map of size 256 × 256 × 1, by keeping the depth-wise argmax of each pixel. The pre-processing of the tiles begins with normalizing the intensity values of the pixels from the range of (0,255) to (0 and 1), this way avoiding the computation of very large numbers.

To overcome the overfitting behavior, we applied online data augmentation, exposing the model to more aspects of the data. Some of the transformations applied include random crops, random rotation and brightness, horizontal and vertical flips, and contrast and gamma shifts. We also applied feature mapping (transfer learning) from the ILSVRC dataset [39] to start from pre-trained weights instead of randomly initialized weights. This way, we ensure a better convergence of the model [40].

We started with a learning rate of 0.01; this value is reduced with a factor of 0.1 when the metrics plateaued for more than 10 epochs (up to a minimum value of $1 \times e^{-5}$). The learning process stopped when the performance stalled for more than 10 epochs or when the network started to display overfitting behavior (a pronounced higher performance on the train set and a lower performance on the test set). The models trained were stored in the open h5 format. In Figure 7, we can observe the behavior displayed by the models during the training process.

To use the models stored in the h5 exchange format, a virtual Python with Conda environment was configured on a Windows OS computer, while installing and configuring the GDAL v3.0.2 [41] and NumPy [42] libraries and dependencies.

The developed scripts rely on NumPy to handle multidimensional arrays of numbers (the initial 3-dimensional arrays of the RGB orthoimages and the final monoband array containing the prediction), to perform the decimation of the orthophotographs from higher spatial resolution (0.25 m) to a lower spatial resolution (0.5 m), to merge the centers of the prediction images (as described in Figure 6), and to perform the thresholding or binarization of the prediction images.

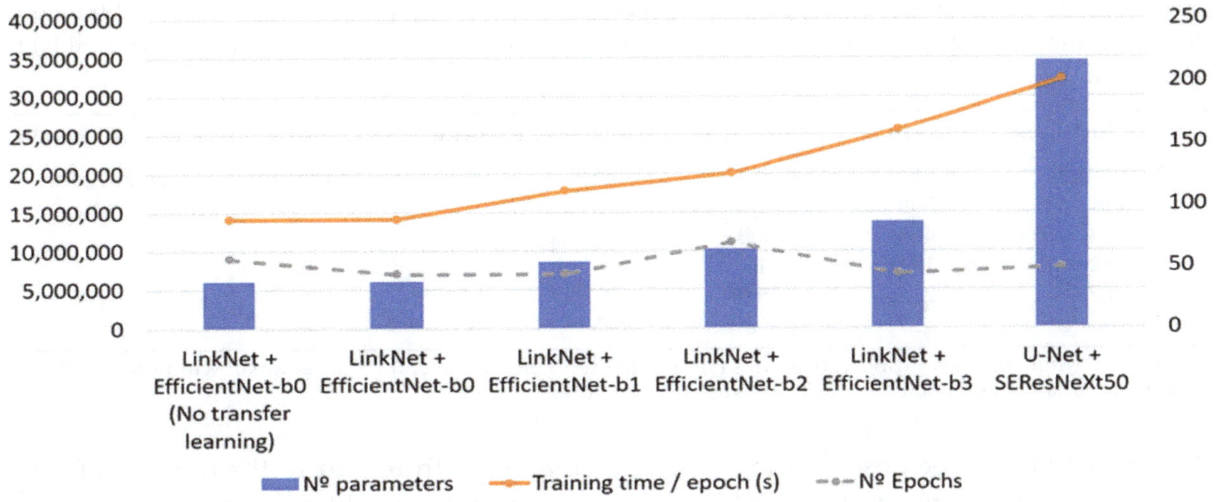

Figure 7. Description of the relation between the training time, the depth of the network and the number of epochs until convergence displayed by the semantic segmentation models during the training process.

The GDAL/ORG libraries were used to open the images stored in ECW format, to create new images both in-memory and in GeoTIFF format, to create vector datasets in memory or files in GeoJSON format, and to generate the polygons that enclose the pixels classified as belonging to the "Wind Turbine Exists" category (by applying the "*ContourGenerate*" function). To reduce the number of false-positives, we used GDAL to calculate the centroid of the feature corresponding to the georeferenced wind turbine, to eliminate the small errors related to the misclassification of other objects as a wind turbine, and to filter objects whose area is smaller than a selected threshold.

4. Results and Discussion

Table 1 shows the results of a systematical evaluation operation of the five test areas (presented in Figure 2) with the semantic segmentation networks trained to extract the positions of the wind turbines and features the number of wind turbines elements identified by the models.

Table 1. Results obtained from applying the methodology proposed in Section 2.1 (related to the semantic segmentation evaluation) to the five test areas selected.

Sheet/Model	Nº of Wind Turbines	LinkNet + EfficientNet-b0 (No Transfer Learning)	LinkNet + EfficientNet-b0	LinkNet + EfficientNet-b1	LinkNet + EfficientNet-b2	LinkNet + EfficientNet-b3	U-Net + SEResNeXt50
H1077	427	815	684	917	836	558	875
H0383	481	1749	1889	2270	1702	1803	932
H0767	278	477	344	980	714	459	391
H0342	136	395	245	622	398	246	293
H0008	598	995	787	982	2042	626	978

The lowest numbers of wind turbines detected in each area highlighted in green. This criterion is selected to minimize the number of false-positives detected by the network. When results are very close (almost the same), several networks are marked as candidates for that area. We can see that the model that shows the most figures highlighted in green is the hybrid network following the LinkNet architecture, with EfficientNet-b3 as the backbone (LinkNet + EfficientNet-b3).

In Table 2, we can find an evaluation of the completeness accuracy of the best performing model (identified in Table 1) in the five test areas.

Table 2. Results obtained from applying the second methodology proposed in Section 2.2 (related to the evaluation of completeness accuracy) to the five test areas with the hybrid network LinkNet + EfficientNet-b3.

Wind Turbines	Original	Detected with Deep Learning			Completeness Original Data	
Map sheet	#	#	False +	False −	Omission	Commission
H0008 (W)	598	626	28	1	9	8
H0383 (N-E)	481	1303	813	3 **	12	0
H0432 (C)	136	246	106	0	4	0
H0767 (E)	278	459	177	0	6	1
H1077 (S)	427	558	215	85 *	6	18

* Different Wind Turbines support structures (similar to high voltage towers). ** Three small wind turbines built next to an urban environment on a different support structure.

Heterogeneity can be observed in the results, mainly with regard to the number of detected false-positives which is generally large. This fact motivates the need for a cartographic reviewing process to eliminate the false-positives from the generated predictions, despite successfully performing joining and vectorizing operations. In the dataset generated for sheet H0383, this number seems excessive since the number of false-positives is greater than the true ones. False-negative values (when the model omits an element of the wind turbine) are acceptable; in two of the analyzed scenes there are none, while in a third the is only one, and. In the remaining two sheets, false-negatives are due to the existence of very small wind turbines (H0383) or with another type of support structure that is not cylindrical (H1077), but rather on the structure similar to a high-tension tower (as seen in Figure 8a,b) or are mounted on three cylinders that form the edges of a pyramid trunk to support the turbine (as seen in Figure 8c).

(a) (b) (c)

Figure 8. Unconventional types of wind turbines identified. In the left (**a**) and center (**b**), the turbine support structures look like high voltage transport pylons, while the example on the right (**c**) shows a structure consisting of three pillars in the shape of a stool.

Regarding the quality of the available dataset, the evaluation of the completeness in the exhaustively revised sheets shows that there are percentages of errors by omission ranging from 0.7% to 2.9%. However, there are no omission errors in three cases and, in the others, it represents 0.3% and 1.3%. If these results are extrapolated to the entire geography of the dataset, it confirms the trend in the proportions of errors by omission derived from the constructions of new wind turbines. At the end of 2019, the sector stated [43] the existence of 20,940 Wind Turbines in Spain. The 16.3% rate of the omission error suggested by this value is not too misaligned with the rates obtained by the model, knowing that the date of several of the orthophotographs used is 2017. We must also consider that the growth of the global windmill farm is not caused by the growth of existing ones, but rather by the creation of new parks (not registered in the dataset).

Regarding the quality of the positional accuracy of the initial data (18,000 features), we found 259 cases where the best performing segmentation model (LinkNet + EfficientNet-b3) has not identified any wind turbines within the 256 × 256 pixel tiles (centered on the coordinates of the objects). In 181 of these cases, a visual inspection confirmed the absence of the wind turbines (a commission error rate of 1.4% of the set, this value being higher than the systematic analysis of five well-distributed areas). On the contrary, in 69 cases, the network generated false-negatives, which represent a rate of 0.38%, within the range of values obtained in the areas studied. However, of those 69 cases, only 22 can be considered false-negative, 45 cases being caused by a special type of wind turbine base (not cylindrical, as can be seen in Figure 8a, as was the case in sheet H1077). For the remaining two cases, the orthophotographs shows different types of turbines (as seen in the Figure 8b). In addition, another nine wind turbines have a low power production capacity; therefore, they are smaller and do not physically resemble normal wind turbines.

By applying the proposed methodology systematically to the 18,000 wind turbines, a set of 18,602 positional relationships was obtained, corresponding to 17,741 different features (18,000 minus the previous 259). This difference in observations is because, in some cases, several turbines were identified in the same tile, all of them being stored. In the subsequent analysis, these cases are filtered (step 6 from Section 2.2), reaching the result of 17,741 turbines correctly identified by the network. The filtering was carried out by applying a distance criterion between the initial coordinates and those obtained from the deep learning evaluation; 861 observations being eliminated this way.

After filtering the data, we obtained a dataset that is 98.56% of the original and we proceeded to characterize the positional error in the planimetric coordinates (X, Y). The results obtained are presented in Table 3. This allows us to conclude that the dataset has positional accuracy characteristics that are a little worse for the Y component of the planimetry (0.5 m), a value that matches the resolution of the images used. In any case, the RMSE (root-mean-square error) is approximately 7 m according to this first analysis.

Table 3. Mean errors, standard deviation, and root-mean-square error (RMSE) in the planimetric coordinates (X, Y).

	Min	Max	Mean	Sigma	RMSE
X	−71.58 m	74.27 m	−0.17 m	6.94 m	6.95 m
Y	−73.89 m	74.48 m	−0.32 m	7.42 m	7.43 m

To determine the main ranges of errors in both components, the histogram of the distances was calculated by grouping the values into 19 binds. We can see that, for approximately 75% of the georeferenced wind turbines, the error in X is between −1 and 5 m and between −3 and 4 m for the Y coordinate. Both histograms are shown in Figure 9.

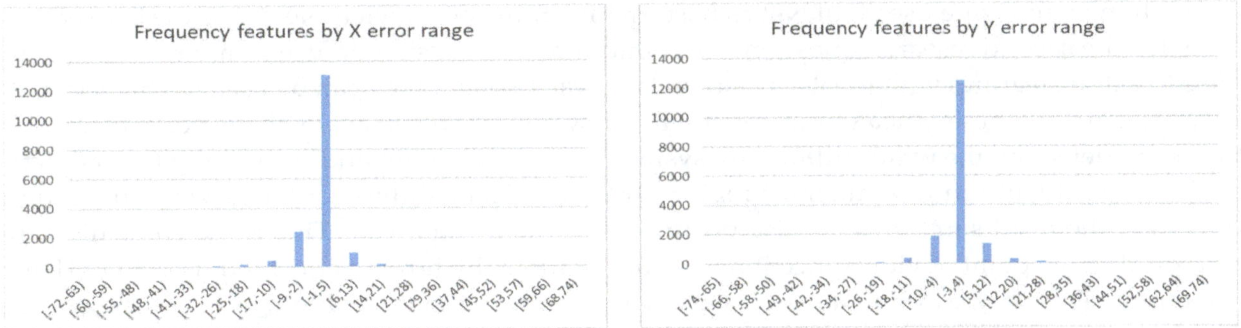

(a) (b)

Figure 9. Histograms of positional errors in the X-coordinate (**a**) and Y-coordinate (**b**) grouped in 19 binds.

5. Conclusions

In this paper, we explored the capacity of different hybrid semantic segmentation models to extract wind turbines in high-resolution aerial orthoimagery. These models were trained on a dataset generated especially for this purpose and their generalization capacity (goodness of the predictions) was evaluated on five different areas of the Spanish territory (regions of 28 km × 19 km, represented with red rectangles in Figure 2).

We also proposed a new strategy to automatically deal with the problem of inaccurate extractions near the boundaries of the images. This strategy is described in Figure 6 and consists of automatically (1) dividing a given tile into four quadrants, (2) extract four secondary images with the same dimensions correspondent to each quadrant from the orthoimage with higher resolution, (3) segment each secondary tile, and (4) only consider their central part in the final segmentation map of the original tile. By applying this strategy, we achieved considerable improvements when extracting wind turbines close to the edges.

Moreover, we proposed a methodology as a workflow that allows the generation of a continuous image with the semantic segmentation predictions and its subsequent vectorization to extract the centers from the bases of the wind turbines. This methodology enabled the comparison of the performance of several hybrid segmentation models in different regions of the Spanish geography and the selection of the one delivering the lowest number of false-positives (in our case, LinkNet + EfficientNet-b3).

We also proposed a methodology as a workflow that has allowed the analysis of the quality in terms of positional accuracy and completeness of an unofficial dataset containing georeferenced wind turbines. In Section 5, we presented the statistical discrepancies of the georeferenced data (Table 3) and the discrepancies in completeness by terms of omission and commission (Table 2). As for the quality of completeness, the values presented are objective, but must be handled with caution, as the date of the latest available orthoimage does not correspond with the latest update of the analyzed dataset. This situation was observed in areas where wind turbines were registered, although they appear under construction in the imagery.

Regarding the geometric quality of the georeferencing of the evaluated dataset, the errors are more dispersive in the X coordinate than the Y coordinate and the values are significantly higher than the resolution of the aerial orthoimagery used 0.5 m (pixel size).

It has been proven that based on the proposed methodology, it is possible to evaluate the quality parameters of datasets whose genealogy and metadata are known using deep learning and semantic segmentation techniques. However, this evaluation has an important cost in the creation of the training datasets of the networks and the tests to select the architecture or hybrid network that generates better results for the application domain (in our case, wind turbines).

The results of the experiments carried out with the six hybrid semantic segmentation networks show that applying Feature Mapping from networks pre-trained with millions of images (e.g., ImageNet Large Scale Visual Recognition Challenge) helps to obtain better results when the training dataset is not very large, as in this case (5140 tiles), as seen in Figure 7 and Table 1. In our case, the best performing segmentation architecture uses LinkNet as base architecture and EfficientNet-b3 as the backbone.

As for the second research question, we found that it is possible to use semantic segmentation models to extract features with greater positional accuracy, so that they can be used to characterize the quality of a dataset whose genealogy is unknown. We were also able to evaluate the performance of the networks in detecting the wind turbines by systematically reviewing and comparing the predictions with the ground-truth in five different regions from the Spanish territory, this way obtaining the goodness of the results in terms of false-positives and false-negatives. This is a great achievement from a productive point of view, since it noticeably reduces the human effort required to update the cartography. This way, a systematic and very laborious revision is avoided; only places where the network has extracted features need to be reviewed.

The methodology proposed to characterize the quality in the studied terms (geometric and completeness) can be applied to any other type of feature that can be modeled as point representation in the cartography (e.g., electric towers, telecommunications antennas). Additionally, it can improve the quality of collaboratively generated datasets (e.g., crowdsourcing) in which quality criteria are not well defined and the quality of each contributor's contributions cannot be quantified.

Author Contributions: Conceptualization, M.-Á.M.-C., C.-I.C. and R.A.; Data curation, M.-Á.M.-C. and J.-J.A.-J.; Formal analysis, M.-Á.M.-C., C.-I.C. and J.-J.A.-J.; Investigation, M.-Á.M.-C., C.-I.C. and R.A.; Methodology, M.-Á.M.-C. and C.-I.C.; Software, M.-Á.M.-C., C.-I.C. and R.A; Supervision, M.-Á.M.-C. and R.A.; Writing—original draft, M.-Á.M.-C. and C.-I.C.; Writing—review and editing, M.-Á.M.-C., C.-I.C., R.A., and J.-J.A.-J. All authors have read and agreed to the published version of the manuscript.

References

1. Ma, L.; Liu, Y.; Zhang, X.; Ye, Y.; Yin, G.; Johnson, B.A. Deep learning in remote sensing applications: A meta-analysis and review. *ISPRS J. Photogramm. Remote Sens.* **2019**, *152*, 166–177. [CrossRef]
2. Stetco, A.; Dinmohammadi, F.; Zhao, X.; Robu, V.; Flynn, D.; Barnes, M.; Keane, J.; Nenadic, G. Machine learning methods for wind turbine condition monitoring: A review. *Renew. Energy* **2019**, *133*, 620–635. [CrossRef]
3. Abedini, F.; Bahaghighat, M.; S'hoyan, M. Wind turbine tower detection using feature descriptors and deep learning. *Facta Univ. Ser. Electron. Energ.* **2020**, *33*, 133–153. [CrossRef]
4. Krizhevsky, A.; Sutskever, I.; Hinton, G.E. ImageNet classification with deep convolutional neural networks. *Commun. ACM* **2017**, *60*, 84–90. [CrossRef]
5. Simonyan, K.; Zisserman, A. Very Deep Convolutional Networks for Large-Scale Image Recognition. *arXiv* **2014**, arXiv:1409.1556.
6. Szegedy, C.; Liu, W.; Jia, Y.; Sermanet, P.; Reed, S.; Anguelov, D.; Erhan, D.; Vanhoucke, V.; Rabinovich, A. Going deeper with convolutions. In Proceedings of the 2015 IEEE Conference on Computer Vision and Pattern Recognition (CVPR), Boston, MA, USA, 7–12 June 2015; pp. 1–9.
7. Ronneberger, O.; Fischer, P.; Brox, T. U-Net: Convolutional Networks for Biomedical Image Segmentation. In Proceedings of the Medical Image Computing and Computer-Assisted Intervention–MICCAI 2015, Munich, Germany, 5–9 October 2015; Navab, N., Hornegger, J., Wells, W.M., Frangi, A.F., Eds.; Springer International Publishing: Cham, Switzerland, 2015; pp. 234–241.
8. Chaurasia, A.; Culurciello, E. LinkNet: Exploiting Encoder Representations for Efficient Semantic Segmentation. In Proceedings of the 2017 IEEE Visual Communications and Image Processing (VCIP), St. Petersburg, FL, USA, 10–13 December 2017; pp. 1–4. [CrossRef]
9. Tan, M.; Le, Q.V. EfficientNet: Rethinking Model Scaling for Convolutional Neural Networks. *arXiv* **2019**, arXiv:1905.11946.
10. Hu, J.; Shen, L.; Sun, G. Squeeze-and-Excitation Networks. In Proceedings of the 2018 IEEE/CVF Conference on Computer Vision and Pattern Recognition, Salt Lake City, UT, USA, 18–23 June 2018; pp. 7132–7141.
11. Xie, S.; Girshick, R.B.; Dollár, P.; Tu, Z.; He, K. Aggregated Residual Transformations for Deep Neural Networks. In Proceedings of the 2017 IEEE Conference on Computer Vision and Pattern Recognition (CVPR 2017), Honolulu, HI, USA, 21–26 July 2017; pp. 5987–5995.
12. Cai, B.; Jiang, Z.; Zhang, H.; Zhao, D.; Yao, Y. Airport Detection Using End-to-End Convolutional Neural Network with Hard Example Mining. *Remote Sens.* **2017**, *9*, 1198. [CrossRef]
13. Zuo, J.; Xu, G.; Fu, K.; Sun, X.; Sun, H. Aircraft Type Recognition Based on Segmentation with Deep Convolutional Neural Networks. *IEEE Geosci. Remote Sens. Lett.* **2018**, *15*, 282–286. [CrossRef]
14. Li, Y.; Fu, K.; Sun, H.; Sun, X. An Aircraft Detection Framework Based on Reinforcement Learning and Convolutional Neural Networks in Remote Sensing Images. *Remote Sens.* **2018**, *10*, 243. [CrossRef]
15. Ding, P.; Zhang, Y.; Deng, W.-J.; Jia, P.; Kuijper, A. A light and faster regional convolutional neural network for object detection in optical remote sensing images. *ISPRS J. Photogramm. Remote Sens.* **2018**, *141*, 208–218. [CrossRef]
16. Alidoost, F.; Arefi, H. A CNN-Based Approach for Automatic Building Detection and Recognition of Roof Types Using a Single Aerial Image. *PFG J. Photogramm. Remote Sens. Geoinf. Sci.* **2018**, *86*, 235–248. [CrossRef]
17. Ma, J.; Wu, L.; Tang, X.; Liu, F.; Zhang, X.; Jiao, L. Building Extraction of Aerial Images by a Global and Multi-Scale Encoder-Decoder Network. *Remote Sens.* **2020**, *12*, 2350. [CrossRef]
18. Chen, Q.; Wang, L.; Wu, Y.; Wu, G.; Guo, Z.; Waslander, S.L. Aerial Imagery for Roof Segmentation: A Large-Scale Dataset towards Automatic Mapping of Buildings. *arXiv* **2018**, arXiv:1807.09532. [CrossRef]

19. Yang, H.L.; Yuan, J.; Lunga, D.; Laverdiere, M.; Rose, A.; Bhaduri, B. Building Extraction at Scale Using Convolutional Neural Network: Mapping of the United States. *IEEE J. Sel. Top. Appl. Earth Obs. Remote Sens.* **2018**, *11*, 2600–2614. [CrossRef]
20. Shorter, N.; Kasparis, T. Automatic Vegetation Identification and Building Detection from a Single Nadir Aerial Image. *Remote Sens.* **2009**, *1*, 731–757. [CrossRef]
21. Li, Y.; Zhang, Y.; Huang, X.; Yuille, A.L. Deep networks under scene-level supervision for multi-class geospatial object detection from remote sensing images. *ISPRS J. Photogramm. Remote Sens.* **2018**, *146*, 182–196. [CrossRef]
22. Dutta, T.; Sharma, H.; Vellaiappan, A.; Balamuralidhar, P. Image Analysis-Based Automatic Detection of Transmission Towers using Aerial Imagery. In *Pattern Recognition and Image Analysis*; Paredes, R., Cardoso, J.S., Pardo, X.M., Eds.; Lecture Notes in Computer Science; Springer International Publishing: Cham, Switzerland, 2015; Volume 9117, pp. 641–651. ISBN 978-3-319-19389-2.
23. Tragulnuch, P.; Chanvimaluang, T.; Kasetkasem, T.; Ingprasert, S.; Isshiki, T. High Voltage Transmission Tower Detection and Tracking in Aerial Video Sequence using Object-Based Image Classification. In Proceedings of the 2018 International Conference on Embedded Systems and Intelligent Technology & International Conference on Information and Communication Technology for Embedded Systems (ICESIT-ICICTES), Khon Kaen, Thailand, 7–9 May 2018; pp. 1–4.
24. Lu, J.; Ma, C.; Li, L.; Xing, X.; Zhang, Y.; Wang, Z.; Xu, J. A Vehicle Detection Method for Aerial Image Based on YOLO. *J. Comput. Commun.* **2018**, *6*, 98–107. [CrossRef]
25. Cao, Y.; Wang, G.; Yan, D.; Zhao, Z. Two Algorithms for the Detection and Tracking of Moving Vehicle Targets in Aerial Infrared Image Sequences. *Remote Sens.* **2016**, *8*, 28. [CrossRef]
26. Malof, J.M.; Bradbury, K.; Collins, L.M.; Newell, R.G.; Serrano, A.; Wu, H.; Keene, S. Image features for pixel-wise detection of solar photovoltaic arrays in aerial imagery using a random forest classifier. In Proceedings of the 2016 IEEE International Conference on Renewable Energy Research and Applications (ICRERA), Birmingham, UK, 20–23 November 2016; pp. 799–803.
27. Yu, J.; Wang, Z.; Majumdar, A.; Rajagopal, R. DeepSolar: A Machine Learning Framework to Efficiently Construct a Solar Deployment Database in the United States. *Joule* **2018**, *2*, 2605–2617. [CrossRef]
28. Shamsolmoali, P.; Zareapoor, M.; Zhou, H.; Wang, R.; Yang, J. Road Segmentation for Remote Sensing Images using Adversarial Spatial Pyramid Networks. *arXiv* **2020**, arXiv:2008.04021. [CrossRef]
29. Dong, R.; Li, W.; Fu, H.; Gan, L.; Yu, L.; Zheng, J.; Xia, M. Oil palm plantation mapping from high-resolution remote sensing images using deep learning. *Int. J. Remote Sens.* **2020**, *41*, 2022–2046. [CrossRef]
30. Marín, A. Wtg_Spain_27052017. Available online: https://amezet.carto.com/builder/fd430f8c-41cb-11e7-876d-0ecd1babdde5/embed (accessed on 5 November 2020).
31. Instituto Geográfico Nacional Plan Nacional de Ortofotografía Aérea. Available online: https://pnoa.ign.es/caracteristicas-tecnicas (accessed on 25 November 2019).
32. De la Fuente Castillo, V.; Díaz-Álvarez, A.; Manso-Callejo, M.-Á.; Serradilla García, F. Grammar Guided Genetic Programming for Network Architecture Search and Road Detection on Aerial Orthophotography. *Appl. Sci.* **2020**, *10*, 3953. [CrossRef]
33. Cira, C.-I.; Alcarria, R.; Manso-Callejo, M.-Á.; Serradilla, F. A Framework Based on Nesting of Convolutional Neural Networks to Classify Secondary Roads in High Resolution Aerial Orthoimages. *Remote Sens.* **2020**, *12*, 765. [CrossRef]
34. Cira, C.-I.; Alcarria, R.; Manso-Callejo, M.-Á.; Serradilla, F. A Deep Learning-Based Solution for Large-Scale Extraction of the Secondary Road Network from High-Resolution Aerial Orthoimagery. *Appl. Sci.* **2020**, *10*, 7272. [CrossRef]
35. Yakubovskiy, P. *Segmentation Models*; GitHub: San Francisco, CA, USA, 2019; Available online: https://github.com/qubvel/segmentation_models (accessed on 16 October 2020).
36. Chollet, F. *Keras*. 2015. Available online: https://github.com/fchollet/keras (accessed on 16 October 2020).
37. Abadi, M.; Barham, P.; Chen, J.; Chen, Z.; Davis, A.; Dean, J.; Devin, M.; Ghemawat, S.; Irving, G.; Isard, M.; et al. TensorFlow: A system for large-scale machine learning. In Proceedings of the 12th USENIX Symposium on Operating Systems Design and Implementation (OSDI '16), Savannah, GA, USA, 2–4 November 2016; p. 21.
38. Kingma, D.P.; Ba, J. Adam: A Method for Stochastic Optimization. *arXiv* **2014**, arXiv:1412.6980.

39. Russakovsky, O.; Deng, J.; Su, H.; Krause, J.; Satheesh, S.; Ma, S.; Huang, Z.; Karpathy, A.; Khosla, A.; Bernstein, M.; et al. ImageNet Large Scale Visual Recognition Challenge. *Int. J. Comput. Vis.* **2015**, *115*, 211–252. [CrossRef]
40. Gupta, S.; Girshick, R.B.; Arbeláez, P.A.; Malik, J. Learning Rich Features from RGB-D Images for Object Detection and Segmentation. In *Proceedings of the 13th European Conference on Computer Vision (ECCV 2014), Zurich, Switzerland, 6–12 September 2014*; Proceedings, Part VII; Fleet, D.J., Pajdla, T., Schiele, B., Tuytelaars, T., Eds.; Springer: Berlin/Heidelberg, Germany, 2014; Volume 8695, pp. 345–360.
41. GDAL/OGR Contributors. *GDAL/OGR Geospatial Data Abstraction Software Library*; Open Source Geospatial Foundation: Chicago, IL, USA, 2020; Available online: https://gdal.org/index.html (accessed on 30 March 2020).
42. Van Der Walt, S.; Colbert, S.C.; Varoquaux, G. The NumPy array: A structure for efficient numerical computation. *Comput. Sci. Eng.* **2011**, *13*, 22–30. [CrossRef]
43. Asociación Empresarial Eólica-Spanish Wind Energy Association-Energía Eólica La eólica en España. Available online: https://www.aeeolica.org/sobre-la-eolica/la-eolica-espana (accessed on 7 November 2020).

8

Wind Turbine Power Curve Upgrades

Davide Astolfi * and Francesco Castellani

Department of Engineering, University of Perugia, Via G. Duranti 93, 06125 Perugia, Italy; francesco.castellani@unipg.it

* Correspondence: davide.astolfi@unipg.it

Abstract: Wind turbine power upgrades have recently become a debated topic in wind energy research. Their assessment poses some challenges and calls for devoted techniques: some reasons are the stochastic nature of the wind and the multivariate dependency of wind turbine power. In this work, two test cases were studied. The former is the yaw management optimization on a 2 MW wind turbine; the latter is a comprehensive control upgrade (pitch, yaw, and cut-out) for 850 kW wind turbines. The upgrade impact was estimated by analyzing the difference between the post-upgrade power and a data-driven simulation of the power if the upgrade did not take place. Therefore, a reliable model for the pre-upgrade power of the wind turbines of interest was needed and, in this work, a principal component regression was employed. The yaw control optimization was shown to provide a 1.3% of production improvement and the control re-powering provided 2.5%. Another qualifying point was that, for the 850 kW wind turbine re-powering, the data quality was sufficient for an upgrade estimate based on power curve analysis and a good agreement with the model result was obtained. Summarizing, evidence of the profitability of wind turbine power upgrades was collected and data-driven methods were elaborated for power upgrade assessment and, in general, for wind turbine performance control and monitoring.

Keywords: wind energy; wind turbines; control and optimization

---

1. Introduction

The wind capacity worldwide is impressively growing and furthermore many multi-megawatt wind turbines have been operating for years. The production optimization therefore has two main directions as regards each single wind turbine: On the one side, diminishing the unavailability time through condition-based maintenance strategies. For example, it is estimated that the unavailability time of a modern wind turbine is currently of the order of 3% [1] and can further diminish. On the other side, the technology update of wind turbines in their operational lifetime has been flourishing in the latest years and has been producing non-negligible improvements of wind kinetic energy conversion efficiency: the assessment and the methodologies for studying these wind turbine power upgrades constitute the topic of the present work.

For completeness, it should be said that production optimization can be conceived also at the wind farm level and there is very interesting scientific research devoted to layout optimization [2–6], wind farm control [7–9], and yaw active control for wake interactions management [10–14].

There are basically two types of wind turbine power upgrades that are currently employed in operating wind turbines: aerodynamic and control upgrades, or possibly a combination of the two. Examples of aerodynamic retrofitting of the blades are installation of vortex generator, passive flow control devices, Gurney flaps and so on [15–23]. Control upgrades typically deal with pitch [24,25], rotor revolutions per minute [26], and yaw management. The increase of production can be achieved also by modifying the wind speed cut-out management, as discussed, for example, in [27–29].

It likely happens that wind farm manufacturers and wind farm owners cooperate as regards to the technology improvement of operating wind turbines with forms of profit sharing of wind turbine power upgrades. This fact has considerably stimulated the high-level analysis of operational data in the industry and the collaboration with academia. Actually, there are several critical points about the assessment of wind turbine power upgrades:

- The wind source is stochastic and it does not make sense to compare the cumulative production before and after a power upgrade.
- It is difficult to account for the multiple dependency of wind turbine power on climate and operating conditions.
- It can be difficult to reliably know the wind conditions on site, in general because nacelle anemometers are mounted behind the rotors of the wind turbines and in particular because cup anemometers might not provide adequate measurement precision.

On these grounds, the power curve study might be a reliable tool for assessing power upgrades only when considerably long datasets are available, in order to avoid the effect of seasonal biases due to the variation of climate conditions on site. If, as commonly happens, wind farm practitioners aim at obtaining an estimate after just few months of upgrade operation, more complex and powerful methods are needed. A certain amount of literature has been flourishing about this problem and some interesting methods have recently been proposed. The common ground is the following idea:

- After the upgrade, of course, the power production is known if operation data are available.
- The production improvement is the difference between the measured production post-upgrade and a simulation of how much the wind turbine would have produced, in the same conditions, if the upgrade did not take place.
- The simulation must be achieved with a model based on pre-upgrade.

Chronologically, the first relevant study is [24]: in that work, a modification of the Gaussian kernel regression method is proposed to account for the multivariate dependency of the power of the wind turbine. Two upgrade test cases are studied: one is aerodynamic (vortex generator installation) and is studied through the analysis of operation data and the other regards the control of the pitch and is studied artificially because the pitch behavior is simulated and data are synthesized accordingly. In [30], another critical point of this kind of problems is discussed in depth: the statistical significance and the dataset dimensionality. The proposed solution is the use of time-resolved operation data, rather than Supervisory Control And Data Acquisition (SCADA) data. The former kind of data actually has sampling time of the order of the second, while the latter kind of data has sampling time of some minutes (typically, ten). In [30], it is shown that, using the time-resolved data, it is possible to obtain results that are similar to the ones from the Kernel-plus of [24], but with a much simpler method: it is the so-called power–power or side-by-side and it is based on the study of the power difference between the target (upgraded) wind turbine and a reference wind turbine, before and after the upgrade of the wind turbine of interest. In [25], three test cases of wind turbine power curve upgrades are considered: pitch angle optimization near the cut-in, vortex generators and passive flow control devices installation, cut-out management optimization. The first two test cases are studied by modeling the pre-upgrade power of the wind turbines of interest using an Artificial Neural Network (ANN) model having as input some operation variables of the nearby wind turbines. A control upgrade, dealing with the rotor revolutions per minute optimization in order to reach the most appropriate induction level, is studied in [26]: in that work, the power-power method is generalized by modeling the power of the upgraded wind turbine through a multivariate linear, employing as input variables some operation parameters of the nearby wind turbines. For other issues regarding this topic, see also [31–33].

On these grounds, the objective of the present work was furnishing further contributions to the topic of wind turbine power curve upgrades assessment. For doing this, two test cases were considered:

- The first test case deals with the yaw management optimization on a 2 MW wind turbine. There is a considerable literature about the potentiality of wind turbine efficiency improvement through the advances in yaw management (see, for example, [34]), but, at this stage, the available studies mainly deal with simulation estimates (for example, recently, in [35], a yaw control strategy based on reinforced learning is designed). To the best of the authors knowledge, the study in this work is the first in the literature that is based on wind turbines in operation.

- The second test case deals with a control upgrade on a 850 kW wind turbine. Since the technology of this kind of device is gradually becoming obsolete, the re-powering on this wind turbine has been more impacting and has dealt with pitch, yaw, and cut-out management optimization. An interesting point about this upgrade is that the measuring chain was improved, through the installation of a sonic anemometer. Furthermore, the wind farm manufacturer arranged a testing period of the upgrade for some months, by alternating half-hour intervals characterized by the operation with the pre- and post-upgrade control logic. Therefore, it was possible to compare the two power curves quite reliably, because the wind speed data have good quality and because the data were collected in the same period and seasonal biases were therefore avoidable. This gives the possibility of verifying the model-based estimate of the production improvement through another, independent, approach.

The two above test cases were studied with particular attention to the methodology. The selected model for the power of the wind turbines of interest was a multivariate linear and it was decided that several operation parameters of the nearby wind turbines could in principle be input variables for the model. This can be considered a generalization of the concept of rotor-equivalent wind speed [36]: the conditions on site can be described, for example, through the blade pitches, the rotor revolutions per minute, and the power output of the wind turbines constituting the wind farm. As discussed in detail throughout the manuscript, remarkable collinearity between the possible covariates of the models was observed and for this reason a principal component regression [37] was employed, differently with respect, for example, to [33], where a stepwise regression algorithm [38] was used for input variables selection for an ordinary least squares regression. This approach is general and does not depend on the test case: therefore, it can be considered a contribution to the methodologies for wind turbine performance control and monitoring.

As regards the selected test cases, the results of this work are that the yaw control optimization on the 2 MW wind turbine provided a production improvement of 1.3% of the AEP; and the 850 kW wind turbine re-powering provided an improvement of 2.5% of the AEP.

The structure of the manuscript is as follows. In Section 2, the test cases and the datasets are described. Section 3 is devoted to the methods: the employed model is discussed in general and implemented in particular for the two selected test cases. In Section 4, the results for the production improvement are collected and discussed. Section 5 is devoted to the conclusions and to some further directions of the present work.

2. The Test Cases and the Datasets

One wind turbine for each test case wind farm underwent the corresponding upgrade (WTG02 in Wind Farm 1 and WTG022 in Wind Farm 2). Actually, the wind farm owner has been adopting the following approach as regards power upgrades: selecting some test wind turbines and, after some months of operation, assessing the impact of the upgrade on the grounds of studies such as the present one. Subsequently, the wind farm owner decides if it is worth extending the upgrade to the other wind turbines in the wind farm.

The employed datasets were obtained from the SCADA collected databases of the wind turbines. Their quality was checked as follows:

- Data were filtered on the request that all wind turbines in the wind farm were productive. This was done using the appropriate operation time counter available in the dataset.
- The quality of the anemometer data was crosschecked overall for each wind turbine through the analysis of the average power curve against the theoretical one and no relevant anomalies were detected.

- The quality of the data for each time step for each wind turbine was crosschecked by comparing the actual power production for the measured nacelle wind speed against the theoretical power curve. If a deviation larger than 30% was detected, the measurement was rejected.

2.1. Test Case 1: Yaw Control Optimization, 2 MW Wind Turbine

The wind farm of interest is composed of six horizontal-axis three-bladed wind turbines having 2 MW of rated power each and the rotor diameter is 92.5 m. The cut-in is 3.5 m/s and the cut-out is 25 m/s. The nominal wind speed is 14.5 m/s.

The layout of the wind farm is reported in Figure 1 and the wind turbine of interest (WTG02) is indicated in red. The wind farm is sited onshore in a gentle terrain in southern Italy. The inter-turbine distances go from the order of 7 rotor diameters (between nearest neighbors) up to the order of 19 rotor diameters.

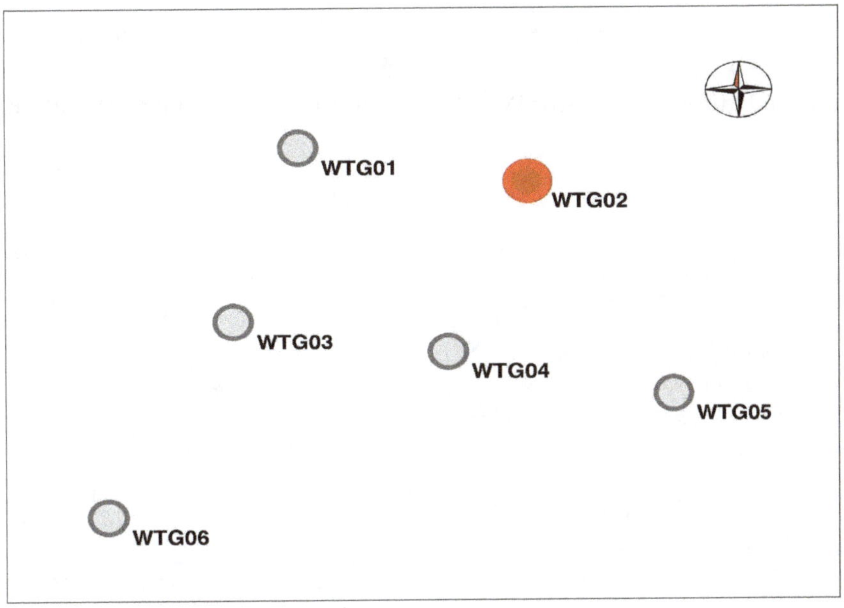

Figure 1. The layout of Wind Farm 1.

The data available were organized in two datasets as follows:

- The first dataset is denoted as D_{bef} and contains the data collected from 1 January 2017 to 20 August 2018. It is a period prior to the yaw control upgrade on turbine WTG02. It is composed of 35,971 data.
- The second dataset is denoted as D_{aft} and contains the data collected from 1 September 2018 to 1 January 2019. It is a period after the control optimization on turbine WTG02. It is composed of 9288 data.

In Figure 2, the normalized autocovariance of the power output of WTG02 is reported as a function of the lag (up to 20) for the D_{bef} dataset. This was done to crosscheck the assumption that each measurement can be considered independent with respect to the others.

The wind direction roses, measured at WTG02, during D_{bef} and D_{aft} are reported in Figure 3 and it arises that the distributions are very similar before and after the upgrade of the WTG02. Therefore, it can be argued that, as far as can be analyzed from the data available, the model formulation and use are not biased by differences in climatology. This is supported also by the fact that the ratio between the average nacelle wind speeds at WTG02 during D_{bef} and D_{aft} is 1.04.

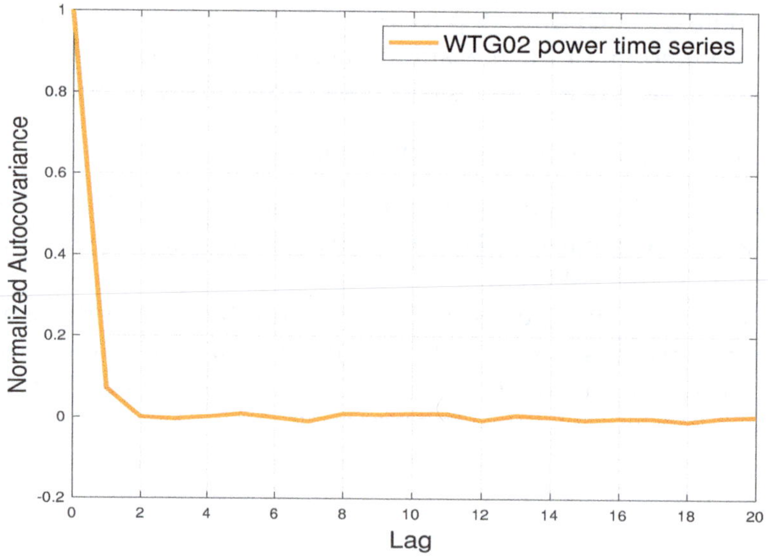

Figure 2. The normalized autocovariance of WTG02 power output (D_{bef} dataset) with maximum lag of 20.

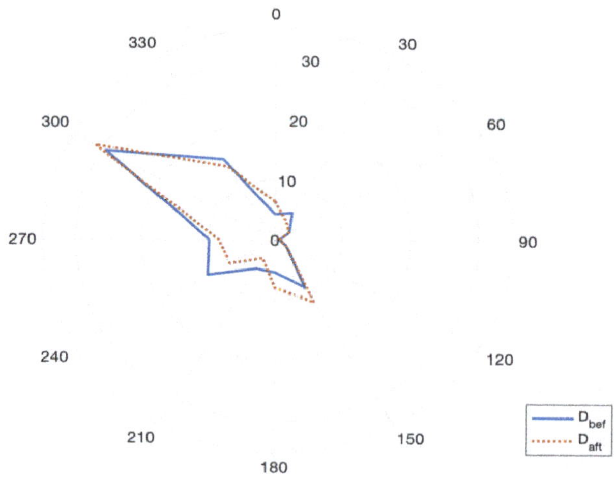

Figure 3. The wind direction rose for Wind Farm 1 during D_{bef} and D_{aft}.

The SCADA collected data have ten minutes of sampling time. The available validated measurements are:

- nacelle wind speed;
- nacelle wind direction;
- nacelle position;
- temperature outside the nacelle;
- active power;
- rotor rotational speed;
- generator rotational speed; and
- reference blade pitch.

The effect of the control upgrade was an improvement of the yaw management, especially for low and moderate wind intensities. This resulted in a decrease of the occurrence of high yaw misalignment. Qualitatively, this was assessed as follows: the yaw misalignment of WTG02 was computed as the difference between the wind direction measured at the nacelle and the position of the nacelle itself. As a pre-requisite, data were filtered on the request that WTG02 was in production. The plot of the yaw misalignment (in degrees) against the power is reported in Figure 4 for the datasets D_{bef} and D_{aft}.

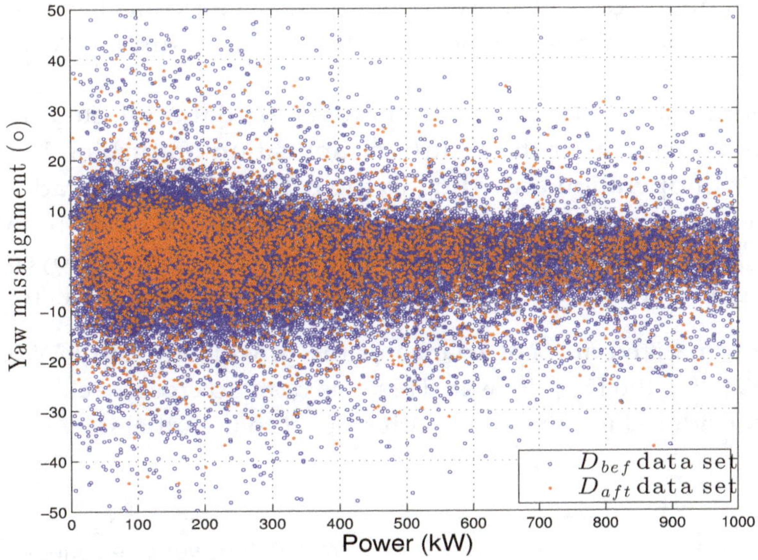

Figure 4. WTG02 yaw angle misalignment as a function of the power, for datasets D_{bef} and D_{aft}.

2.2. Test Case 2: Control Re-Powering, 850 kW Wind Turbine

The wind farm of interest is composed of twenty-three horizontal-axis three-bladed wind turbines having 850 kW of rated power each and the rotor diameter is 58 m. The cut-in is 3 m/s and the cut-out is 20 m/s. The nominal wind speed is 12.5 m/s.

The layout of the wind farm is reported in Figure 5 and the wind turbine of interest (WTG022) is indicated in red. The wind farm is sited onshore in a gentle terrain in northern France. The inter-turbine distances go from the order of four rotor diameters (between nearest neighbors) to the order of 100 rotor diameters. The wind direction rose on-site is quite uniform.

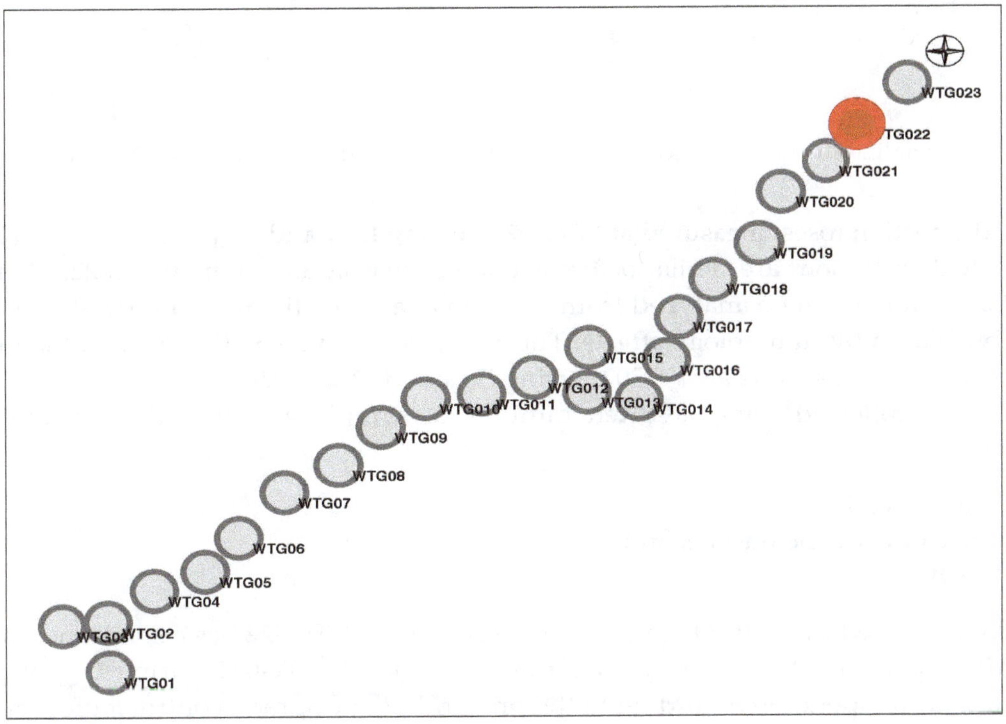

Figure 5. The layout of Wind Farm 2.

The data available were organized into two datasets as follows:

- The first dataset is denoted as D_{bef} and contains the data collected from 1 February 2018 to 20 August 2018. It is a period prior to the control upgrade on turbine WTG022. It is composed of 15,353 data.
- The second dataset is denoted as D_{aft} and contains the data collected from 24 August 2018 to 1 April 2019. It is a period after the control optimization on turbine WTG022. In this period, half-hour intervals of operation according to the pre- and post-upgrade logic were alternated. The former subset is indicated as D_{aft}^{non-up} and is composed of 4245 data. The latter subset is indicated as D_{aft}^{up} and is composed of 4265 data. Only D_{aft}^{up} is employed for the model-based estimate of the upgrade, while both subsets are employed for the power curve study in Section 4.2.

In Figure 6, the normalized autocovariance of the power output of WTG022 is reported as a function of a lag up to 20, for the D_{bef} dataset. This was done to crosscheck the assumption that each measurement can be considered independent with respect to the others.

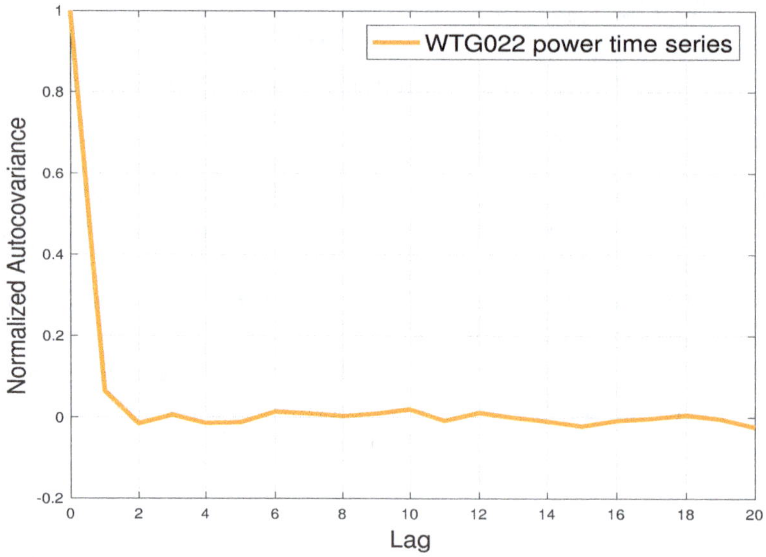

Figure 6. The normalized autocovariance of WTG022 power output (D_{bef} dataset) with maximum lag of 20.

The wind direction roses, measured at WTG022, during D_{bef} and D_{aft} are reported in Figure 7 and it arises that the distributions are similar before and after the upgrade of the WTG022. Therefore, it can be argued that, as far as can be analyzed from the data available, the model formulation and use are not remarkably biased by climatology effects. This is supported also by the fact that the ratio between the average nacelle wind speeds at WTG02 during D_{bef} and D_{aft} is 0.96.

The SCADA collected data have ten minutes of sampling time. The available validated measurements are:

- nacelle wind speed;
- temperature outside the nacelle; and
- active power.

It should be noticed that, after the upgrade intervention, WTG022 has a sonic anemometer at the nacelle. As discussed in detail in Section 4.2, the operation of WTG022 during D_{aft} was as follows: half-hour intervals of operation according to the pre- and post-upgrade control logic were alternated. Therefore, to assess the upgrade using the techniques proposed in Section 3, only the data in D_{aft} characterized by operation according to the upgraded control were selected. Instead, for the power curve study of Section 4.2, all data in D_{aft} were used, after dividing them according to the pre or post upgrade behavior.

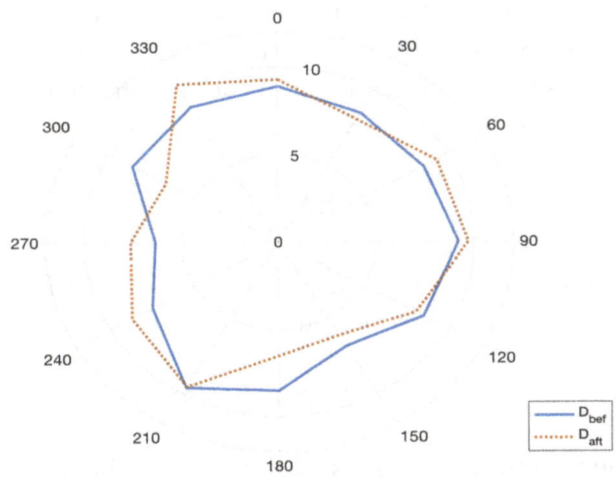

Figure 7. The wind direction rose for Wind Farm 1 during D_{bef} and D_{aft}.

3. The Methods

This section presents the formulating of a reliable model for the pre-upgrade power of the wind turbines of interest (WTG02 in Wind Farm 1 and WTG022 in Wind Farm 2). Section 4 is devoted in detail to the use of these models for the performance improvement estimate. For the moment, it is important to recall that a good model for the power of the wind turbines was needed because it was trained with pre-upgrade data and validated against a pre-upgrade dataset; the upgrade was quantified by simulating through the adopted model how the post-upgrade power would have been if the upgrade did not take place. In other words, the performance improvement was elaborated from how the residuals between measurements and model estimates changed after the upgrade with respect to before.

As anticipated in Section 1, the critical point was selecting the model type and the input variables. The discussion in [33], in relation to the work in [25], indicates that a linear model can be adequate for this objective. In other words, the general sense is that it is possible to approximate reliably the power of a wind turbine as a linear function of operation variables measured at the nearby wind turbines in the farm. This makes sense, not only by a statistical point of view, but also by the point of view of wind energy practice: actually, since a wind turbine acts as a filter to the wind fluctuations, the blade pitch, the rotor revolutions per minute and the active power of a wind turbine can likely be used for accounting for the on-site wind conditions [36].

The possible variables fed as input to the model are those indicated in Section 2.1 for Test Case 1 and those indicated in Section 2.2 for Test Case 2, for all the wind turbines in the wind farms except the upgraded ones. The decision of excluding the variables of the upgraded wind turbines as input variables to the model was motivated by the fact that the wind sensors might change after the upgrade (as in Test Case 2, see Section 2.2), or the upgrade might affect the measuring chain of the wind conditions (as discussed in [25,33]), or in general the relation between the power and the control (pitch, rotor revolutions per minute, etc) might change as a consequence of the upgrade. Therefore, since for the employed method one must assume that the input variables to the model are "probes" of the external conditions whose behavior does not change after the upgrade of the wind turbine of interest, it is straightforward that the variables of the upgraded wind turbine can only be the target (i.e., the output) of the model.

Similar to Astolfi et al. [33], a linear model was considered adequate for the objectives of this work. The critical point is the input variables selection: Tables 1 and 2 indicate that the possible covariates of the model can be highly correlated and this would lead to a non-optimal standard linear regression. On these grounds, Principal Component Regression (PCR) [39] was selected for this study. The use of this method for control and monitoring purposes in wind energy has been growing [40]. The procedure is as follows. Let $\boldsymbol{Y}_{n,1} = (y_i, \ldots, y_n)^T$ be the vector of measured output and $\boldsymbol{X}_{n,p} = (x_i, \ldots, x_n)^T$ be the matrix of covariates. n is the number of observations and p is the number of covariates. In the following, it is assumed that $\boldsymbol{X}$ is normalized such that each covariate has zero mean.

The standard least squares regression poses that

$$\boldsymbol{Y} = \boldsymbol{X}\boldsymbol{\beta} + \boldsymbol{\epsilon}, \tag{1}$$

where $\boldsymbol{\beta}$ is the vector of regression coefficients that must be estimated from the data and $\boldsymbol{\epsilon}$ is a vector of random errors. The ordinary least squares estimate of $\boldsymbol{\beta}$ is given by

$$\boldsymbol{\beta}_{ols} = \left(\boldsymbol{X}^T\boldsymbol{X}\right)^{-1}\boldsymbol{X}^T\boldsymbol{Y} \tag{2}$$

The principal component estimate of $\boldsymbol{\beta}$ is obtained as follows. The principal component transformation of the covariates matrix can easily be expressed in terms of the singular value matrix factorization. Therefore, let

$$\boldsymbol{X} = \boldsymbol{U}\boldsymbol{\Delta}\boldsymbol{V}^T \tag{3}$$

be the singular value decomposition of $\boldsymbol{X}$. This means that the columns of $\boldsymbol{U}$ and $\boldsymbol{V}$ are orthonormal sets of vectors denoting the left and right singular vectors of $\boldsymbol{X}$ and $\boldsymbol{\Delta}$ is a diagonal matrix, whose elements are the singular values of $\boldsymbol{X}$. This allows decomposing $\boldsymbol{X}\boldsymbol{X}^T$ as:

$$\boldsymbol{X}\boldsymbol{X}^T = \boldsymbol{V}\boldsymbol{\Lambda}\boldsymbol{V}^T, \tag{4}$$

where $\boldsymbol{\Lambda} = diag\left(\lambda_1, \ldots, \lambda_p\right)$ and $\lambda_1 \geq \cdots \geq \lambda_p \geq 0$.

$\boldsymbol{X}\boldsymbol{V}_i$ is the ith principal component and $\boldsymbol{V}_i$ is the ith loading corresponding to the ith principal value λ_i. Therefore, $\boldsymbol{W} = \boldsymbol{X}\boldsymbol{V}$ can be viewed as a new covariates data matrix and the principal component regression basically is an ordinary least squares regression between $\boldsymbol{Y}$ and $\boldsymbol{W}$. A powerful aspect of the principal component regression is that the decomposition in Equation (4) indicates a sort of regularization scheme: namely, the matrix $\boldsymbol{W}$ can be truncated including a desired number of principal components. This regularization addresses the problem of multicollinearity of covariates, because, when two or more covariates are highly correlated, $\boldsymbol{X}$ tends to lose its full rank and this implies that $\boldsymbol{X}\boldsymbol{X}^T$ has some eigenvalues tending to 0: excluding the principal components associated to the smallest eigenvalues λ_i means regularizing the covariates matrix in order that it has full rank.

Finally, the PCR estimate of $\boldsymbol{\beta}$ is given as

$$\boldsymbol{\beta}_{PCR} = \boldsymbol{V}\left(\boldsymbol{W}^T\boldsymbol{W}\right)^{-1}\boldsymbol{W}^T\boldsymbol{Y}, \tag{5}$$

where it is assumed that the matrices can be truncated to a desired number of columns, i.e., principal components.

The selection of an adequate number of principal components for the regression is performed through K-fold cross-validation [41]. D_{bef} is divided randomly in two fractions: $(K-1)/K$ of the data are used for training and the remaining $1/K$ are used for validation. K was selected to be 10 for this study. The training data are employed for estimating $\boldsymbol{\beta}$ through principal component regression (Equation (5)) and the model estimate of the validation data is given by

$$\hat{\boldsymbol{Y}}_{valid} = \boldsymbol{X}_{valid}\boldsymbol{\beta}_{PCR} \tag{6}$$

For each fold selection, the root mean square error is used as a metric for the goodness of the regression: it is given in general by

$$RMSE = \sqrt{\frac{1}{n_{valid}} \sum_{i=1}^{n_{valid}} (\hat{y}_i - y_i)^2}, \tag{7}$$

where n_{valid} is the number of rows of $\mathbf{X}_{valid}$. The $RMSE$ values are subsequently averaged on the folds selection and, therefore, for a given number j of principal components included in the regression, one can obtain a unique metric $RMSE_j$ for estimating the quality of the regression. The final selection of the number of principal components to be kept is performed as follows: given $RMSE_j$, the error estimate for the model with j principal components, if $RMSE_j - RMSE_{j+1} < 10\%$, k is selected. It should be noticed that, as discussed in Section 4, the results for both test cases do not depend sensibly on this choice, as long as a certain minimum number of principal components are included in the model.

A test can be formulated for inquiring the statistical significance of the fact the performance of the wind turbine of interest has changed after the upgrade. One can pose that the output can be modeled through a linear model before and after the upgrade and inquire to whether there has been a structural change, i.e. if the linear models before and after the upgrade are different. Suppose therefore that

$$\begin{cases} y_{bef} = \mathbf{X}_{bef}\boldsymbol{\beta}_{bef} + \boldsymbol{\epsilon}_{bef} \\ y_{aft} = \mathbf{X}_{aft}\boldsymbol{\beta}_{aft} + \boldsymbol{\epsilon}_{aft}, \end{cases} \tag{8}$$

where $\mathbf{X}$ is the matrix of explanatory variables, $\boldsymbol{\beta}$ is the vector of regression coefficients and $\boldsymbol{\epsilon}$ are vectors of random errors.

The hypothesis test about the structural change of the regression regards the null hypothesis:

$$H_0 : \boldsymbol{\beta}_{bef} = \boldsymbol{\beta}_{aft}. \tag{9}$$

This is known as the Chow test and is based on the fact that, indicated with RSS the residuals sum of squares between measurements and model estimates, the quantity

$$F_{chow} = \frac{RSS_{bef} - RSS_{aft}}{RSS_{aft}} \frac{N - 2K - 2}{K} \tag{10}$$

is distributed as $F(K, N - 2K - 2)$, where K is the number of covariates and N is the number of data points.

Practically, the Chow test is performed as follows. The covariates matrices and the output vectors before and after the upgrade are vertically juxtaposed to form a total covariates matrix $\mathbf{X}_{TOT}$ and a total output vector $\mathbf{Y}_{TOT}$. The test is performed with the assumption that the break point where the structural change can happen occurs when the data before the upgrade end and the data after the upgrade start.

3.1. Test Case 1

In Table 2, some sample results are reported for supporting the selection of the principal component regression: the correlation coefficients between the rotor rotational speeds of WTG01 and WTG03–WTG06 are reported. These covariates were selected because the rotor basically acts as a filter, smoothing the fluctuations caused by the turbulence, and it is therefore likely that in a wind farm the rotor speeds of nearby wind turbines are highly collinear.

Table 1. Matrix of model input variables correlation coefficients.

WTG	01	03	04	05	06
01	1	0.82	0.90	0.89	0.79
03	0.82	1	0.86	0.80	0.94
04	0.90	0.86	1	0.91	0.85
05	0.89	0.80	0.91	1	0.80
06	0.79	0.94	0.85	0.80	1

The structure of the model for the test case of interest was selected as follows. The output $\boldsymbol{Y}$ is the power of WTG02; the covariates matrix $\boldsymbol{X}$ was selected to be composed of power, rotor rotational speed, generator rotational speed, blade pitches, nacelle position and ambient temperature at each wind turbine of the wind farm, except WTG02. Therefore, if one considers the filtered D_{bef} dataset, $\boldsymbol{Y}$ is a vector of 25,044 data and $\boldsymbol{X}$ is a matrix having 24,055 rows and 30 columns (six variables for five wind turbines).

The results for the model K-fold cross-validation are reported in Figure 8 and, with the criterion exposed in Section 3, five principal components were selected. It should be noticed that a sensitivity analysis was carried and it was observed that the results do not change substantially by including more than five principal components.

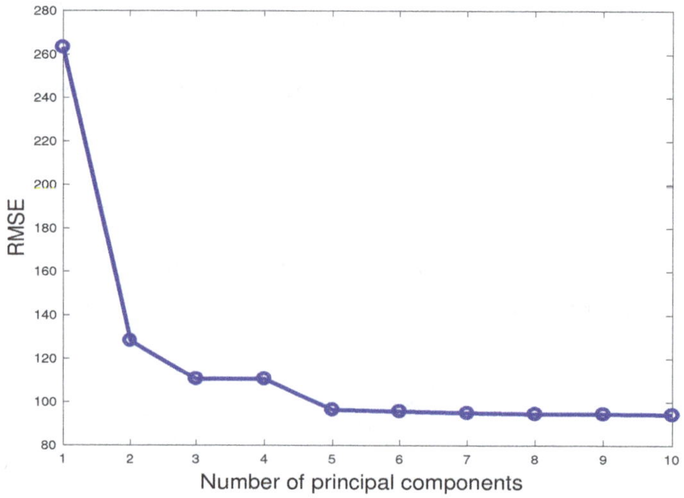

Figure 8. Average $RMSE$ as a function of the number of principal components included in the regression.

As regards the Chow Test, the matrix X_{TOT} is composed of 31,392 rows (25,044 before upgrade and 6348 after upgrade) and 30 columns, and the vector Y_{TOT} is composed of 31392 elements. The break point position for the Chow test is 25,044 and the computed p-value is lower than 10^{-32}. This clearly indicates that the linear relation between covariates matrix and the target has a structural change after the upgrade of WTG02.

3.2. Test Case 2

In Table 2, the correlation coefficients between some sample covariates are reported. The powers of WTG018 WTG021 and WTG023 was selected for reporting in Table 2: these covariates was selected for readability of the table and mostly because those wind turbines are the nearest to the target WTG022 and therefore those covariates are likely to be selected for a standard least squares regression. The remarkably high values reported in Table 2 support the selection of the principal component regression as model type.

Table 2. Matrix of model input variables correlation coefficients.

WTG	018	019	020	021	023
018	1	0.94	0.93	0.92	0.87
019	0.94	1	0.93	0.93	0.90
020	0.93	0.93	1	0.95	0.90
021	0.92	0.93	0.95	1	0.93
023	0.97	0.90	0.90	0.93	1

The structure of the model for the test case of interest was selected as follows. The output $\boldsymbol{Y}$ is the power of WTG022. The covariates matrix $\boldsymbol{X}$ is composed of the available validated measurements: nacelle wind speed, power and ambient temperature at each wind turbine of the wind farm, except WTG022. Therefore, if one considers the filtered D_{bef} dataset, $\boldsymbol{Y}$ is a vector of 15,353 data and $\boldsymbol{X}$ is a matrix having 15,353 rows and 66 columns (three variables for 22 wind turbines).

The results for the model *K*-fold cross-validation are reported in Figure 9 and, through the criterion exposed in Section 3, the number of selected principal components results to be six. It should be noticed that the results do not change significantly by including more than six principal components in the model.

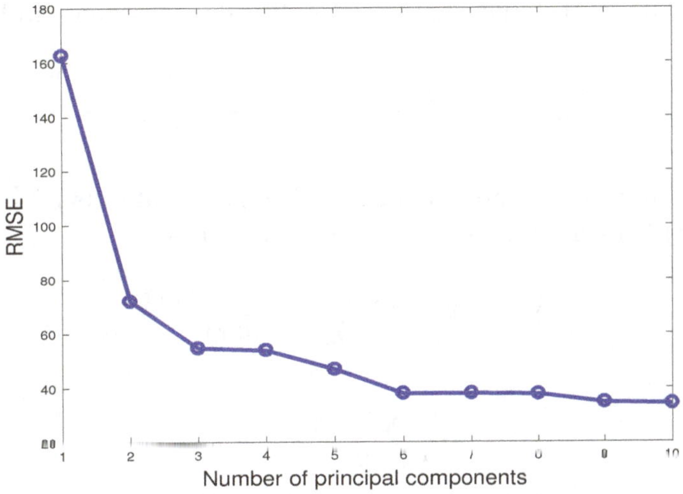

Figure 9. Average *RMSE* as a function of the number of principal components included in the regression.

As regards the Chow Test, the matrix X_{TOT} is composed of 19,618 rows (15,353 before upgrade and 4265 after upgrade) and 66 columns, the vector Y_{TOT} is composed of 19,618 elements. The break point position for the Chow test is 15,353 and the computed p-value is lower than 10^{-32}. This clearly indicates that the linear relation between covariates matrix and the target has a structural change after the upgrade of WTG022.

4. The Results

The procedure for assessing the upgrade Was based on the following idea. After an upgrade installation, through the SCADA collected data, it is possible to know the power production of the upgraded wind turbine. To estimate the impact of the upgrade, one should know how much the wind turbine would have produced if the upgrade did not take place. The most reliable and practical way to obtain this kind of estimate is through a data-driven model, based on the pre-upgrade datasets. A reliable model was achieved (Section 3) and it was used for the upgrade assessment presented below.

The procedure is as follows. The datasets available were organized in this way:

- D_{bef} was randomly divided in two subsets: D0 ($\frac{2}{3}$ of the data) and D1 ($\frac{1}{3}$ of the data). D0 was used for training the model and constructing the weight matrix $\boldsymbol{W}$ and D1, the pre-upgrade dataset, was employed for validating the model.
- D_{aft}, the post-upgrade dataset, was employed for estimating the power upgrade. For notation consistency, it is also referred to equivalently as D2.

Notice that, for Test Case 2, the dataset D^{up}_{aft} was employed as D2.

The residuals between the measurement y and the simulation $\hat{y}$, for the datasets D1 and D2, were studied. The focus was in how the residuals varied after the upgrade. Therefore, consider Equation (11) with $i = 1, 2$.

$$R(\boldsymbol{x}_i) = y(\boldsymbol{x}_i) - \hat{y}(\boldsymbol{x}_i). \tag{11}$$

A Student's t-test was performed to inquire if there was any statistically significant change in the turbine output after the upgrade. The t statistic is computed as

$$t = \frac{\bar{R}_2 - \bar{R}_1}{\sigma_R \sqrt{\frac{1}{N_1} + \frac{1}{N_2}}}. \tag{12}$$

In Equation (12), N_1 and N_2 are the numbers of data in, respectively, D1 and D2; $\bar{R}_1$ and $\bar{R}_2$ are the average residuals in datasets D1 and D2 respectively; and σ_R is given in Equation (13):

$$\sigma_R = \sqrt{\frac{(N_1 - 1)\, S_2^2 + (N_2 - 1)\, S_2^2}{N_1 + N_2 - 2}}, \tag{13}$$

where S_1 and S_2 are the standard deviations of the residuals in datasets D1 and D2, respectively.

As regards the upgrade estimate, for $i = 1, 2$, one computes

$$\Delta_i = 100 * \frac{\sum_{x \in Di} (y(\boldsymbol{x}) - \hat{y}(\boldsymbol{x}))}{\sum_{x \in Di} y(\boldsymbol{x})} \tag{14}$$

and

$$\delta_i = \frac{1}{N_i} \sum_{x \in Di} y(\boldsymbol{x}_i) - \hat{y}(\boldsymbol{x}_i), \tag{15}$$

where N_i is the number of data points in the datasets D1 and D2, respectively. Notice that, if the model is reliable, one should have that $\delta_1 \simeq 0$ and $\Delta_1 \simeq 0$, differently with what should happen as regards δ_2 and Δ_2 if the upgrade is really effective. Finally, the quantity

$$\Delta = \Delta_2 - \Delta_1 \tag{16}$$

can be taken a percentage estimate of the production improvement. In the case the datasets D1 and D2 are characterized by considerably different y distributions, it might be appropriate to take this into account by renormalizing Equation (16): a reasonable correction factor can be the ratio between the y averages in datasets D2 and D1.

The above procedure can be repeated several times to synthesize experiment repetition: at each run of the model, a different D0 (training set) and therefore D1 (pre-upgrade validation dataset) can be selected. Notice that this basically corresponds to repeating the K-fold cross-validation. The difference with respect to the procedure described in Section 3 is that in this case the model structure was always the same and it was exactly the one selected on the grounds of the discussion in Section 3. The way the pre-upgrade data were divided was also changed with respect to Section 3. The selection of D0 and D1 actually corresponded to $K = 33.\bar{3}$. This was done because it agrees with most of the rule of thumbs for data partition for this kind of tasks and because, with this selection, the dimensions of D1 and D2 (the post-upgrade dataset) have the same order of magnitude.

Therefore, the Δ estimate varied at each run of the model, because the training data changed and, therefore Δ_1 and Δ_2 changed. In principle, it could be possible to select randomly a subset of D2 as post-upgrade simulation dataset, but in this work this choice was avoided. The reason was that typically D2 is shorter than D0 and D1, because for practical reasons an upgrade is assessed as soon as possible with good reliability. The above bootstrap technique therefore allowed having several estimates of Δ with the same data: the final estimate was always the average and it is presented below with its standard deviation. This corresponds to the procedure of Section 3 with J repetitions: for this part of the study, J was selected based on when the Δ average and standard deviation became fairly stable. It was observed that $J = 30$ is sufficient for this task.

4.1. Test Case 1

Since the effect of the upgrade regards especially the low-moderate wind intensity, data were filtered on the request that the power of WTG02 is less than 1 MW. After this further filter, the number of data was 25,044 for D_{bef} and 6348 for D_{aft}.

The t-statistic (Equation (12)) was computed to be of the order of 10^{-10} and this indicates that the probability that the upgrade was ineffective was correspondingly low.

Table 3 reports the results for the average (over the J model runs) Δ_i and δ_i with $i = 1, 2$ (Equations (14) and (15)). From these results, it arises that the upgrade could be detected as an average absolute increase of 13.5 kW in the difference between WTG02 power measurements and model estimates. Notice that the average value of the residuals for datasets D1 was extremely low (0.1 kW) and, correspondingly, the average estimate of Δ_1 (Equation (16)), i.e. the percentage error on the cumulative production, was extremely low as well. This indicates that the model was particularly reliable as regards the simulation of the pre-upgrade behavior of the WTG02.

In Figure 10, the plot of $R(x_1)$ and $R(x_2)$ on a sample model run is reported. The data were averaged in power production intervals, whose amplitude was 5% of the rated. From this plot, the effect of the upgrade can be read as an increase of the difference between the WTG02 power measurements and the WTG02 power model estimates.

Table 3. Average absolute and percentage residuals between measurement and model estimation.

Residual	δ (kW)	Δ (%)
$R(x_1)$	0.1	0.009
$R(x_2)$	13.5	4.1

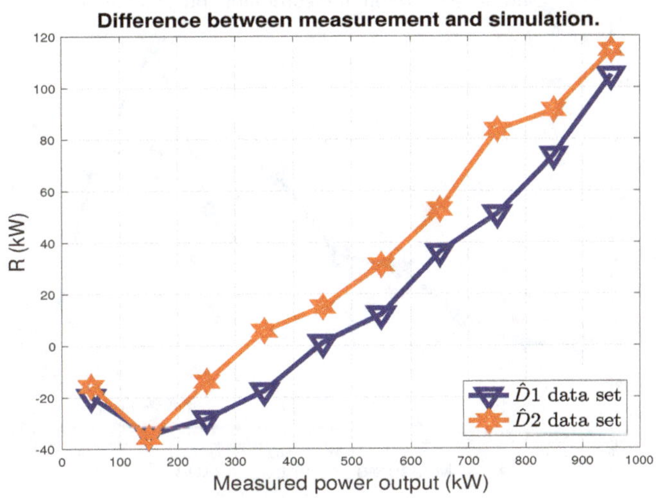

Figure 10. The average difference R between power measurement y and estimation $\hat{y}$ (Equation (11)). Datasets: D1 and D2. Sample run of the model.

From the results in Table 3 and Equation (14), the average production improvement was estimated as $\Delta = 4.3\%$, with a standard deviation of 0.4%: in other words, with the proposed method it was computed that WTG02, during the dataset D2, produced below 1 MW, the 4.1% more than it would have done if the upgrade had not been adopted. A reference long-term power or wind speed distribution can be employed to estimate how much this corresponds in terms of annual energy production and the average result is $\Delta_{AEP} = 1.3 \pm 0.1\%$. This result is consistent with the test case studies in [25]: the order of magnitude of the impact of multi-megawatt wind turbine control optimization can typically be estimated as 1% of the AEP. It is interesting notice that, to the best of the authors knowledge, this is the first estimate in the literature based on operation data of the impact of yaw control optimization.

4.2. Test Case 2

The t-statistic (Equation (12)) was computed to be of the order of 10^{-15} and this indicates that the probability that the upgrade was ineffective was correspondingly low.

Table 4 reports the results for the average (over the J model runs) Δ_i and δ_i with $i = 1, 2$ (Equations (14) and (15)). It arises that the upgrade could be detected as an average absolute increase of 3.9 kW in the difference between WTG022 power measurements and model estimates. The average value of the residuals for datasets D1 was very low (0.2 kW) and correspondingly, the average estimate of Δ_1 (Equation (16)), i.e. the percentage error on the cumulative production, was extremely low as well. This indicates that the model was reliable as regards the simulation of the pre-upgrade behavior of the WTG022.

Table 4. Average absolute and percentage residuals between measurement and model estimation.

Residual	δ (kW)	Δ (%)
$R(x_1)$	0.2	0.009
$R(x_2)$	3.9	2.5

In Figure 11, the plot of $R(x_1)$ and $R(x_2)$ on a sample model run is reported. The data were averaged in power production intervals, whose amplitude was 10% of the rated. From this plot, the effect of the upgrade can be read as an increase of the difference between the WTG022 power measurements and the WTG022 power model estimates, especially for moderately low wind intensities and approaching rated power.

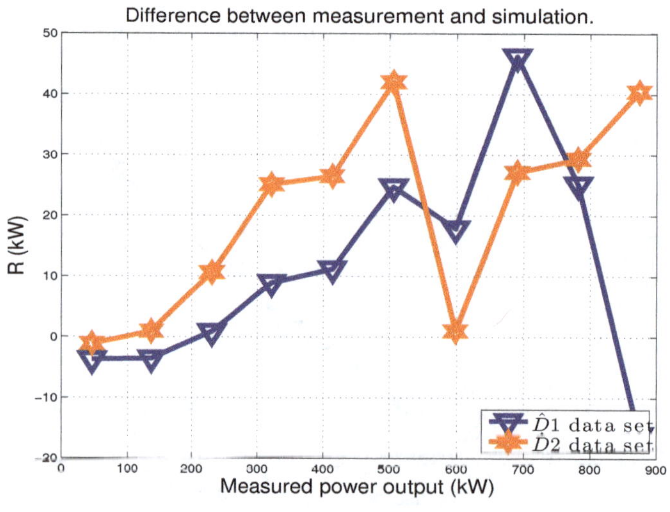

Figure 11. The average difference R between power measurement y and estimation $\hat{y}$ (Equation (11)). Datasets: D1 and D2. Sample run of the model.

From the results in Table 3 and Equation (14), the average production improvement was estimated as $\Delta = 2.5\%$, with a standard deviation of 0.2%: in other words, with the proposed method it was computed that WTG022, during the dataset D2, produced 2.5% more than it would have done if the upgrade had not been adopted.

Power Curve Analysis

As anticipated in Section 2, the post-upgrade operation during dataset D_{aft} was as follows: half-hour intervals were alternated, during which WTG022 was operating, respectively, according to the pre- and post-upgrade control logic. This was done to assess practically in real time the effect of the upgrade. With these data available and taking into account that, during D_{aft}, a sonic anemometer was collecting data at WTG022 nacelle, it was reasonable to study the power curve.

In Figure 12, the two power curves measured during D_{aft} are reported. Data were averaged in wind speed intervals having 0.5 m/s of amplitude. In Figure 13, the difference between these two curves is plotted. In Figure 13, it can interestingly be observed that the upgraded operation mode indeed lost performance around 10 m/s: the same situation was observed from the residuals presented in Figure 11. Since this study was performed with only few months of data in D_{aft}, it is plausible to expect that this situation was adjusted in the following, to obtain a performance improvement along the whole power curve.

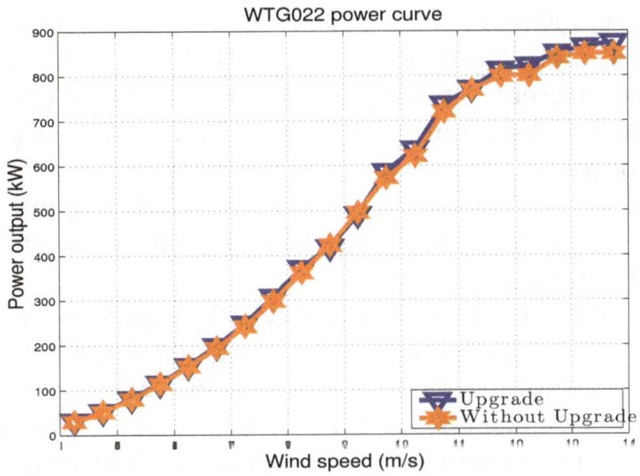

Figure 12. WTG022 power curve during D_{aft}.

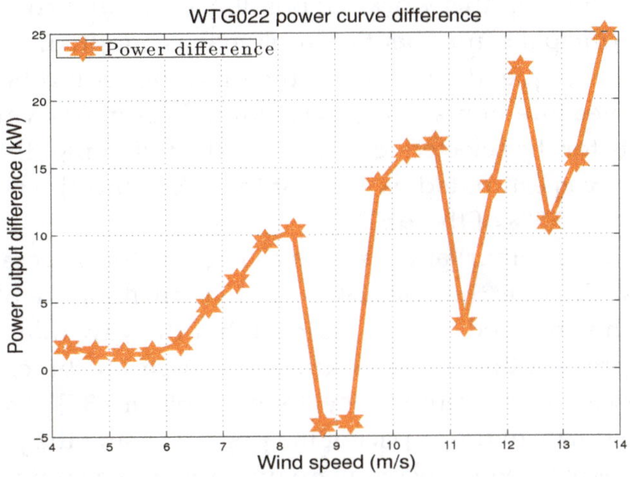

Figure 13. Difference between the WTG022 power curves of Figure 12.

The production improvement was estimated as follows: the power curve, according to the pre-upgrade logic in Figure 12, and the power difference in Figure 13, were weighted against the wind distribution during the whole D_{aft} dataset. The ratio between these two quantities provided an estimate of how much the production would have improved during D_{aft} if the power curve was always the improved one, with respect to the production that would have been obtained if the power curve was always the non-improved one. The improvement computed in this way amounted to 2.3% of the production. Even though it was computed with a different approach, it is interesting to notice that it agreed fairly well with the estimate reported in Section 4.2.

5. Conclusions

In this study, two test cases of wind turbine power curve upgrades were analyzed: the common ground between them is that the upgrade regards the control of the wind turbines. The difference between the two test cases is that one wind turbine (Test Case 1) has a quite recent technology (it is a 2 MW wind turbine) and the control upgrade deals only with one aspect (the management of the yaw); the other wind turbine under investigation (Test Case 2) belongs instead to a less recent technology (the rated power is 850 kW) and the upgrade has consequently involved several aspects of the control (yaw, pitch, and cut-out) and included the update of the anemometer sensors at the nacelle.

Despite being organized as a test case discussion, this study was strongly characterized by a methodological approach. Actually, the point with the study of wind turbine power curve upgrades is that it is difficult to assess them reliably using operation data analysis techniques such as the power curve, because of the multivariate dependence of the power of a wind turbine on climate conditions and working parameters. The problem of wind turbine power curve upgrades study therefore translates into the following question: how can the power of a wind turbine be modeled reliably? It is evident that the answer to this question can be exploited for several problems regarding the control and monitoring of wind turbines and, in general, of complex systems. As regards wind turbines, for example, similar approaches are employed in [42] for the study of how much the pitch misalignment impacts on the performance.

The turning point for the present study was practically adopting the other wind turbines in the wind farm as probes of on-site conditions. This somehow generalized the concept of rotor-equivalent wind speed, discussed, for example, in [36]: since the wind turbine acts as a filter, some working parameters such as active power, blade pitches, rotor or generator revolutions per minute can robustly describe the wind farm at the micro-scale level. Therefore, the idea of this study was modeling the power of the wind turbines of interest, according to their pre-upgrade behavior, as a linear function of the wind and operation conditions measured at the nearby wind turbines: this can basically be considered a generalization of the so-called power–power method, adopted, for example, in [30]. Since for the test cases considered in this work the possible covariates for a linear model displayed a remarkable collinearity, a principal component regression was adopted.

Using this modeling technique, the impact of the upgrades could be elaborated from how the residuals between power measurements and power model estimates vary after the upgrade with respect to before. The results for the selected test cases are the following: the yaw control optimization on the 2 MW wind turbine was estimated as 1.3% of the AEP; and the re-powering on the 850 kW wind turbine was estimated as 2.5% of the AEP.

There are at least two other remarkable aspects as regards the selected test cases. To the best of the authors knowledge, Test Case 1 is the first assessment in the literature of yaw control optimization using operation data and the obtained results indicate that the yaw management optimization is a promising direction for improving the power production of wind turbines. It is therefore valuable to push forward this line of research, as recently done, for example, in [35]. As regards Test Case 2, it was possible to obtain another estimate of the impact of the upgrade using the power curve study. Actually, with the re-powering, the anemometer sensors were updated and a sonic anemometer was installed. Furthermore, in the post upgrade period examined for this study, the operation of the wind turbine was alternated: half an hour according to the pre-upgrade logic and half an hour according to the post-upgrade logic, and so on. The quality of the data and the fact that they were collected in the same period (avoiding seasonal biases) allowed studying the power curve reliably and the improvement estimate was shown to be in good agreement with the computation from the multivariate model.

There are several further directions of the present work. Currently, some test cases are at study for which a linear model is not adequate, probably because of complex climatology conditions on site. Therefore, it is planned to investigate nonlinear approaches for this kind of studies and to inquire what site characteristics call for nonlinearity. An interesting development is the use of the methods of this work for other control and monitoring issues related to wind turbine operation: for example, monitoring the effect of blade pitches re-alignment according to the technique proposed in [43], or monitoring the operation of the wind turbines [40]. Furthermore, a very promising direction of the studies about wind turbine power curve upgrades is the use of time-resolved data, having sampling time of the order of second: this kind of data have considerable potentiality for performance control and monitoring [44], but their time scale calls for more advanced time-series analysis [45].

Author Contributions: Conceptualization, D.A. and F.C.; Data curation, D.A.; Formal analysis, D.A.; Investigation, D.A., F.C.; Methodology, D.A.; Project administration, F.C.; Software, D.A.; Supervision, F.C., Validation, D.A.; Writing—original draft, D.A.; and Writing— review adn editing, F.C.

Acknowledgments: The authors thank Ludovico Terzi, technology manager of Renvico.

References

1. Tchakoua, P.; Wamkeue, R.; Ouhrouche, M.; Slaoui-Hasnaoui, F.; Tameghe, T.A.; Ekemb, G. Wind turbine condition monitoring: State-of-the-art review, new trends, and future challenges. *Energies* **2014**, *7*, 2595–2630. [CrossRef]
2. Parada, L.; Herrera, C.; Flores, P.; Parada, V. Wind farm layout optimization using a Gaussian-based wake model. *Renew. Energy* **2017**, *107*, 531–541. [CrossRef]
3. Shakoor, R.; Hassan, M.Y.; Raheem, A.; Wu, Y.K. Wake effect modeling: A review of wind farm layout optimization using Jensen's model. *Renew. Sustain. Energy Rev.* **2016**, *58*, 1048–1059. [CrossRef]
4. Feng, J.; Shen, W.Z. Solving the wind farm layout optimization problem using random search algorithm. *Renew. Energy* **2015**, *78*, 182–192. [CrossRef]
5. Sorkhabi, S.Y.D.; Romero, D.A.; Yan, G.K.; Gu, M.D.; Moran, J.; Morgenroth, M.; Amon, C.H. The impact of land use constraints in multi-objective energy-noise wind farm layout optimization. *Renew. Energy* **2016**, *85*, 359–370. [CrossRef]
6. Chen, Y.; Li, H.; He, B.; Wang, P.; Jin, K. Multi-objective genetic algorithm based innovative wind farm layout optimization method. *Energy Convers. Manag.* **2015**, *105*, 1318–1327. [CrossRef]
7. Park, J.; Law, K.H. Cooperative wind turbine control for maximizing wind farm power using sequential convex programming. *Energy Convers. Manag.* **2015**, *101*, 295–316. [CrossRef]
8. Park, J.; Law, K.H. A data-driven, cooperative wind farm control to maximize the total power production. *Appl. Energy* **2016**, *165*, 151–165. [CrossRef]
9. Wang, F.; Garcia-Sanz, M. Wind farm cooperative control for optimal power generation. *Wind Eng.* **2018**, *42*, 547–560. [CrossRef]
10. Gebraad, P.; Thomas, J.J.; Ning, A.; Fleming, P.; Dykes, K. Maximization of the annual energy production of wind power plants by optimization of layout and yaw-based wake control. *Wind Energy* **2017**, *20*, 97–107. [CrossRef]
11. Gebraad, P.; Teeuwisse, F.; Van Wingerden, J.; Fleming, P.A.; Ruben, S.; Marden, J.; Pao, L. Wind plant power optimization through yaw control using a parametric model for wake effects—A CFD simulation study. *Wind Energy* **2016**, *19*, 95–114. [CrossRef]
12. Fleming, P.A.; Ning, A.; Gebraad, P.M.; Dykes, K. Wind plant system engineering through optimization of layout and yaw control. *Wind Energy* **2016**, *19*, 329–344. [CrossRef]

13. Campagnolo, F.; Petrović, V.; Bottasso, C.L.; Croce, A. Wind tunnel testing of wake control strategies. In Proceedings of the IEEE American Control Conference (ACC), Boston, MA, USA, 6–8 July 2016; pp. 513–518.
14. Fleming, P.; Annoni, J.; Shah, J.J.; Wang, L.; Ananthan, S.; Zhang, Z.; Hutchings, K.; Wang, P.; Chen, W.; Chen, L. Field test of wake steering at an offshore wind farm. *Wind Energy Sci.* **2017**, *2*, 229–239. [CrossRef]
15. Barlas, T.K.; Van Kuik, G. Review of state of the art in smart rotor control research for wind turbines. *Prog. Aerosp. Sci.* **2010**, *46*, 1–27. [CrossRef]
16. Tsai, K.C.; Pan, C.T.; Cooperman, A.M.; Johnson, S.J.; Van Dam, C. An innovative design of a microtab deployment mechanism for active aerodynamic load control. *Energies* **2015**, *8*, 5885–5897. [CrossRef]
17. Fernández-Gámiz, U.; Marika Velte, C.; Réthoré, P.E.; Sørensen, N.N.; Egusquiza, E. Testing of self-similarity and helical symmetry in vortex generator flow simulations. *Wind Energy* **2016**, *19*, 1043–1052. [CrossRef]
18. Aramendia, I.; Fernandez-Gamiz, U.; Ramos-Hernanz, J.A.; Sancho, J.; Lopez-Guede, J.M.; Zulueta, E. Flow Control Devices for Wind Turbines. In *Energy Harvesting and Energy Efficiency*; Springer: Berlin, Germany, 2017; pp. 629–655.
19. Fernandez-Gamiz, U.; Zulueta, E.; Boyano, A.; Ansoategui, I.; Uriarte, I. Five megawatt wind turbine power output improvements by passive flow control devices. *Energies* **2017**, *10*, 742. [CrossRef]
20. Fernandez-Gamiz, U.; Gomez-Mármol, M.; Chacón-Rebollo, T. Computational Modeling of Gurney Flaps and Microtabs by POD Method. *Energies* **2018**, *11*, 2091. [CrossRef]
21. Gutierrez-Amo, R.; Fernandez-Gamiz, U.; Errasti, I.; Zulueta, E. Computational Modelling of Three Different Sub-Boundary Layer Vortex Generators on a Flat Plate. *Energies* **2018**, *11*, 3107. [CrossRef]
22. Fernandez-Gamiz, U.; Errasti, I.; Gutierrez-Amo, R.; Boyano, A.; Barambones, O. Computational Modelling of Rectangular Sub-Boundary Layer Vortex Generators. *Appl. Sci.* **2018**, *8*, 138. [CrossRef]
23. Aramendia, I.; Fernandez-Gamiz, U.; Zulueta, E.; Saenz-Aguirre, A.; Teso-Fz-Betoño, D. Parametric Study of a Gurney Flap Implementation in a DU91W (2) 250 Airfoil. *Energies* **2019**, *12*, 294. [CrossRef]
24. Lee, G.; Ding, Y.; Xie, L.; Genton, M.G. A kernel plus method for quantifying wind turbine performance upgrades. *Wind Energy* **2015**, *18*, 1207–1219. [CrossRef]
25. Astolfi, D.; Castellani, F.; Terzi, L. Wind Turbine Power Curve Upgrades. *Energies* **2018**, *11*, 1300. [CrossRef]
26. Astolfi, D.; Castellani, F.; Berno, F.; Terzi, L. Numerical and Experimental Methods for the Assessment of Wind Turbine Control Upgrades. *Appl. Sci.* **2018**, *8*, 2639. [CrossRef]
27. Bossanyi, E.; King, J. Improving wind farm output predictability by means of a soft cut-out strategy. In Proceedings of the European Wind Energy Conference and Exhibition EWEA, Copenhagen, Denmark, 16–19 April 2012; Volume 2012.
28. Petrović, V.; Bottasso, C.L. Wind turbine optimal control during storms. *J. Phys. Conf. Ser.*. **2014**, *524*, 012052. [CrossRef]
29. Astolfi, D.; Castellani, F.; Lombardi, A.; Terzi, L. About the extension of wind turbine power curve in the high wind region. *J. Solar Energy Eng.* **2019**, *141*, 014501. [CrossRef]
30. Hwangbo, H.; Ding, Y.; Eisele, O.; Weinzierl, G.; Lang, U.; Pechlivanoglou, G. Quantifying the effect of vortex generator installation on wind power production: An academia-industry case study. *Renew. Energy* **2017**, *113*, 1589–1597. [CrossRef]
31. Astolfi, D.; Castellani, F.; Terzi, L. A SCADA data mining method for precision assessment of performance enhancement from aerodynamic optimization of wind turbine blades. *J. Phys. Conf. Ser.* **2018**, *1037*, 032001. [CrossRef]
32. Terzi, L.; Lombardi, A.; Castellani, F.; Astolfi, D. Innovative methods for wind turbine power curve upgrade assessment. *J. Phys. Conf. Ser.* **2018**, *1102*, 012036. [CrossRef]
33. Astolfi, D.; Castellani, F.; Fravolini, M.L.; Cascianelli, S.; Terzi, L. Precision Computation of Wind Turbine Power Upgrades: An Aerodynamic and Control Optimization Test Case. *J. Energy Resour. Technol.* **2019**, *141*, 051205. [CrossRef]
34. Song, D.; Yang, J.; Fan, X.; Liu, Y.; Liu, A.; Chen, G.; Joo, Y.H. Maximum power extraction for wind turbines through a novel yaw control solution using predicted wind directions. *Energy Convers. Manag.* **2018**, *157*, 587–599. [CrossRef]
35. Saenz-Aguirre, A.; Zulueta, E.; Fernandez-Gamiz, U.; Lozano, J.; Lopez-Guede, J.M. Artificial Neural Network Based Reinforcement Learning for Wind Turbine Yaw Control. *Energies* **2019**, *12*, 436. [CrossRef]

36. Wagner, R.; Cañadillas, B.; Clifton, A.; Feeney, S.; Nygaard, N.; Poodt, M.; St Martin, C.; Tüxen, E.; Wagenaar, J. Rotor equivalent wind speed for power curve measurement—Comparative exercise for IEA Wind Annex 32. *J. Phys. Conf. Ser.* **2014**, *524*, 012108. [CrossRef]
37. Green, W.H. *Econometric Analysis*; Maxwell Macmillan International: New York City, NY, USA, 1997.
38. Pope, P.; Webster, J. The use of an F-statistic in stepwise regression procedures. *Technometrics* **1972**, *14*, 327–340. [CrossRef]
39. Frank, L.E.; Friedman, J.H. A statistical view of some chemometrics regression tools. *Technometrics* **1993**, *35*, 109–135. [CrossRef]
40. Pozo, F.; Vidal, Y. Wind turbine fault detection through principal component analysis and statistical hypothesis testing. *Energies* **2016**, *9*, 3. [CrossRef]
41. Refaeilzadeh, P.; Tang, L.; Liu, H. Cross-validation. In *Encyclopedia of Database Systems*; Springer: Berlin, Germany, 2009; pp. 532–538.
42. Astolfi, D. A Study of the Impact of Pitch Misalignment on Wind Turbine Performance. *Machines* **2019**, *7*, 8. [CrossRef]
43. Elosegui, U.; Egana, I.; Ulazia, A.; Ibarra-Berastegi, G. Pitch angle misalignment correction based on benchmarking and laser scanner measurement in wind farms. *Energies* **2018**, *11*, 3357. [CrossRef]
44. Gonzalez, E.; Stephen, B.; Infield, D.; Melero, J.J. Using high-frequency SCADA data for wind turbine performance monitoring: A sensitivity study. *Renew. Energy* **2019**, *131*, 841–853. [CrossRef]
45. Rodrigues, P.M.; Salish, N. Modeling and forecasting interval time series with threshold models. *Adv. Data Anal. Classif.* **2015**, *9*, 41–57. [CrossRef]

9

Wind Turbine Control using Nonlinear Economic Model Predictive Control over All Operating Regions

Xiaobing Kong [1], Lele Ma [1], Xiangjie Liu [1,*], Mohamed Abdelkarim Abdelbaky [1,2] and Qian Wu [1]

1 The State Key Laboratory of Alternate Electrical Power System with Renewable Energy Sources, North China Electric Power University, Beijing 102206, China; kongxiaobing@ncepu.edu.cn (X.K.); 1172127008@ncepu.edu.cn (L.M.); m_abdelbaky@ncepu.edu.cn (M.A.A.); wuqian@ncepu.edu.cn (Q.W.)

2 Electrical Power and Machines Engineering Department, Faculty of Engineering, Cairo University, Giza 12613, Egypt

* Correspondence: liuxj@ncepu.edu.cn

Abstract: With the gradual increase in the installed capacity of wind turbines, more and more attention has been paid to the economy of wind power. Economic model-predictive control (EMPC) has been developed as an effective advanced control strategy, which can improve the dynamic economy performance of the system. However, the variable-speed wind turbine (VSWT) system widely used is generally nonlinear and highly coupled nonaffine systems, containing multiple economic terms. Therefore, a nonlinear EMPC strategy considering power maximization and mechanical load minimization is proposed based on the comprehensive VSWT model, including the dynamics of the tower and the gearbox in this paper. Three groups of simulations verify the effectiveness and reliability/practicability of the proposed nonlinear EMPC strategy.

Keywords: variable-speed wind turbine; tower fatigue; drive-shaft torsion; nonlinear economic-model predictive control

---

1. Introduction

As a consequence of energy shortages around the world, environmental protection requirements, and the higher cost of traditional power, renewable energy sources are receiving a great deal of attention worldwide nowadays. Wind energy becomes the dominant renewable energy source due to its plentifulness, cleanness, and economic advantages. By the ending of 2018, the cumulative installed capacity of wind is estimated to be more than 221 GW in China, with an increase of 19 GW and 25.9 GW in 2017 and 2018 respectively, accounting for 31.9% and 32.8% of global overall wind power capacity [1]. Therefore, enhancing the efficiency and economy of wind energy conversion systems has become an important issue to handle the growing need for energy.

Numerous control schemes have been applied in variable-speed wind turbine (VSWT) controller designs to enhance efficiency. Generally, PID controllers are widely used in the pitch control of wind turbine systems, which only focuses on decreasing the blades' mechanical loads within region three, as shown in Figure 1 [2]. A feedback gain-scheduled PI controller was designed to maximize the captured power and mitigate the mechanical loads, where the fixed-phase and gain-margins are obtained through a frequency response analysis [3]. In [4], a gain-scheduled L1 adaptive optimal controller is proposed for the VSWT control system, where the nonconvex optimization problem is solved by using a genetic algorithm based on a single linearized model of the wind-turbine (WT) system at different operating points. To obtain the rated generator power and reduce the mechanical load, a nonlinear-PI controller with an extended state-observer was developed in order to control the blade pitch angle [5]. Although these PI controllers have led to the effective steady-state performance of VSWTs, the distortion of the transient performance according to variations in the wind speed is still a drawback. Additionally, PI/PID controllers offer good performance, but they still are not optimal [6].

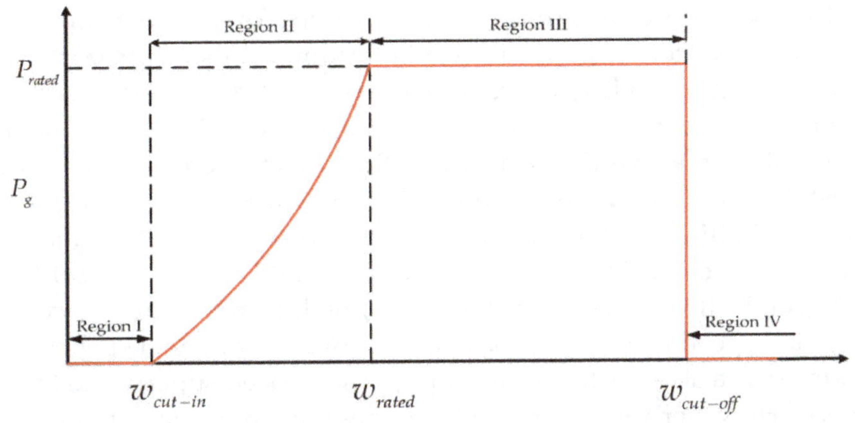

Figure 1. VSWT operation regions.

However, owing to multiple constraints, multiple variables, the highly coupled nonaffine basis of the VSWT systems, and the stochastic shape of wind speed input, high-performance wind turbine control (WTC) has become a challenging problem [7]. Various nonlinear control schemes have been applied to the WTC problem. In [8], an artificial neural network technique is proposed for the pitch control in region three, where the VSWT dynamics are modeled using a back-propagation learning algorithm. In [9], a fuzzy logic pitch controller is proposed for WTC, considering the generator power and speed as control inputs. In [10], a multiple feedback pitch controller is proposed for the nacelle fore-aft speed and generator power regulation. This controller is designed by employing a trade-off between pitch angle fluctuations and output power fluctuations. In [11], an adaptive neural pitch controller is proposed for the WTC problem. An online, two-layer neural network model is proposed for estimating the unknown wind turbine aerodynamics. Recently in [12], a T-S fuzzy modeling method with nonlinear consequents is proposed for the VSWT control problem. However, the wind turbine actuator constraints are not considered in the controller designs presented in these approaches, which may cause a wind up phenomenon and significant performance deterioration if the actuator inputs reach saturation limits.

Among all the control strategies, model predictive control (MPC) is emerging as a powerful candidate for WTC thanks to its capacity to handle all the system constraints directly, such as physical limitations and operational constraints [13]. Classical MPC schemes have been used for WTC to ensure that the maximum power is captured in region two, and the rated power is tracked in region three. In [14,15], a linear MPC strategy is proposed based on a single linear state-space model to deal with the actuator constraints. Although these approaches have good performance concerning certain operating points and system constraints handling, no special attention has been paid to either the strong nonlinearity of the system or the mechanical load mitigation to enhance the economy of the VSWT system.

Some MPC techniques have been utilized by taking the wind turbine nonlinearity into account. In [16–18], a multivariable MPC is proposed for VSWT pitch control in region three, using Takagi-Sugeno (T-S) fuzzy models. In [19], a nonlinear MPC (NMPC) strategy is investigated for VSWT control for both regions two and three by tuning the penalty parameters. In [20,21], multiple model predictive controllers are proposed based on multiple linearized time-invariant state-space models, guaranteeing the actuator limits in all operation regions. However, it is obvious that the controller switching results in more difficulties. The authors take the load of WT into account as one of completing penalties based on a simplified nonlinear WT model. The shortcoming of these control schemes is the difficulty of adequately tuning the tracking weights; furthermore, economic factors are not considered in the design of the controller.

To enhance the dynamic economy, the economic MPC (EMPC) was developed for the VSWT system. Generally, the EMPC uses a measure of the system performance directly as an economic cost function [22]. However, the economic cost function may not be a positive definite with regard to any tracking trajectory in order to guarantee the stability of the operating points. In [23], the closed-loop stability of the system is guaranteed via terminal constraints depending on the strong duality assumption. In addition, in [24], a Lyapunov-based EMPC is constituted using two separate operation modes to make the closed-loop system ultimately bounded within a small region. In the context of the WTC problem, the EMPC objective function directly considers VSWT energy generation maximization against the actuator fatigue minimization, as opposed to tracking some reference states and inputs. The proposed NEMPC in this paper has a unique weight in the objective function, while the weights must be chosen carefully for every input and state in the classical NMPC strategy [19]. In addition, in [25,26], an EMPC strategy has been designed for wind turbine control systems for all regions. In [26], the wind speed is produced based on the Van der Hoven spectrum [27], in which the wind speed is considered as a slowly-changing average wind speed superposed by a rapidly-changing turbulence wind speed. However, these EMPC approaches pay no special attention to the mitigation of mechanical loads on the tower in the economic objective function. Additionally, different treatment of the random wind speed can be considered, consisting of summing the mean wind speed, which follows a two-parameter Weibull distribution, and the turbulent wind speed, which follows a zero-mean Gaussian, white noise distribution [28].

In this study, a nonlinear EMPC (NEMPC) strategy is proposed for all the operating regions of the VSWT system. A nonlinear VSWT model, which considers the tower and gearbox dynamics, is established. The NEMPC optimization problem considers all the actuator constraints (pitch angle, and torque constraints with their rate of change constraints) and the hard constraints (rotor speed, generator speed, and electrical power). The economic cost function seeks the maximum VSWT generated power against the competing penalties regardless of wind input, in order to achieve the best economic operation, as well as the fatigue load mitigation of key mechanical structures, including both drive-shaft torsion and tower fore-aft motion. This control strategy can provide potential improvements in the closed-loop performance, and satisfy the economy in contrast with classical WTC strategies. Finally, in the simulations, random wind speed is considered by summing the mean wind speed and the turbulent wind speed, which follow a two-parameter Weibull distribution and a zero-mean Gaussian white noise distribution, respectively.

The paper is organized as follows. The nonlinear modeling of the VSWT system is established in Section 2. The proposed NEMPC strategy is derived in Section 3. In Section 4, the simulation results for a 5-MW VSWT system are demonstrated, with comparisons between the classical NMPC strategy and the proposed NEMPC strategy. Also, in this section, a typical wind turbine benchmark simulator FAST is used to test the validity and practicability of the proposed NEMPC strategy. Lastly, Section 5 provides conclusions.

2. Wind Turbine Modeling

Wind-energy conversion systems consist of a windmill, a gearbox, and a generator, as demonstrated in Figure 2. In this section, nonlinear VSWT modeling is investigated. To establish this nonlinear model, the aerodynamics, tower, drive train, and the generator should be considered. In the following subsections, these subsystems are discussed separately. Finally, a nonlinear state-space model is established.

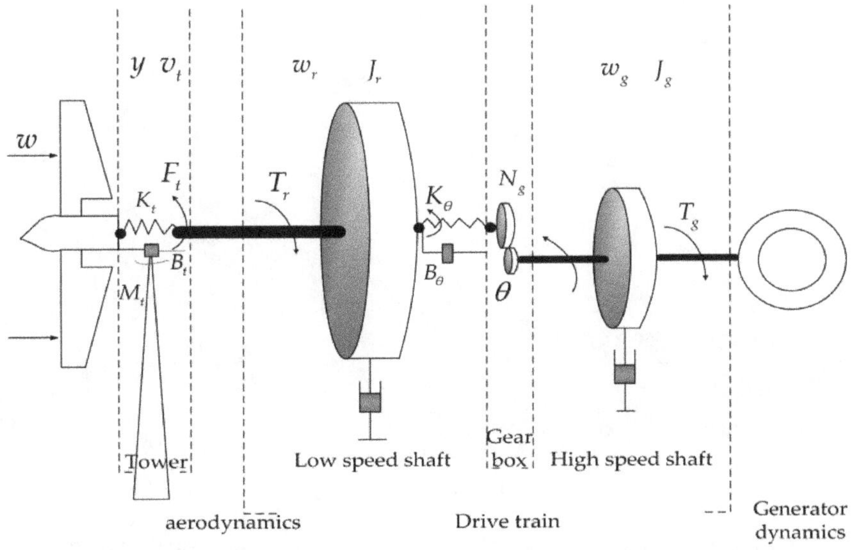

Figure 2. The mechanical structure of the VSWT generation system.

2.1. *Rotor Aerodynamics*

The airflow on VSWT blades causes aerodynamic torque, which represents one of the causes of VSWT nonlinearities. The aerodynamic torque T_r can be stated as below [29]:

$$T_r = \frac{1}{2}\rho\pi R^2 \frac{v_r^3}{w_r} C_p(\lambda(v_r, w_r), \beta) \tag{1}$$

The aerodynamic power that the VSWT extracts is expressed by [29]:

$$P^A = \frac{1}{2}\rho\pi R^2 v_r^3 C_p(\lambda(v_r, w_r), \beta) \tag{2}$$

where ρ is the air density, v_r represents the effective wind speed on the turbine rotor [30], R is the rotor radius of VSWT, w_r is the VSWT rotor angular speed, $C_p(\lambda(v_r, w_r), \beta)$ is the aerodynamic power coefficient, and β is the blade pitch angle. The tip speed ratio λ is defined as [30]:

$$\lambda(w_r, v_r) = \frac{Rw_r}{v_r} \tag{3}$$

The following approximate expression of the aerodynamic power coefficient derived from [18] is widely used:

$$C_p(\lambda, \beta) = c_1(\frac{c_2}{\Lambda} - c_3\beta - c_4) \times e^{-\frac{c_5}{\Lambda}} + c_6\lambda \tag{4}$$

where $1/\Lambda = 1/(\lambda + 0.08\beta) - 0.035/(1 + \beta^3)$, and the coefficients $c_i(i = 1, 2, \cdots, 6)$ depend on the blade shape and its aerodynamic performance [18,29]. In this paper, NREL 5-MW VSWT system is discussed; the six chosen coefficient values are [18]: $c_1 = 0.5176$, $c_2 = 116$, $c_3 = 0.4$, $c_4 = 5$, $c_5 = 21$, $c_6 = 0.0068$. This nonlinear function $C_P(\lambda, \beta)$ is indicated in Figure 3.

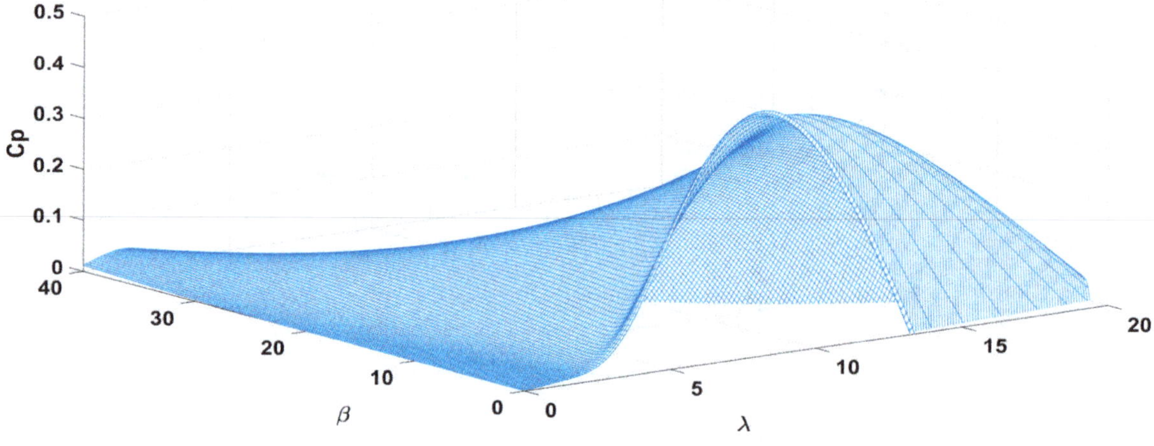

Figure 3. The aerodynamic power coefficient for 5-MW VSWT.

2.2. Tower Dynamics

The blade vibration results in the tower deflection. Thus, the thrust force on a tower top over the nacelle, which causes undesired nodding for the tower and fatigue loads on the VSWT, must be considered [31]:

$$F_t = \frac{1}{2}\rho\pi R^2 {v_r}^2 C_t(\lambda(v_r, w_r), \beta) \tag{5}$$

where $C_t(\lambda(v_r, w_r), \beta)$ is the thrust coefficient, which is the nonlinear function in terms of blade pitch angle β and tip speed ratio λ. By adopting a polynomial fitting algorithm based on the data-driven from a typical wind turbine benchmark simulator, the thrust coefficient can be derived as:

$$\begin{aligned} C_t(\lambda, \beta) = \; & 0.08698 - 0.003371\beta - 0.053272\lambda + 0.06499\lambda^2 - 0.003096\beta\lambda^2 - 0.009575\lambda^3 \\ & + 0.0002103\beta\lambda^3 + 0.0005667\lambda^4 - 5.24\times10^{-6}\beta\lambda^4 - 1.199\times10^{-5}\lambda^5 \end{aligned} \tag{6}$$

where $C_t(\lambda, \beta)$ represents a nonlinear function, as indicated in Figure 4.

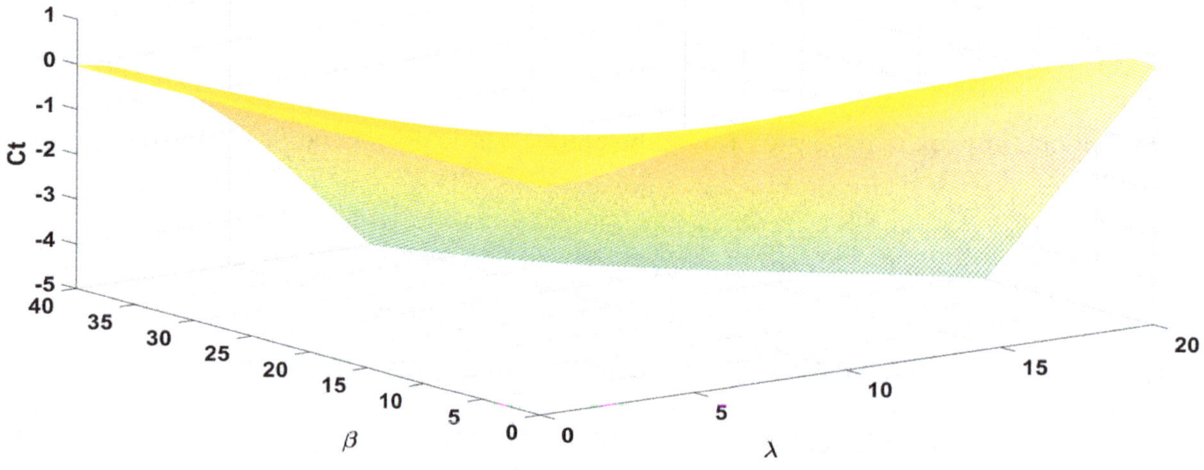

Figure 4. The thrust coefficient for 5-MW VSWT.

Thrust force towards the tower top over the nacelle causes tower fore-aft motion. The dynamics of tower fore-aft motion can be stated as a simplified second-order differential equation:

$$M_t\ddot{y} + B_t\dot{y} + K_t y = F_t \tag{7}$$

where M_t represents the mass model, B_t and K_t are the structural damping and structural stiffness coefficients of the tower model respectively, y represents the tower top displacement, and $\dot{y}$ is the tower deflection rate.

Tower fore-aft motion causes a nonnegligible effect on effective wind speed. The tower deflection rate is defined as $v_t = \dot{y}$. Thus, the effective (relative) wind speed v_r on the turbine rotor can be expressed with the normal wind speed w and the tower deflection as follows [30]:

$$v_r = w - v_t \tag{8}$$

The above formula shows that the actual wind speed acting on the blade is a relative wind speed, which is the difference between the natural wind speed w and the displacement speed of the tower v_t. It can be imagined that when the tower moves forward and backward in the horizontal direction with the wind speed, the wind speed acting on the blade is not exactly equal to the wind speed directly given by the natural environment. This formula makes the wind speed acting on the blade more accurate.

In Figure 2, w represents the nature wind speed towards the blade, which cannot be replaced by the relative (effective) wind speed v_r because the relative (effective) wind speed is just a virtual physical variable which is defined for the aerodynamic torque and power, not a real physical variable. Therefore, the relative (effective) wind speed v_r cannot appear before the blade, replacing wind speed w in Figure 2.

2.3. Drive Train

The wind mechanical energy is transformed through a drive train to the electrical generator. The drive train is rigidly coupled in the rotor side with a flexible connection in the generator side, as demonstrated in Figure 2. The shaft torsion represents the torque of the drive train flexible shaft, which is a vital factor affecting economic performance. Oscillations in the shaft torsion cause damage to the turbine components. The shaft torsion is replaced by the absolute angular position of the generator and rotor; its derivative is [17]:

$$\dot{\theta} = w_r - \frac{w_g}{N_g} \tag{9}$$

where θ is the shaft torsional angle, w_g is the generator angular speed, and N_g is the gearbox ratio.

Assuming that the low-speed shaft is one spring and one damper, the two-mass model is used to describe the drive train from a low- to a high-speed shaft through a gearbox as follows [17,26]:

$$J_r\dot{w}_r = T_r - K_\theta\theta - B_\theta\dot{\theta} \tag{10}$$

$$J_g\dot{w}_g = -T_g + \frac{K_\theta}{N_g}\theta + \frac{B_\theta}{N_g}\dot{\theta} \tag{11}$$

where J_r is the inertia of the VSWT rotor and the low-speed shaft, J_g is the inertia of the generator, K_θ and B_θ represent the stiffness and damping of the drive train, T_g is the generator torque.

2.4. Generator

The electrical generator power P^E can be derived as [17]:

$$P^E = \eta_g w_g T_g \tag{12}$$

where η_g represents the generator and power efficiency. Ignoring the losses, the derivation between P^A (the mechanical power) and P^E stems notably from the energy tentatively stored as the rotor kinetic energy [25].

2.5. The Nonlinear State-Space Model of VSWT

Based on Equations (1)–(12), the nonlinear dynamics of the VSWT system can be transformed into the following nonlinear state-space form:

$$\begin{aligned} \dot{x} &= f_c(x,u,w) \\ y &= g_c(x,u,w) \end{aligned} \tag{13}$$

where $g_c(x,u,w) = \begin{bmatrix} w_g \\ \eta_g w_g T_g \end{bmatrix}$, $f_c(x,u,w) = \begin{bmatrix} -\frac{B_\theta}{J_g N_g^2} w_g + \frac{B_\theta}{J_g N_g} w_r + \frac{K_\theta}{J_g N_g}\theta - \frac{1}{J_g} T_g \\ \frac{B_\theta}{J_r N_g} w_g - \frac{B_\theta}{J_r} w_r - \frac{K_\theta}{J_r}\theta + \frac{\pi}{2J_r}\rho_a R_r^2 \frac{(w-v_t)^3}{w_r} C_p(\lambda,\beta) \\ -\frac{w_g}{N_g} + w_r \\ v_t \\ -\frac{K_t}{M_t} y - \frac{B_t}{M_t} v_t + \frac{\pi}{2M_t}\rho_a R_r^2 (w-v_t)^2 C_t(\lambda,\beta) \end{bmatrix}$.

Define the state variables, input variables, and output variables as: $x = \begin{bmatrix} w_g & w_r & \theta & y & v_t \end{bmatrix}^T$, $u = \begin{bmatrix} T_g & \beta \end{bmatrix}^T$, $y_o = \begin{bmatrix} w_g & P^E \end{bmatrix}^T$. Then, the discrete wind turbine system model (13) can be obtained by forth stage Runge–Kutta (RK4) method with the sampling time T,

$$\begin{aligned} x(k+1) = f(x(k),u(k),w(k)) &= x(k) + \tfrac{T}{6}(k_1 + 2k_2 + 2k_3 + k_4) \\ y(k) &= g(x(k),u(k),w(k)) \end{aligned} \tag{14}$$

where $k_1 = f_c(x(k),u(k))$, $k_2 = f_c(x(k) + \frac{T}{2}k_1, u(k))$, $k_3 = f_c(x(k) + \frac{T}{2}k_2, u(k))$, $k_4 = f_c(x(k) + Tk_3, u(k))$, $g(x(k),u(k),w(k)) = g_c(x,u,w)$.

3. NEMPC Strategy for VSWT Control

EMPC for WTC seeks the optimal operation between the maximization of the generated power and the minimization of the cost related to care and maintenance. In a real wind farm, the oscillatory transient of tower deflection and shaft torsion can create microcracks in the materials, which can lead to component failure and increase maintenance costs. Therefore, not only generated power, pitch-angle and generator-torque, but also shaft torsion angle and tower displacement are considered in the economic cost function for the reduction of fatigue loads in this paper.

3.1. Economic Cost Function

In this paper, economic indexes are defined as follows. Firstly, the generator power must be considered in the economic index l_{e1} to capture the maximum power [19]:

$$l_{e1}(x,u) = -\lambda_1 P^E \tag{15}$$

Secondly, the shaft torsional angle and tower displacement must be considered in the economic index l_{e2} to reduce the fatigue of the tower structure caused by the tower deflection [19] and the gearbox load caused by the drive-shaft torsion [26]:

$$l_{e2}(x,u) = \lambda_2\theta^2 + \lambda_3 v_t^2 \tag{16}$$

Finally, the pitch-angle and generator-torque must be considered in the economic indexes l_{e3} and l_{e4}, respectively, to smoothen the control performance and reduce the fluctuations of the output electric power [19]:

$$l_{e3}(x,u) = \lambda_4\Delta\beta^2 + \lambda_5\beta^2 \tag{17}$$

$$l_{e4}(x,u) = \lambda_6 \Delta T_g^2 \tag{18}$$

where λ_1, λ_2, λ_3, λ_4, λ_5, and λ_6 are weighted coefficients.

In designing the NEMPC for the VSWT system, the economic cost function can be obtained as:

$$l_e(x,u) = l_{e1} + l_{e2} + l_{e3} + l_{e4} \tag{19}$$

which aims to decrease the load and fatigue on the wind power system and smooth the generator torque input and the output power while capturing the maximal power.

3.2. The Operational and Physical Constraints

Due to the electrical limitations of the actuator electronics and the safety requirements of the control process, a set of physical constraints must be fulfilled during operation in VSWT. Thus, the following set of constraints must be taken into consideration along the predictive/control horizon:

$$w_{g\min} \le w_g \le w_{g\max} \quad w_{r\min} \le w_r \le w_{r\max} \quad \theta \ge \theta_{\min} \tag{20}$$

$$0 \le T_g \le T_{g\max} \quad -\Delta T_{g\max} \le \Delta T_g \le \Delta T_{g\max} \tag{21}$$

$$\beta_{\min} \le \beta \le \beta_{\max} \quad -\Delta\beta_{\max} \le \Delta\beta \le \Delta\beta_{\max} \tag{22}$$

$$0 \le P^E \le P^E{}_{\max} \tag{23}$$

where the maximum rotor speed $w_{r\max}$, generator speed $w_{g\max}$, and torque $T_{g\max}$ are set slightly higher than their rated values $w_{r\ rated}$, $w_{g\ rated}$, and $T_{g\ rated}$. The definition and values of related parameters are listed in Table A2 in Appendix A [19,32].

3.3. The Optimization Problem for the Proposed NEMPC Strategy

Based on the above nonlinear model and cost function in Sections 3.1 and 3.2, the optimization problem for the proposed NEMPC strategy can be summed up as follows:

$$\begin{aligned}
&\min J = \sum_{i=1}^{N_P} l_e(x(k+i-1|k), u(k+i-1|k)) \\
&s.t. \\
&x(k+i|k) = f(x(k+i-1|k), u(k+i-1|k), w(k+i-1)), i = 1, \cdots, N_P, x(k|k) = x(k) \\
&y(k+i-1|k) = g(x(k+i-1|k), u(k+i-1|k), w(k+i-1)) \\
&\Delta u(k+i|k) = u(k+i|k) - u(k+i-1|k) \\
&equation(20) - equation(23)
\end{aligned} \tag{24}$$

Define $\mathbf{X} = \begin{bmatrix} x(k+1|k)^T & x(k+2|k)^T & \cdots & x(k+N_P|k)^T \end{bmatrix}^T$, $\mathbf{U} = \begin{bmatrix} u(k|k)^T & u(k+1|k)^T & \cdots & u(k+N_P-1|k)^T \end{bmatrix}^T$, $\mathbf{\Delta U} = \begin{bmatrix} \Delta u(k|k)^T & \Delta u(k+1|k)^T & \cdots & \Delta u(k+N_P-1|k)^T \end{bmatrix}^T$, $\mathbf{P^E} = \begin{bmatrix} P^E(k|k) & \cdots & P^E(k+N_P-1|k) \end{bmatrix}^T$, where N_P is the predictive/control horizon.

Define $\Phi = \begin{bmatrix} \mathbf{X} & \mathbf{U} & \mathbf{\Delta U} & \mathbf{P^E} \end{bmatrix}^T$ as the new optimal variables set. Then, the optimization problem (24) can be rewritten as:

$$\begin{aligned}
&\min J \\
&s.t. \\
&Aeq\Phi = bu(k-1), \\
&Ceq_i(\Phi) = \begin{bmatrix} x(k+i|k) - f(x(k+i-1|k), u(k+i-1|k), w(k+i-1)) \\ \hat{P}^E(k+i-1|k) - g(x(k+i-1|k), u(k+i-1|k), w(k+i-1)) \end{bmatrix}_{i=1,\cdots,N_P,\hat{x}(k|k)=x(k)} = 0 \\
&\Phi_{\min} \le \Phi \le \Phi_{\max}
\end{aligned} \tag{25}$$

$$Aeq = \begin{bmatrix} 0 & I & -S & 0 \end{bmatrix}_{2N_P\times 10N_P}, b = \begin{bmatrix} I & I & \cdots & I \end{bmatrix}^T_{2N_P\times 2N_P}$$

where
$$S = \begin{bmatrix} I & 0 & \cdots & 0 \\ I & I & & 0 \\ \vdots & \vdots & \ddots & \\ I & I & \cdots & I \end{bmatrix},$$

$$\Phi_{\min} = \begin{bmatrix} X_{\min} & U_{\min} & \Delta U_{\min} & \hat{P}^E_{\min} \end{bmatrix}, \Phi_{\max} = \begin{bmatrix} X_{\max} & U_{\max} & \Delta U_{\max} & \hat{P}^E_{\max} \end{bmatrix}$$

$$X_{\min} = \begin{bmatrix} x_{\min} \\ x_{\min} \\ \vdots \\ x_{\min} \end{bmatrix}, X_{\max} = \begin{bmatrix} x_{\max} \\ x_{\max} \\ \vdots \\ x_{\max} \end{bmatrix}, U_{\min} = \begin{bmatrix} u_{\min} \\ u_{\min} \\ \vdots \\ u_{\min} \end{bmatrix}, U_{\max} = \begin{bmatrix} u_{\max} \\ u_{\max} \\ \vdots \\ u_{\max} \end{bmatrix},$$

$$\Delta U_{\min} = \begin{bmatrix} \Delta u_{\min} \\ \Delta u_{\min} \\ \vdots \\ \Delta u_{\min} \end{bmatrix}, \Delta U_{\max} = \begin{bmatrix} \Delta u_{\max} \\ \Delta u_{\max} \\ \vdots \\ \Delta u_{\max} \end{bmatrix}, \hat{P}^E_{\min} = \begin{bmatrix} 0 \\ 0 \\ \vdots \\ 0 \end{bmatrix}, \hat{P}^E_{\max} = \begin{bmatrix} P^E_{\max} \\ P^E_{\max} \\ \vdots \\ P^E_{\max} \end{bmatrix}$$

$$x_{\min} = \begin{bmatrix} w_{g\ \min} & w_{r\ \min} & \theta_{\min} & y_{\min} & v_{t\ \min} \end{bmatrix}^T, u_{\min} = \begin{bmatrix} T_{g\ \min} & \beta_{\min} \end{bmatrix}^T, \Delta u_{\min} = \begin{bmatrix} \Delta T_{g\ \min} & \Delta\beta_{\min} \end{bmatrix}^T$$
$$x_{\max} = \begin{bmatrix} w_{g\ \max} & w_{r\ \max} & \theta_{\max} & y_{\max} & v_{t\ \max} \end{bmatrix}^T, u_{\max} = \begin{bmatrix} T_{g\ \max} & \beta_{\max} \end{bmatrix}^T, \Delta u_{\max} = \begin{bmatrix} \Delta T_{g\ \max} & \Delta\beta_{\max} \end{bmatrix}^T.$$

The optimization problem (25) can be solved by using the interior point method, which is realized by using IPOT solver in MATLAB [33]. Then, the optimal solution of the control inputs (pitch angle and the generator torque) at the current time can be extracted and implemented.

4. Simulations Results

Three simulations are performed based on a typical 5-MW NREL VSWT system to validate the proposed NEMPC strategy. The corresponding parameters of the 5-MW NREL VSWT system are listed in Table A1 in the Appendix A [32]. The corresponding constraints are shown in Table A2 [19,32]. In the simulations, the weights in the cost function (15)–(18) contain: $\lambda_1 = 20, \lambda_2 = 100, \lambda_3 = 5 * 10^8, \lambda_4 = 1, \lambda_5 = 0, \lambda_6 = 0.001$. The maximum generator speed, rotor speed, and generator torque are all arbitrarily set at 5% above their rated values, i.e.,$w_{g\max} = 1.05 w_{g\ rated}$, $w_{r\max} = 1.05 w_{r\ rated}$, $T_{g\max} = 1.05 T_{g\ rated}$. For the purpose of simplicity, the maximum electric power is chosen as $P^E_{\max} = P^E_{rated}$.

For comparison purposes, the classical NMPC strategy based on [19] is also constituted, which has the same model (13) and constraints (20)–(23) as the proposed NEMPC strategy in this paper. However, the tracking objective function (not economic cost function) is used in the classical NMPC, which can be described as:

$$l_{mpc}(x,u) = \sum_{i=1}^{N_P} \left[Q_\Omega \left(w_r(k+i-1|k) - w_{r\ ref}(w) \right)^2 + Q_\beta (\beta(k+i-1|k) - \beta_{ref}(w))^2 \right] \quad (26)$$

where $Q_\Omega = 3.4 \times 10^7$, $Q_\beta = 4.6 \times 10^8$ for the wind speed region II, $Q_\Omega = 3.4 \times 10^5$, $Q_\beta = 4.6 \times 10^5$ for wind speed region III, $\beta_{ref}(w)$ is the pitch angle reference and $w_{r\ ref}(w)$ is the rotor speed reference, which can be derived as:

$$w_{r\ ref}(w) = \begin{cases} w_r^*(w) & if \quad w_r^*(w) \in [w_{r\ cut-in}, w_{r\ rated}] \\ w_{r\ cut-in} & if \quad w_r^*(w) < w_{r\ cut-in} \\ w_{r\ rated} & if \quad w_r^*(w) > w_{r\ rated} \end{cases}, \beta_{ref}(w) = \begin{cases} \beta^*(w) & if \quad w \le w_{rated} \\ \beta_{AR}(w) & if \quad w > w_{rated} \end{cases} \quad (27)$$

β^* and λ^* in Equation (27) are the optimal pitch angle and tip ratio in wind speed region II as demonstrated in Figure 1, can be derived via maximizing the power coefficient C_P:

$$\beta^*, \lambda^* = \max_{\beta,\lambda} C_P(\beta, \lambda) \tag{28}$$

Then, the optimal rotor speed w_r^* can be obtained:

$$w_r^*(w) = \frac{\lambda^* w}{R} \tag{29}$$

$\beta_{AR}(w)$ in Equation (27) can be solved from: $\frac{1}{2}\rho A C_P\left(\beta_{AR}(w), \frac{R w_{r\ rated}}{w}\right) w^3 = P_{rated}$.

The predictive/control horizon is set as $N_P = 14$ for both classical NMPC and the proposed NEMPC in the following three groups of simulations.

4.1. Gradient Normal Wind

As indicated in Figure 5, a gradient normal wind varying between 7.5 m/s and 15 m/s is used in the simulation. The initial condition is set as $x_0 = \begin{bmatrix} 93.54 & 0.96 & 1.85 \times 10^{-3} & 0.21 & 0 \end{bmatrix}^T$, *and* $t_0 = 0$. All the input variables, state variables, and output variables adopting both the proposed NEMPC and the classical NMPC strategies are shown as Figure 6a–h. During the time period from 0 s to 50 s, all the dynamic trajectories are almost the same when using the proposed NEMPC and the classical NMPC strategies, because the chosen initial state is just the optimal solution for both controllers under the initial 7.5 m/s normal wind speed.

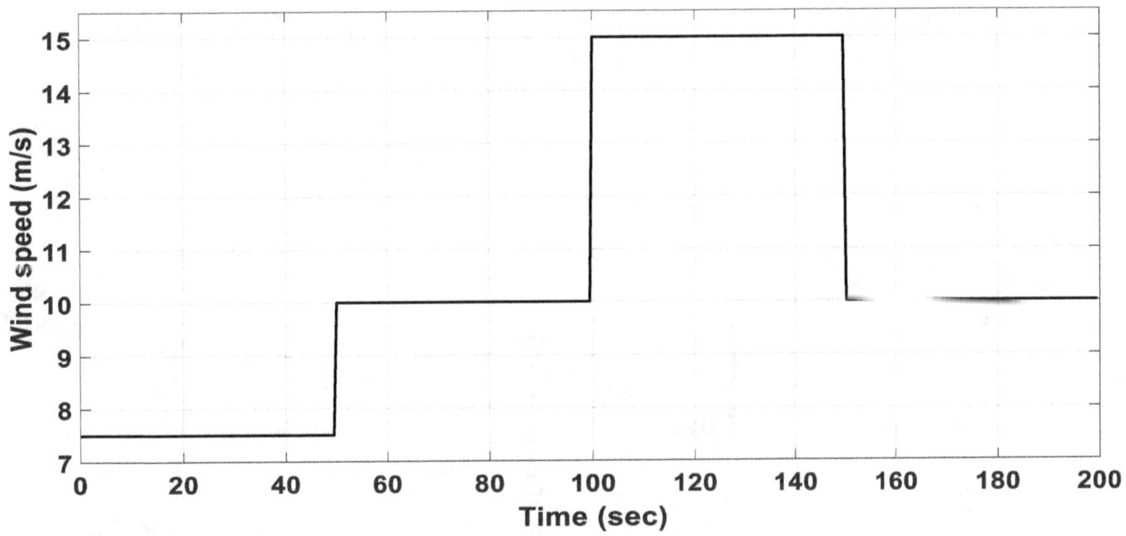

Figure 5. Gradient wind speed.

At time instant 50 s, the normal wind speed steps from 7.5 m/s to 10 m/s, which is below the rated wind speed. Obviously, the VSWT system remains working in region II by using the classical NMPC strategy, which aims to follow the maximum power point and keeps the pitch angle zero. The proposed NEMPC strategy aims to attain a trade-off between the power maximization and turbine fatigue minimization by minimizing the economic cost function (19). Therefore, the pitch-angle is not always kept at zero, while the vibrations of the shaft torsion angle, the tower displacement, and the generation torque are much slighter when using the proposed NEMPC strategy, as shown in Figure 6c–f. No constraints are violated due to the constraints in Equations (20)–(23), which are imposed in both the classical NMPC strategy and the proposed NEMPC strategy, as seen in Figure 6.

At time instant 100 s, the normal wind speed steps from 10 m/s to 15 m/s, which is above the rated wind speed. Therefore, the working region of the VSWT system changes from region II to region

III. The aim of the classical NMPC changes from the maximization power capturing to rated power capturing. The generator speed, rotor speed, generator torque, and generator power are changed to the rated values. Simultaneously, the pitch angle is enabled. As shown in Figure 6c–f, the vibration of the shaft torsional angle, the pitch angle, and especially the tower displacement, is much smaller when using the proposed NEMPC strategy than when using the classical NMPC strategy, due to θ, v_t and β being considered in the economic cost function (19).

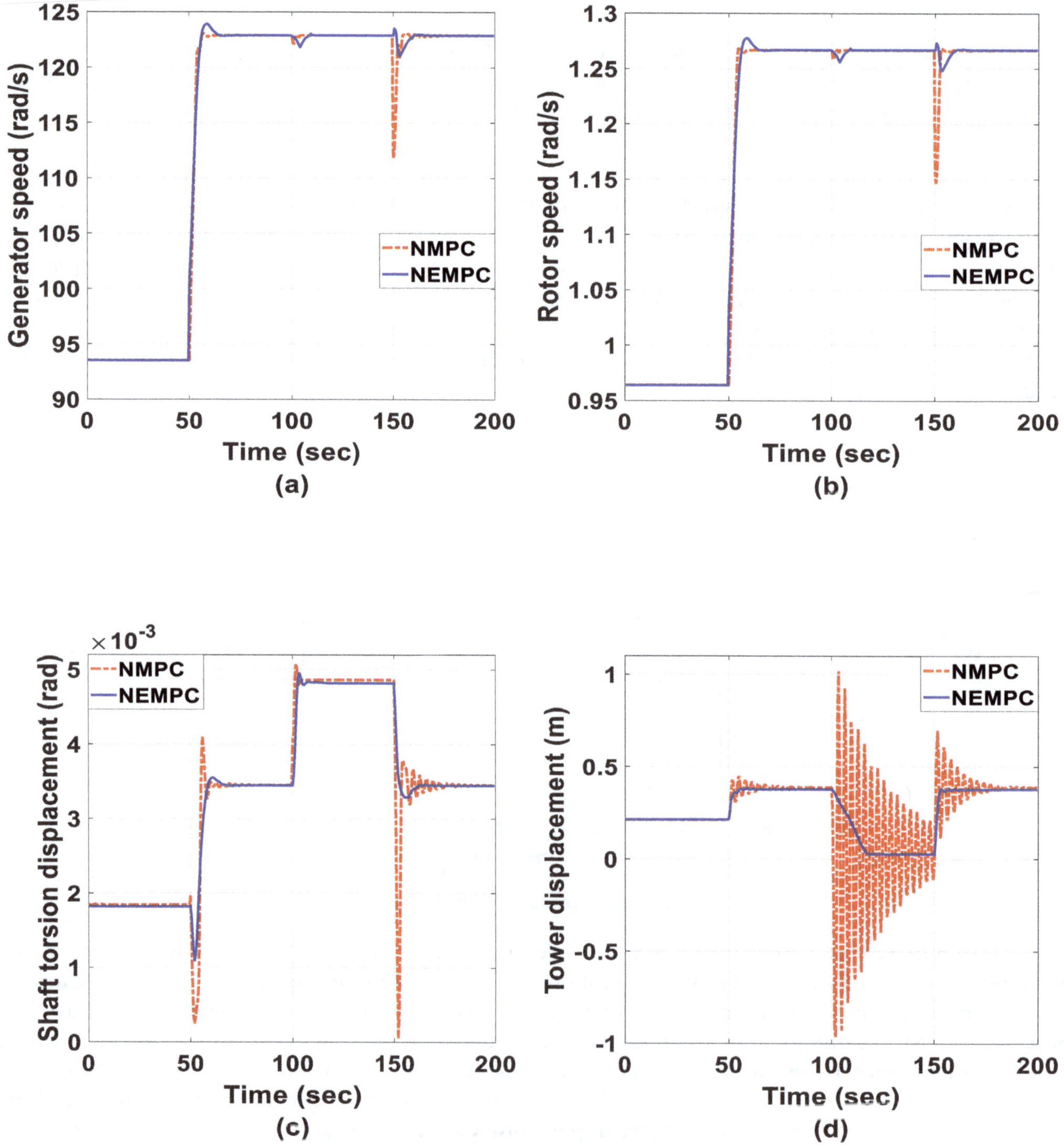

Figure 6. *Cont.*

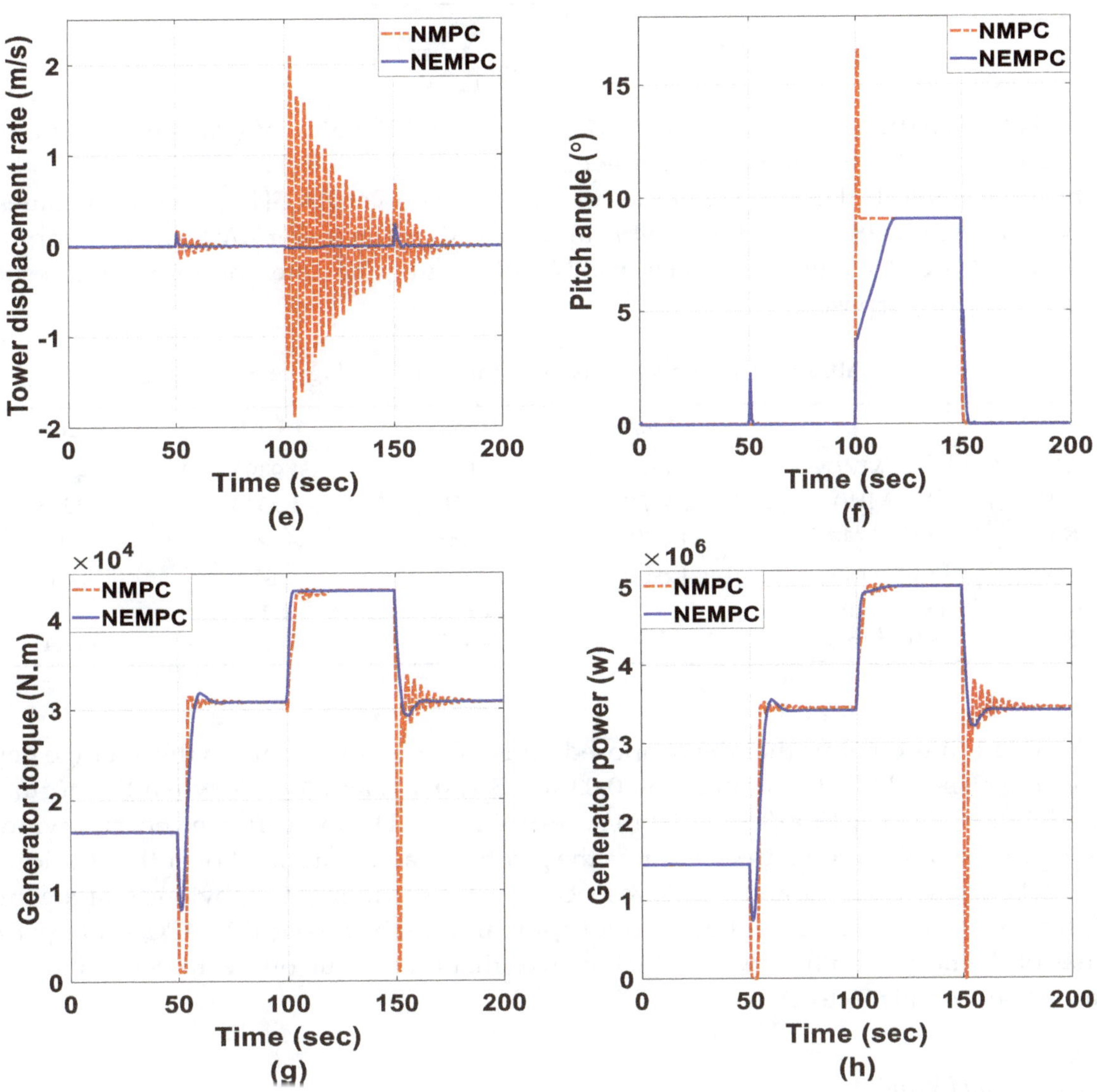

Figure 6. (**a**) Generator speed; (**b**) Rotor speed; (**c**) Shaft torsional angle; (**d**) Tower displacement; (**e**) Tower displacement rate; (**f**) Pitch angle; (**g**) Generator torque; (**h**) Generator power.

At time instant 150 s, the normal wind speed decreases from 15 m/s to 10 m/s. Thus, the wind turbine works back to region II. The pitch angle is disabled, the generator speed, rotor speed, and generator torque are performed to follow the maximum power point by using the classical NMPC strategy. From 6b–h, it is obvious that the vibration of each state variable is much more severe when using the classical NMPC strategy compared with the proposed NEMPC strategy. This is because the shaft load and the tower fatigue are not considered in the tracking objective function (26) for the classical NMPC strategy, while they are emphasized in the economic cost function (19) for the proposed NEMPC strategy. Vibrations in the shaft torsional angle, the tower displacement, and the tower displacement rate are harmful to wind turbine systems. Thus, the proposed NEMPC strategy is much better than the classical NMPC strategy in this respect.

From a more intuitive point of view, numerical comparisons have been made to illustrate the difference between these two controllers. Define the root mean square (RMS) values of mechanical loads as follows:

$$RMS_{shaft} = \sqrt{\frac{1}{Nsim}\sum_{k=1}^{Nsim} \theta^2(k)} \tag{30}$$

$$RMS_{tower} = \sqrt{\frac{1}{Nsim}\sum_{k=1}^{Nsim} v_t^2(k)} \tag{31}$$

where $Nsim$ is the length of the simulation, RMS_{shaft} is the RMS value of gearbox load on the shaft, RMS_{tower} is the RMS value of fatigue on the tower.

As shown in Table 1, the average values of generator power $AVG(P_g)$, the RMS values of the gearbox load on shaft RMS_{shaft}, and the RMS values of fatigue on tower RMS_{tower} by adopting both the proposed NEMPC strategy and the classical NMPC strategy at different normal wind speed are listed for comparison purposes.

Table 1. Data analysis based on the results in Figure 6.

	7.5 m/s	10 m/s	15 m/s	10 m/s
$AVG(P_g^{NEMPC})$ (MW)	1.4575	3.1873	4.9493	3.4317
$AVG(P_g^{NMPC})$ (MW)	1.4570	3.1714	4.9343	3.3323
RMS_{shaft}^{NEMPC}($\times10^{-3}$ rad)	1.820	3.288	4.784	3.416
RMS_{shaft}^{NMPC} ($\times10^{-3}$ rad)	1.846	3.300	4.851	3.476
RMS_{tower}^{NEMPC} ($\times10^{-2}$ m/s)	0	1.643	1.10	3.31
RMS_{tower}^{NMPC} ($\times10^{-2}$ m/s)	0	4.483	76.58	16.44

From Table 1, it is obvious that the proposed NEMPC strategy enhances the average generator power over the classical NMPC strategy by 0.03%, 0.5%, 0.3%, and 2.98%, when the normal wind speeds are 7.5 m/s, 10 m/s, 15 m/s, and 10 m/s, respectively. During different normal wind speed periods, the proposed NEMPC strategy reduces the gearbox load on the shaft over the classical NMPC strategy by 1.41%, 0.36%, 1.38%, and 1.73%, respectively. Furthermore, it is obvious that the proposed NEMPC strategy reduces the tower fatigue a lot compared to the classical NMPC strategy from the RMS values of the fatigue on the tower, which proves that the fatigue on the tower can't be ignored with regards to economic factors.

4.2. Stepwise Normal Wind

In general, 11.4 m/s represents the rated wind-speed, which is used to divide the wind speed into region II and III. Due to the different control objectives in regions II and III, a switching controller is widely used in the actual control of VSWT, which is a great challenge in controller design. To achieve the various operational requirements, the objective function of the classical NMPC strategy switches frequently according to wind speed operating regions. Thus, the rated wind speed 11.4 m/s is actually a very important speed in practical controller designs for VSWT systems.

The proposed NEMPC strategy in this paper adopts a unique economic cost function (19) which does not change according to the normal wind speed between the cut-in and cut-off wind speeds. Suppose the stepwise wind speed is initially set to 10.6 m/s, then steps to 12 m/s with an increment of 0.2 m within 25 s (from region II to region III), as shown in Figure 7a. As indicated in Figure 7b, it's obvious that the pitch angle has a quick and small overshoot, before quickly returning to zero before t = 125 s. After 125 s, the fluctuations of the pitch angle accumulate around a certain optimal value when the proposed NEMPC strategy is used. At 125 s, the wind speed changes from 11.4 m/s to 11.6 m/s. This means that the special wind speed is around 11.4 m/s due to the economic objective function (19), above which the stable pitch angle no longer remains at zero. Thus, these simulation results prove that a wind speed of around 11.4 m/s was set to divide the region II and III, which is reasonable for considering the economic operation of the VSWT system.

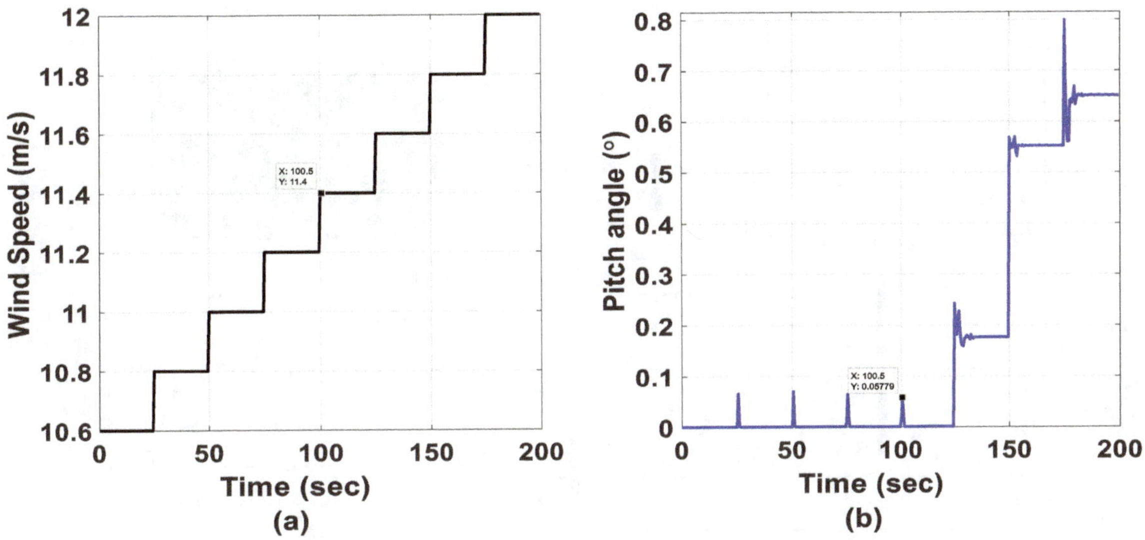

Figure 7. (**a**) Stepwise wind speed; (**b**) Pitch angle.

4.3. Randomly Varying Wind Speed Based on the FAST Simulator

The NREL FAST is an inclusive aeroelastic simulator of two- and three-bladed horizontal-axis WTs [34]. To validate the practicality and validity of the proposed NEMPC strategy, the simulation is then tested on NREL FAST. In this group of simulations, a FAST subroutine was written into the MATLAB S-function, as indicated in Figures 8 and 9.

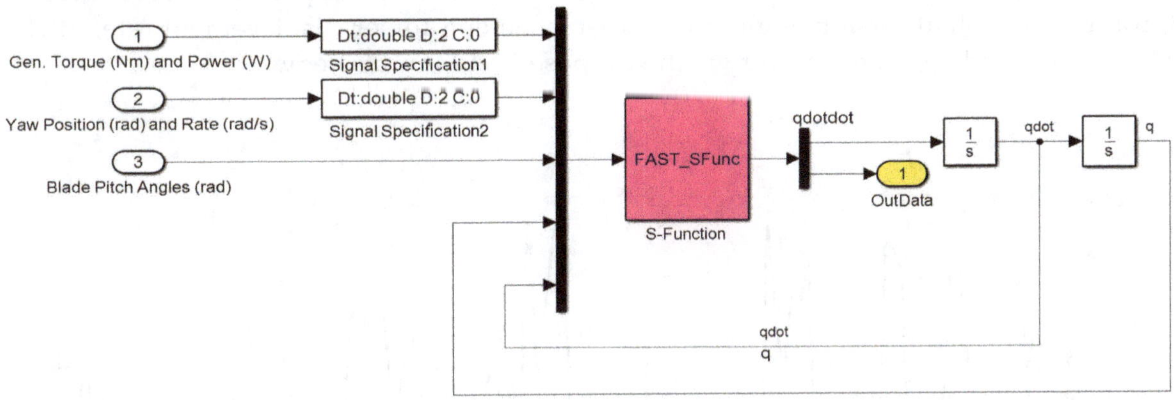

Figure 8. The FAST S-Function Block.

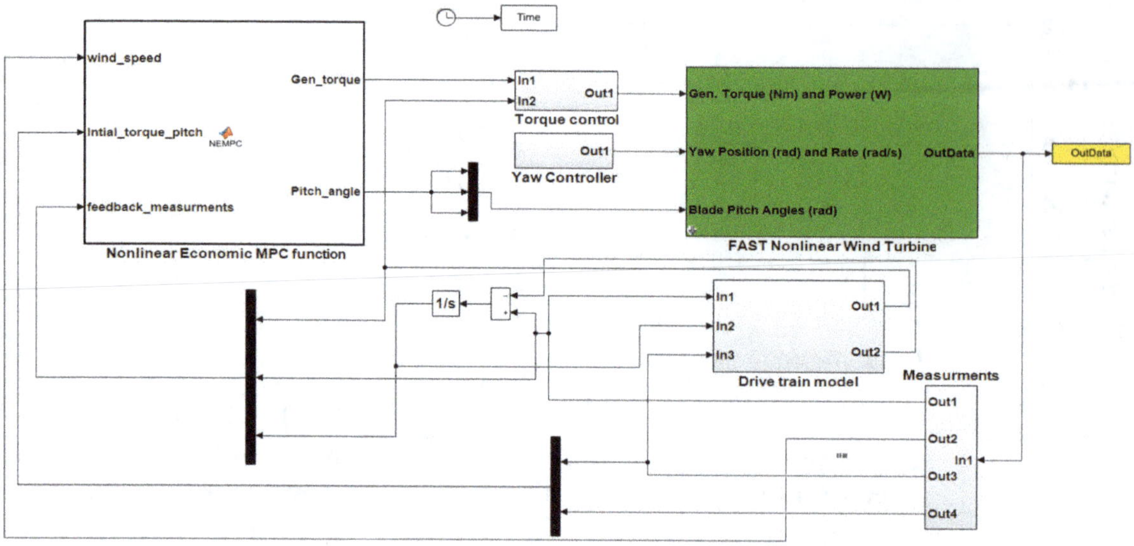

Figure 9. Simulation structure.

The normal wind speed w is always modeled as the combination of a slowly-changing mean wind-speed w_m and a rapidly-changing turbulent wind-speed w_d:

$$w = w_m + w_d \tag{32}$$

where we suppose that w_m follows the two-parameter Weibull distribution through a scale parameter $s1 = 10$ and shape parameter $s2 = 14$, and w_d follows zero-mean Gaussian white noise distribution with standard variation $\sigma = 1.899$ [28].

Then, a group of 600 s random normal wind speed data generated by the FAST simulator as indicated in Figure 10 is adopted in Case 3. The initial condition is set as $x_0 = \begin{bmatrix} 99.77 & 1.03 & 2.1\times10^{-3} & 0.25 & 0 \end{bmatrix}^T$, $t_0 = 0$. Figure 11a–h indicate the performances of the generator speed, rotor speed, shaft torsion angle, tower displacement, tower displacement rate, pitch angle, generator torque, and generator power via the proposed NEMPC strategy.

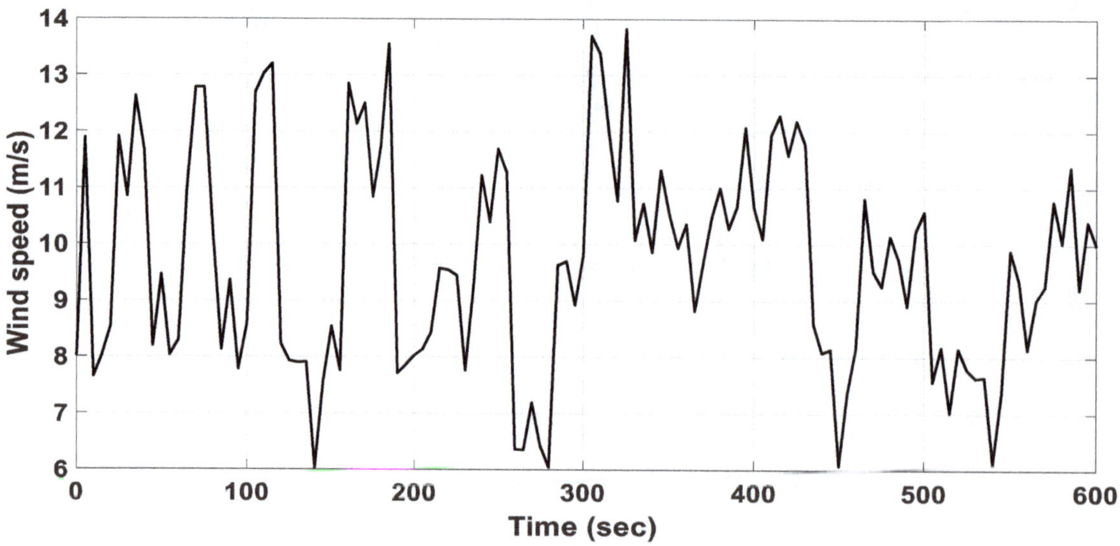

Figure 10. Wind speed.

From a more intuitive viewpoint, it can be observed that the performances of the proposed NEMPC strategy outperform the classical NMPC strategy. More specifically, the classical NMPC strategy shows more noticeable fluctuations in the performance of generator power, shaft torsion displacement, tower displacement, generator torque, and pitch angle when the wind speed has a sudden increase around its rated wind speed, i.e., 11.4 m/s, compared with the proposed NEMPC scheme.

Meanwhile, the classical NMPC strategy shows more successive fluctuations in the performance of generator power, shaft torsion displacement, tower displacement, generator torque, and pitch angle when the normal wind speed has a sudden drop around the rated wind speed, compared with the proposed NEMPC strategy.

As shown in Figure 11d–e, the tower displacement and the tower displacement by using the classical NMPC strategy fluctuates more severely according to the wind speed variation compared with the proposed NEMPC strategy, due to the economic index l_{e2} related with the tower displacement considered in the economic cost function (19) for the proposed NEMPC. It is similar in the performances of the generator power, shaft torsion displacement, generator torque, and pitch angle, as shown in Figure 10. It can be concluded that the proposed NEMPC strategy decreases the operating costs while maintaining energy utilization, compared with the classical NMPC strategy. This is especially obvious when the wind speed varies around the rated wind speed. From Table 2, it is obvious that the proposed NEMPC strategy enhances the average generator power over the classical NMPC strategy by 0.67%, 1.05%, 0.28%, and 1.75%, when the average wind speeds are 9.7 m/s, 9.4 m/s, 10.8 m/s, and 9 m/s, respectively.

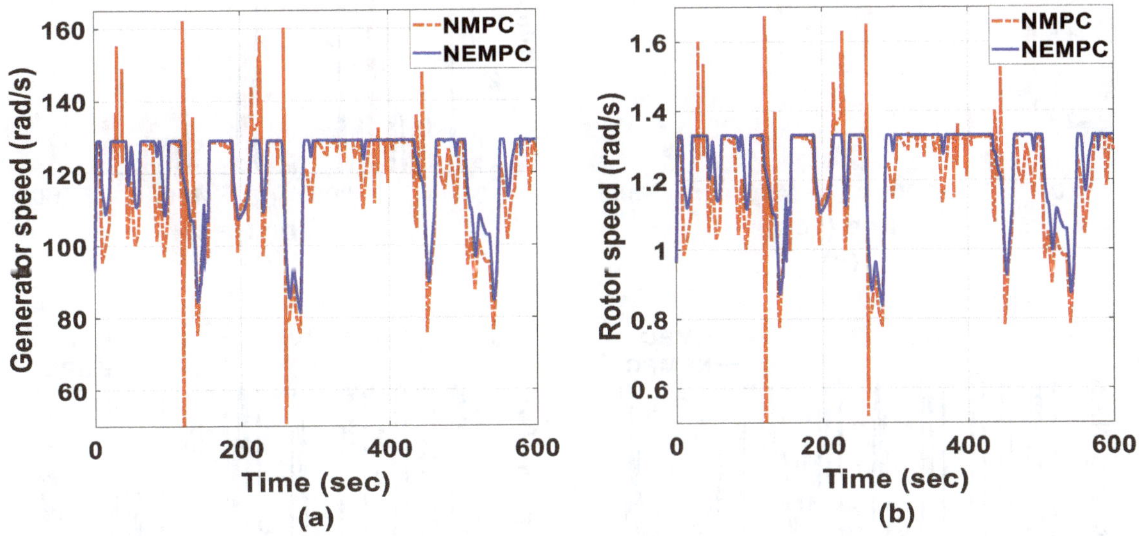

Figure 11. *Cont.*

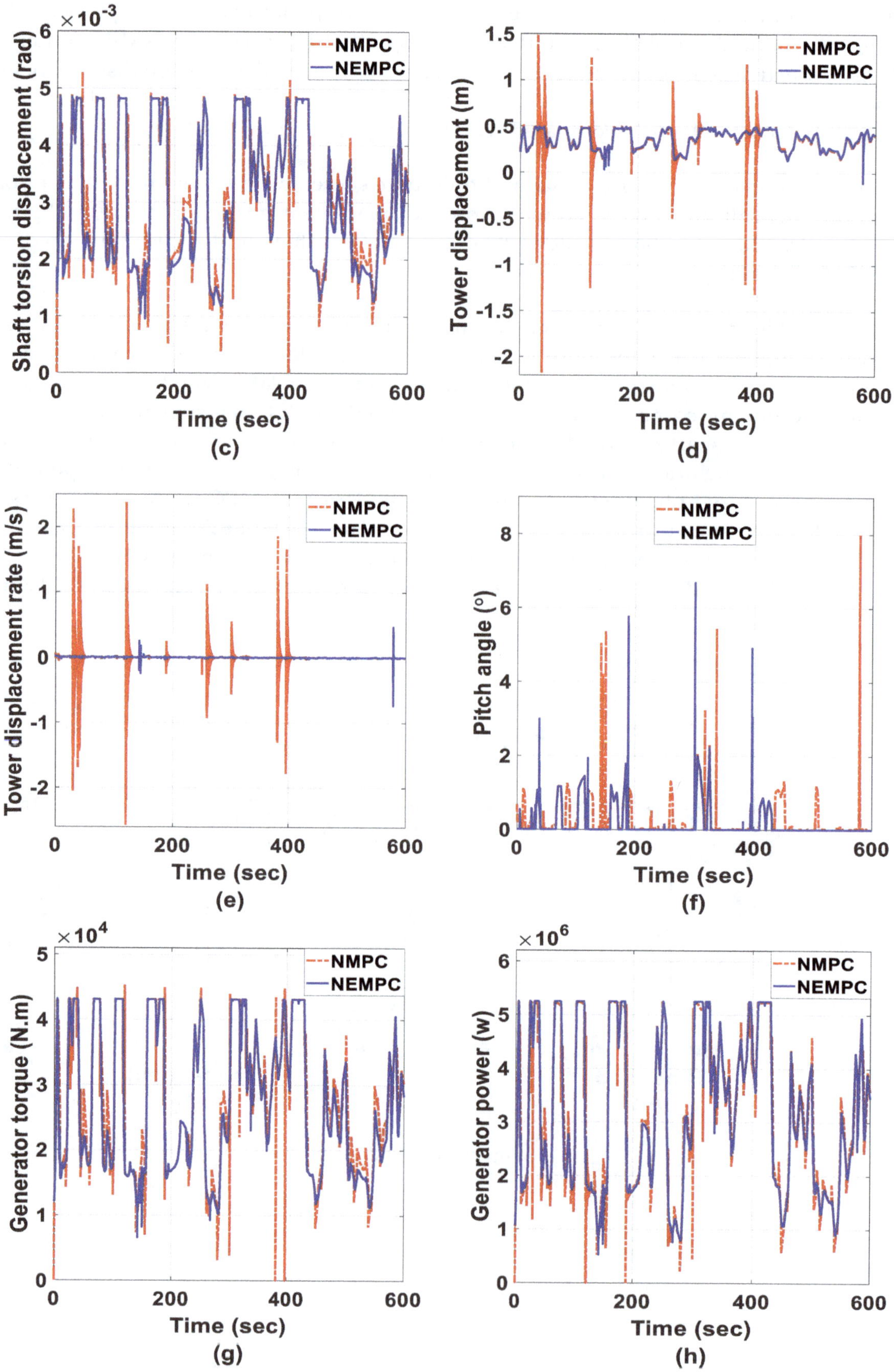

Figure 11. (**a**) Generator speed; (**b**) Rotor speed; (**c**) Shaft torsional angle; (**d**) Tower displacement; (**e**) Tower displacement rate; (**f**) Pitch angle; (**g**) Generator torque; (**h**) Generator power.

Table 2. Data analysis based on the results in Figure 11.

	9.7 m/s	9.4 m/s	10.8 m/s	9 m/s
$AVG(P_g^{NEMPC})$ (MW)	3.1162	2.9598	4.1231	2.5662
$AVG(P_g^{NMPC})$ (MW)	3.095	2.9291	4.1116	2.5221
RMS_{shaft}^{NEMPC}($\times 10^{-3}$ rad)	3.234	3.092	3.913	2.581
RMS_{shaft}^{NMPC} ($\times 10^{-3}$ rad)	3.365	3.215	3.939	2.761
RMS_{tower}^{NEMPC} ($\times 10^{-2}$ m/s)	36.38	35.80	42.59	33.40
RMS_{tower}^{NMPC} ($\times 10^{-2}$ m/s)	39.44	35.83	43.22	32.86

During different average wind speed periods, the proposed NEMPC strategy reduces the gearbox load on the shaft over the classical NMPC strategy by 4.05%, 3.96%, 0.68%, and 6.96%, respectively. Furthermore, it is obvious that the proposed NEMPC strategy reduces the tower fatigue a lot compared to the classical NMPC strategy from the RMS values of the fatigue on the tower. Therefore, this group of simulation-based FAST verifies the economy of the proposed NEMPC strategy.

It is well known that the computational burden of nonlinear MPC strategies increases exponentially according to the incensement of the predictive horizon. Various predictive horizons have been considered to investigate the computational burden by using classical NMPC and the proposed NEMPC strategies. The relative computational times of these two controllers are indicated in Figure 12. From Figure 12, it is obvious that the computational time of the proposed NEMPC is much smaller than the classical NMPC with the same predictive horizon. Simultaneously, the computational time of the classical NMPC increases more rapidly according to the incensement of the predictive horizon, compared with the proposed NEMPC strategy.

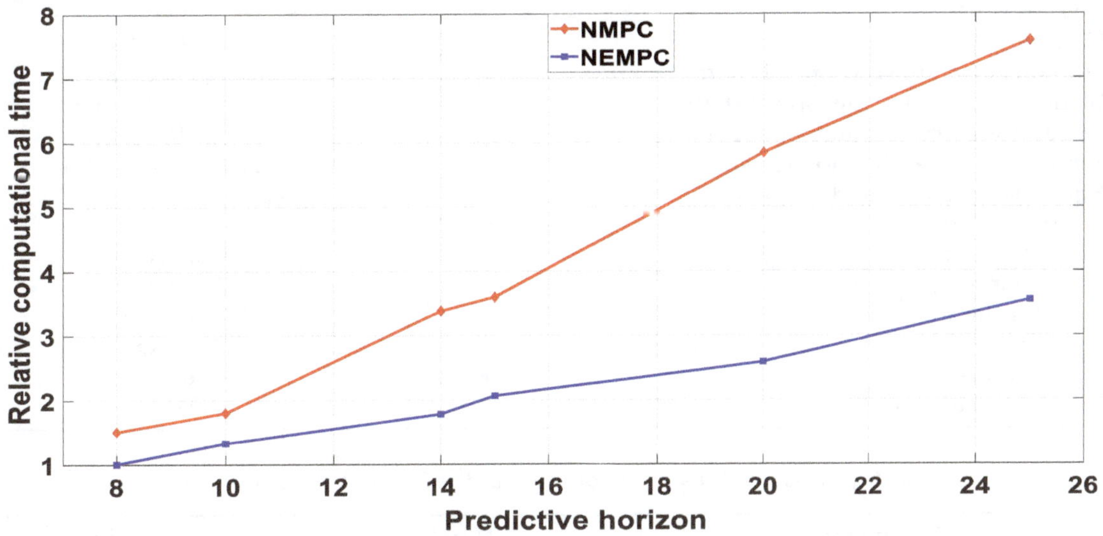

Figure 12. Relative computational times for various predictive horizons.

5. Conclusions

In this paper, the tower displacement and shaft torsional angle are taken into consideration in the development of a nonlinear model for a 5-MW VSWT system. Based on this comprehensive model, a NEMPC strategy for the VSWT system is proposed to increase the dynamic economy, i.e., through power generation maximization and fatigue load minimization during operation. Three groups of simulations for comparisons with the classical NMPC strategy demonstrate the effectiveness of the proposed NEMPC strategy. The simulation under random wind speeds based on the FAST simulator verified the reliability practicability of the proposed NEMPC strategy.

The contributions of this study compared with other current studies can be outlined as follows: Firstly, the tower dynamics are focused on in the modeling process for the VSWT with flexible connections on the shaft, and the tower displacement is considered as a part of the economic objective function, which can reduce tower fatigue. Secondly, all the related operational constraints and physical constraints in the VSWT system with flexible connections on the shaft are considered in the proposed NEMPC strategy in detail. Finally, simulations based on the FAST simulator, which is a realistic WT simulator for testing the system's validity and practicability, were performed to show the practical implementation of the proposed controller using a turbulence wind speed profile. A future research topic is the design of a robust EMPC strategy taking into account the uncertainty of wind speed.

Author Contributions: Conceptualization, X.K. and M.A.A.; methodology, X.K.; software, M.A.A. and Q.W.; validation, X.K., M.A.A., and Q.W.; formal analysis, X.K.; investigation M.A.A. and Q.W.; resources, L.M. and X.L.; data curation, X.K.; writing—original draft preparation, M.A.A. and Q.W.; writing—review and editing, X.K., L.M. and X.L.; visualization, X.K.; supervision, L.M. and X.L.; project administration, L.M. and X.L.; funding acquisition, X.K. All authors have read and agreed to the published version of the manuscript.

Funding: This work was supported in part by the National Nature Science Foundation of China under Grant 61603134, 61673171, U1709211, 61533013, and 61833011, and in part by Fundamental Research Funds for the Central Universities under Grant 2019QN041, 2017MS033, and 2017ZZD004.

Conflicts of Interest: The authors declare no conflict of interest.

Appendix A

Table A1. The parameters in the 5-MW VSWT system.

Parameter	Symbol	Value
Number of blades	-	3
Rotor radius	-	63 m
Hub height	-	84.3 m
Hub inertia on the low-speed shaft	J_H	115,926 kg·m^2
Blade inertia on the low-speed shaft	J_B	11,776,047 kg·m^2
Generator inertia on high-speed shaft	J_g	534.116 kg·m^2
Equivalent drive shaft torsion spring	K_θ	867,637,000 N·m/rad
Equivalent drive shaft torsion damping	B_θ	6,215,000 N·m/(rad/s)
Gearbox ratio	N_g	97
tower fore-aft inertia	M_t	430,000 kg
tower damping	B_t	17,600 N·s/m
tower stiffness	K_t	1,770,000 N/m
Air density	ρ_a	1.2231 kg/m^3
Rotor radius	R_r	63 m
Generator efficiency	η_g	94.4%

Table A2. The values of the bounds in the 5-MW VSWT system.

Parameter	Symbol	Value
Rated generator speed, Min. generator speed	$\omega_{g\ rated}$, $\omega_{g\ min}$	122.9096, 70.1622
Rated rotor speed, Min. rotor speed	$\omega_{r\ rated}$, $\omega_{r\ cut-in}$	1.2671, 0.7226
Min. shaft torsion angle	θ_{min}	0
Rated torque, Min. torque	$T_{g\ rated}$, $T_{g\ min}$	43,093.55, 0
Max. torque rate, Min. pitch, Max. pitch	ΔT_{gmax}, β_{min}, β_{max}	15,000, 0, 90
Max. pitch rate, Rated electrical power	$\Delta\beta_{min}$, P^E_{rated}	8, 5,000,000

References

1. Data, E.S.T. What is the proportion of China's global renewable energy power generation installed data in 2018? *Polaris Power Network: Polaris Power Network News*, 12 April 2019. Available online: http://news.bjx.com.cn/html/20190412/974382.shtml (accessed on 2 May 2019).

2. Bossanyi, E.A. Wind Turbine Control for Load Reduction. *Wind Energy* **2003**, *6*, 229–244. [CrossRef]
3. Nam, Y. Control System Design. In *Wind Turbines*; IntechOpen: Rijeka, Croatia, 2011.
4. Jafarnejadsani, H.; Pieper, J. Gain-Scheduled ℓ_1-Optimal Control of Variable-Speed-Variable-Pitch Wind Turbines. *IEEE Trans. Control Syst. Technol.* **2015**, *23*, 372–379. [CrossRef]
5. Ren, Y.; Li, L.; Brindley, J.; Jiang, L. Nonlinear PI control for variable pitch wind turbine. *Control Eng. Pract.* **2016**, *50* (Suppl. C), 84–94. [CrossRef]
6. El-Shafei, M.A.K.; El-Hawwary, M.I.; Emara, H.M. Implementation of fractional-order PID controller in an industrial distributed control system. In Proceedings of the 2017 14th International Multi-Conference on Systems, Signals & Devices (SSD), Marrakech, Morocco, 28–31 March 2017; pp. 713–718.
7. Kusiak, A.; Zhang, Z.; Verma, A. Prediction, operations, and condition monitoring in wind energy. *Energy* **2013**, *60*, 1–12. [CrossRef]
8. Yilmaz, A.S.; Özer, Z. Pitch angle control in wind turbines above the rated wind speed by multi-layer perceptron and radial basis function neural networks. *Expert Syst. Appl.* **2009**, *36*, 9767–9775. [CrossRef]
9. Van, T.L.; Nguyen, T.H.; Lee, D.C. Advanced Pitch Angle Control Based on Fuzzy Logic for Variable-Speed Wind Turbine Systems. *IEEE Trans. Energy Convers.* **2015**, *30*, 578–587. [CrossRef]
10. Wakui, T.; Yoshimura, M.; Yokoyama, R. Multiple-feedback control of power output and platform pitching motion for a floating offshore wind turbine-generator system. *Energy* **2017**, *141*, 563–578. [CrossRef]
11. Jiao, X.; Meng, W.; Yang, Q.; Fu, L.; Chen, Q. Adaptive Continuous Neural Pitch Angle Control for Variable-Speed Wind Turbines. *Asian J. Control* **2019**, *21*. [CrossRef]
12. Moodi, H.; Bustan, D. Wind turbine control using T-S systems with nonlinear consequent parts. *Energy* **2019**, *172*, 922–931. [CrossRef]
13. Lio, W.H.; Rossiter, J.; Jones, B.L. A review on applications of model predictive control to wind turbines. In Proceedings of the 2014 UKACC International Conference on Control (CONTROL), Loughborough, UK, 9–11 July 2014; pp. 673–678.
14. Henriksen, L.C. *Model Predictive Control of Wind Turbines*; DTU Informatics: Lyngby, Denmark, 2010.
15. Henriksen, L.C.; Hansen, M.H.; Poulsen, N.K. Wind turbine control with constraint handling: A model predictive control approach. *IET Control Theory Appl.* **2012**, *6*, 1722–1734. [CrossRef]
16. Bououden, S.; Chadli, M.; Filali, S.; El Hajjaji, A. Fuzzy model based multivariable predictive control of a variable speed wind turbine: LMI approach. *Renew. Energy* **2012**, *37*, 434–439. [CrossRef]
17. Abdelbaky, M.A.; Liu, X.; Jiang, D. Design and implementation of partial offline fuzzy model-predictive pitch controller for large-scale wind-turbines. *Renew. Energy* **2020**, *145*, 981–996. [CrossRef]
18. Abdelbaky, M.A.; Liu, X.; Kong, X. Wind Turbines Pitch Controller using Constrained Fuzzy-Receding Horizon Control. In Proceedings of the 2019 Chinese Control And Decision Conference (CCDC), Nanchang, China, 3–5 June 2019; pp. 236–241.
19. Gros, S.; Vukov, M.; Diehl, M. A real-time MHE and NMPC scheme for wind turbine control. In Proceedings of the 52nd IEEE Conference on Decision and Control, Florence, Italy, 10–13 December 2013; pp. 1007–1012.
20. Soliman, M.; Malik, O.P.; Westwick, D.T. Multiple model multiple-input multiple-output predictive control for variable speed variable pitch wind energy conversion systems. *IET Renew. Power Gener.* **2011**, *5*, 124–136. [CrossRef]
21. Song, D.; Yang, J.; Dong, M.; Joo, Y.H. Model predictive control with finite control set for variable-speed wind turbines. *Energy* **2017**, *126*, 564–572. [CrossRef]
22. Rawlings, J.B.; Mayne, D.Q. *Model Predictive Control: Theory and Design*; Nob Hill Publishing: Madison, WI, USA, 2009.
23. Diehl, M.; Amrit, R.; Rawlings, J.B. A Lyapunov Function for Economic Optimizing Model Predictive Control. *IEEE Trans. Autom. Control* **2011**, *56*, 703–707. [CrossRef]
24. Heidarinejad, M.; Liu, J.; Christofides, P.D. Economic model predictive control of nonlinear process systems using Lyapunov techniques. *AIChE J.* **2012**, *58*, 855–870. [CrossRef]
25. Gros, S.; Schild, A. Real-time economic nonlinear model predictive control for wind turbine control. *Int. J. Control* **2017**, *90*, 2799–2812. [CrossRef]
26. Cui, J.; Liu, S.; Liu, J.; Liu, X. A Comparative Study of MPC and Economic MPC of Wind Energy Conversion Systems. *Energies* **2018**, *11*, 3127. [CrossRef]
27. Munteanu, I.; Bratcu, A.I.; Cutululis, N.-A.; Ceanga, E. *Optimal Control of Wind Energy Systems: Towards a Global Approach*; Springer Science & Business Media: New York, NY, USA, 2008.

28. Usta, I.; Arik, I.; Yenilmez, I.; Kantar, Y.M. A new estimation approach based on moments for estimating Weibull parameters in wind power applications. *Energy Convers. Manag.* **2018**, *164*, 570–578. [CrossRef]
29. Bianchi, F.; De Battista, H.; Mantz, R.J. *Wind Turbine Control Systems: Principles, Modelling and Gain Scheduling Design*; Springer: London, UK, 2007; p. 208.
30. Imran, R.; Hussain, D.; Chowdhry, B. Parameterized Disturbance Observer Based Controller to Reduce Cyclic Loads of Wind Turbine. *Energies* **2018**, *11*, 1296. [CrossRef]
31. Hur, S.-H. Modelling and control of a wind turbine and farm. *Energy* **2018**, *156*, 360–370. [CrossRef]
32. Jonkman, J.; Butterfield, S.; Musial, W.; Scott, G. *Definition of a 5-MW Reference Wind Turbine for Offshore System Development*; National Renewable Energy Laboratory (NREL): Golden, CO, USA, 2009.
33. Wächter, A.; Biegler, L.T. On the implementation of an interior-point filter line-search algorithm for large-scale nonlinear programming. *Math. Program.* **2006**, *106*, 25–57. [CrossRef]
34. Jonkman, J.M.; Buhl, M.L., Jr. *FAST User's Guide-Updated August 2005*; National Renewable Energy Laboratory (NREL): Golden, CO, USA, 2005.

10

Dynamic Stall of a Vertical-Axis Wind Turbine and its Control using Plasma Actuation

Lu Ma, Xiaodong Wang *, Jian Zhu and Shun Kang

Key Laboratory of Power Station Energy Transfer Conversion and System, Ministry of Education, North China Electric Power University, Beijing 102206, China; 1162102037@ncepu.edu.cn (L.M.); 1111170919@ncepu.edu.cn (J.Z.); Kangs@ncepu.edu.cn (S.K.)

* Correspondence: Wangxd@ncepu.edu.cn

Abstract: In this paper, a dynamic stall control scheme for vertical-axis wind turbine (VAWT) based on pulsed dielectric-barrier-discharge (DBD) plasma actuation is proposed using computational fluid dynamics (CFD). The trend of the wind turbine power coefficient with the tip speed ratio is verified, and the numerical simulation can describe the typical dynamic stall process of the H-type VAWT. The tangential force coefficient and vorticity contours of the blade are compared, and the regular pattern of the VAWT dynamic stall under different tip speed ratios is obtained. Based on the understanding the dynamic stall phenomenon in flow field, the effect of the azimuth of the plasma actuation on the VAWT power is studied. The results show that the azimuth interval of the dynamic stall is approximately 60° or 80° by the different tip speed ratio. The pulsed plasma actuation can suppress dynamic stall. The actuation is optimally applied for the azimuthal position of 60° to 120°.

Keywords: DBD plasma actuation; dynamic stall; vertical-axis wind turbine; active flow control

---

1. Introduction

Dielectric-barrier-discharge (DBD) plasma actuation is an active flow control technology developed in the mid-1960s [1]. The actuator is a plasma-generating device, which is composed of two electrodes, dielectric between the electrodes, and an external high voltage power source. When the actuator is working, the air near the electrode is ionized to ions and electrons. Due to the asymmetry of the electrode arrangement, the ions move in an orientated direction within the electric field, generating a body force that drives the neutral gas molecules to produce a tangential jet, achieving boundary layer flow control. DBD plasma actuation has good performance in boundary layer flow separation control [2], dynamic stall vortex control [3], and so on [4].

The dynamic stall phenomenon can be removed by unsteady pitch control of the blade and the boundary layer separation [5]. Dynamic stall is an inherent feature of VAWT. The angle of attack and relative speed of a blade are changing dramatically during each VAWT rotation cycle. The boundary layer separates and a leading edge vortex is formed on the suction side. This causes the VAWT dynamic stall. Dynamic stall will bring about VAWT structure vibration, noise, and efficiency reduction [6]. VAWT dynamic stall is a complex unsteady flow phenomenon. When the blade moves, the relative velocity and direction are time varying, and the blade is affected by the upstream wake. Although researchers are extremely interested in using flow control methods to improve wind turbine performance [7–10], few researchers have addressed the problem of VAWT dynamic stall. Previous works have only focused on the relationship between the lift/drag and the angle of attack [11], or simply flow field [12,13]. However, there is still a need for investigating the correspondence between the vortex motion and the tangential force of a blade.

There is a considerable amount of research on plasma dynamic stall control for an oscillating airfoil or a flat plate. There is still little knowledge about plasma control of VAWT. In 2006, Post & Corke [3] placed the actuator at the leading edge of NACA0015 airfoil, and the effect of DBD plasma for flow separation and dynamic stall control on an oscillating airfoil was studied experimentally. It was found that the plasma actuation was able to suppress separation and delay stall, and the pulsed actuation was more advantageous. Sato et al. [14] studied the lift enhancement of an oscillating plate by plasma actuation through wind tunnel experiments and numerical simulations. They pointed out that plasma actuation could effectively suppress the flow separation on the flat plate during the up-stroke motion. Phan & Shin [15] using numerical method investigated the impact on the aerodynamic performance at different plasma actuation parameters and actuation positions for an oscillating NACA0012 airfoil. They believed that plasma could increase the lift of the airfoil and suggested to consider the energy consumption of the actuator to achieve the optimal control authority. Greenblatt et al. [16,17] placed a plasma actuator on the leading edge of a small vertical-axis wind turbine and conducted related research. The actuator is placed at the leading edge in a pulsed actuation mode. The results showed that after applying control, the power of the wind turbine increased and the fluctuation decreased, because plasma actuation reduced the leading edge vortex. They also argued that plasma actuation is only effective on the upwind side; however, it was not clear what specific azimuthal interval should be applied actuation. It is valuable and practical to formulate a scientific and economic control strategy to reduce the excited energy consumption and increase the power of the wind turbine.

Two methods can be used to study dynamic stall phenomena: numerical methods [18,19] and wind tunnel experimental methods [20]. Due to the complexity of the VAWT dynamic stall phenomenon, wind tunnel experiments are difficult to capture the full detailed flow field. Therefore, the numerical simulation method is used to study the VAWT dynamic stall problem. The load of VAWT can be calculated using the empirical formula model [21] and the theoretical model [22]. The former is calculated formula fitted by the experimental data, which is easy to implement and low in cost; the latter is mainly solved by the CFD method, and a complex flow field with high kinetic energy and shearing can be obtained, but its more time consuming. Buchner et al. [23] used a theoretical model to study the dynamic stall phenomenon of VAWT and found that the tip speed ratio was less than 3, which significantly reduced the power of VAWT. Most of the dynamic stall effects occur in the low tip speed ratio (less than 3), while the VAWTs actually operate mainly within it, which highlights the industrial relevance of the dynamic stall problem.

A suitable actuator simplified model needs to be selected for numerical simulation of DBD plasma flow control. Shyy et al. [24], Massines et al. [25], Suzen et al. [26], and Abdollahzadeh et al. [27–29] proposed their actuator simplified models based on different assumptions, respectively. The most widely used is the phenomenological model proposed by Shyy et al., which loads the plasma reduced body force into the fluid momentum equation as a source term. The advantages of the Shyy's model are simplicity, fast response, ease of application, and so on. However, the oversimplified model severely overestimates the body forces generated by the plasma actuator. Therefore, the experimental correction of the Shyy's model must be performed before the numerical calculation. The following literature use the corrected Shyy's model for numerical simulation studies [30,31]. Considering the complex unsteadiness of the VAWT dynamic stall and the accuracy of the phenomenological model, in this paper, the modified Shyy's model is selected for numerical simulation.

In this paper, the unsteady CFD method is used to study the dynamic stall phenomenon of VAWT at different tip speed ratios, and the azimuthal position corresponding to the start and end of dynamic stall is found. The modified plasma phenomenological model is used to analyze the influence of plasma actuation on VAWT in different azimuthal intervals. An optimal control strategy for dynamic stall of a vertical-axis wind turbine based on plasma actuation is proposed.

2. Computational Model and Numerical Method

2.1. H-Type VAWT Model

In this paper, a two-dimensional model is used for numerical simulation. The three-blade H-type VAWT model computational domain is shown in Figure 1, which is consistent with the reference [22,32]. *D* represents the diameter of the wind turbine. Inlet, outlet, upper, and lower boundary are 2.5*D*, 4.5*D*, and 2*D* from the VAWT center, respectively.

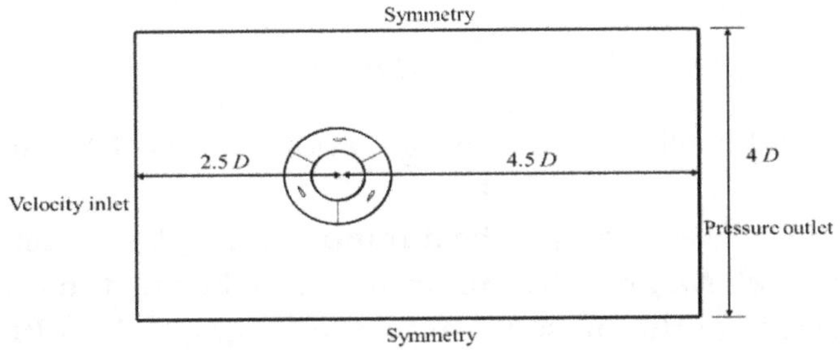

Figure 1. Computational domain and boundary conditions.

The radius of VAWT is $R = 0.3$ m. The blade adopts the NACA0022 airfoil with a chord length $c = 0.1$ m. The inlet velocity is $U_\infty = 5.07$ m/s. The rotational speed $\omega = 16.9$~50.7 rad/s, are studied, corresponding to the tip speed ratio $\lambda = 1$ to 3. θ is the azimuth, and the position where the blade 1 is located in Figure 2 is defined as $\theta = 0°$. $\theta = 0°$~$180°$ is the upwind side, and $\theta = 180°$~$360°$ is the downwind side.

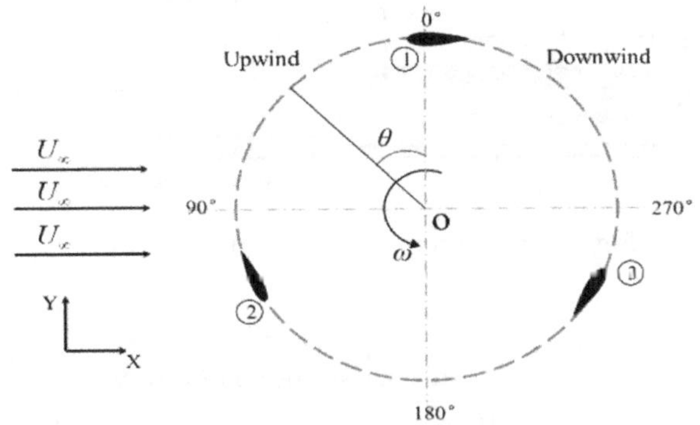

Figure 2. Position of each blade of VAWT.

The plasma actuator is placed at 5~90% *c* from the leading edge, and the purple triangle represents the plasma region (The actuator is placed at 30% *c*), as shown in Figure 3.

Figure 3. Position of the plasma actuator on a blade.

2.2. Plasma Actuator Model

The Shyy's model [24] simplifies the complex body force generated by plasma into linear electric field force. The basic mechanism of the actuator is shown in Figure 4.

Ignoring the secondary factors, assuming that the plasma density ρ_c is constant and the electric field is linearly distributed, the electric field distribution in the OAB is obtained as

$$\left|\vec{E}\right| = E_0 - k_1 x - k_2 y \quad (1)$$

where $E_0 = 1.2 \times 10^4$ kV/m is the maximum electric field strength, and the coefficients $k_1 = 1.8 \times 10^9$ kV/m^2 and $k_2 = 3.6 \times 10^9$ kV/m^2 are obtained from the electric field distribution [30].

The body force induced by the plasma actuator can be expressed as electric field force

$$\vec{F} = \vec{E}\rho_c e\alpha \quad (2)$$

where $e = 1.602 \times 10^{-19}$ C [24] is elementary charge, and $\rho_c = 1 \times 10^{17}$ m^3 [24]. The $\alpha = 0.3$ is the correction factor [30].

The duty cycle is: $D_{tc} = T_d/T$, where T_d is the duration of the plasma actuation and T is the period.

The original Shyy's model is a great oversimplification; hence, the model used in this paper is calibrated by experiment [33]. In the present study, the height and width of the plasma region in Figure 4 are OA = 2.5 mm and OB = 5 mm [34,35]. In Equation (2), The electric field strength is multiplied by the correction factor of 0.3. Figure 5 is the comparison between the computational and experimental [33] streamwise velocity distribution, where u_p is the plasma reduced velocity, and $u_{p,max} = 5.9$ m/s is the maximum plasma reduced velocity. The ordinate represents the distance from the wall. It can be seen from Figure 5 that the corrected model has a good agreement with the experimental values.

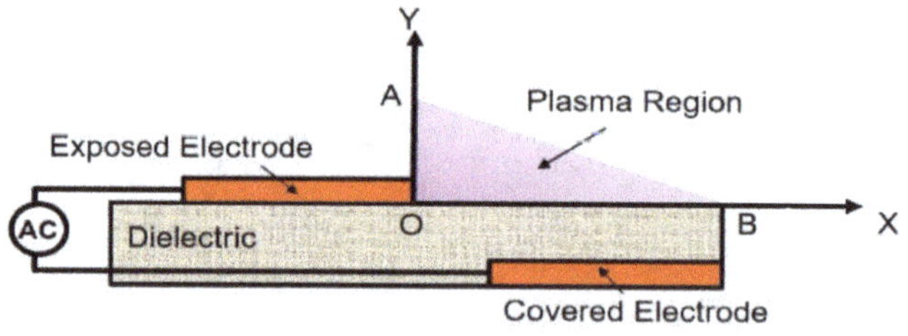

Figure 4. Sketch of the plasma actuator.

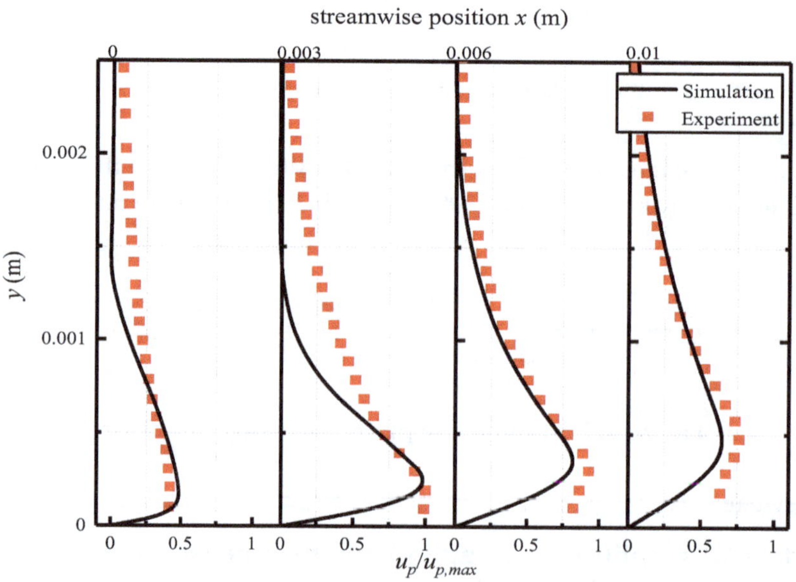

Figure 5. Velocity distribution at different streamwise positions.

2.3. Numerical Method and Boundary Conditions

Figure 6 shows the global mesh of the computational domain. The structured grid is generated, and the total number of cells is 300,000. The sliding mesh is used to establish the rotation domain and the number of cells is 260,000. The grids number along the circumference of a blade is 500. The first cell height of the blade is 0.01 mm, which guarantees $y^+ < 1$, as shown in Figure 7.

Inlet is velocity inlet. Outlet is pressure outlet. Upper and lower boundary are symmetry. The blade surface is a non-slip wall. The electric field force generated by the plasma actuation is loaded as a body force source term though the user-defined function.

Numerical simulations are calculated using ANSYS FLUENT. The turbulence model is k-ω SST model. The blades rotate 0.2 degrees per time step.

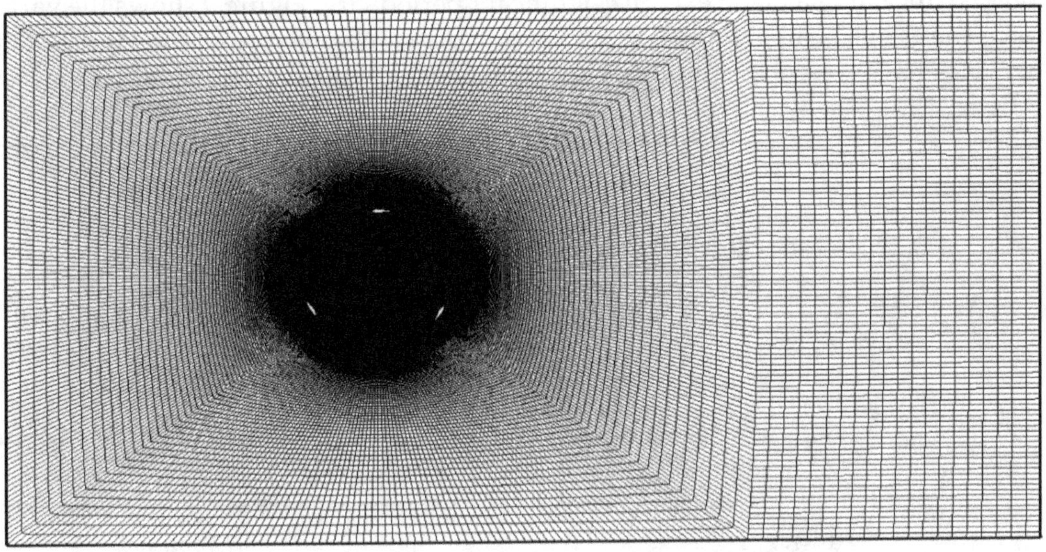

Figure 6. Global mesh of computational domain.

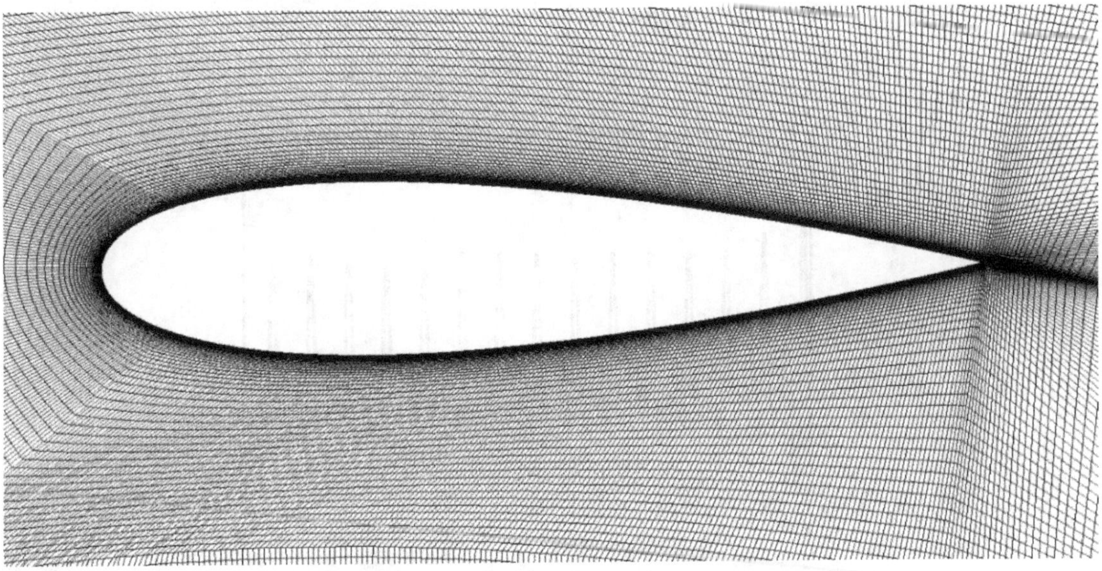

Figure 7. Mesh around the blade.

2.4. CFD Validation

In Figure 8a, grid sensitivity verification is performed on base-line case ($U_\infty = 5.07$ m/s, $\lambda = 2.15$) of three different numbers of cells. As a constant, the first cell height of the blade is 0.01 mm, which guarantees $y^+ < 1$. The grids number along the circumference of a blade are 500, 250, and 125, respectively. As can be seen from the figure, the difference between the three grids is small. Considering the requirements of the plasma model for the grid, the fine mesh (260,000 cells) is selected.

VAWT dynamic stall is a complex unsteady flow phenomenon; hence, the time step sensitivity verification is necessary. Figure 8b shows the verification of three different the time step sizes on base-line case. It can be seen that large time step size ($\Delta t = 0.0003$ s) makes the tangential force larger. Since the frequency of the plasma pulsed actuation is high, the minimum time step size is selected. When the time step size is $\Delta t = 0.0001$ s, the blades rotate 0.2 degrees per time step.

It is important and necessary to choose a suitable turbulence model for numerical simulation. A large number of scholars [36–39] have used k-ω SST model to solve VAWT problems. Therefore, this paper does not perform additional verification on the turbulence model.

The convergence criterions are that all the parameters was set to 10^{-4} at each time step and the tangential force coefficient of a blade exhibits significant periodicity. Figure 9 shows the variation of the tangential force coefficient with azimuth over 15 cycles. As can be seen from the Figure 9, the solution of the last 10 cycles have converged. In this paper, we use the time average result of the last 10 cycles.

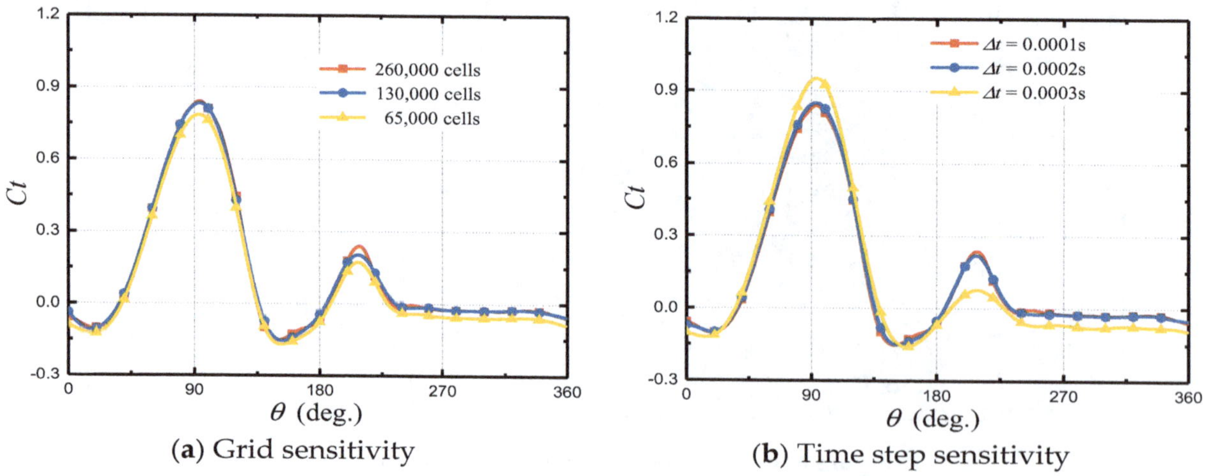

Figure 8. CFD validation: the tangential force coefficient with azimuth.

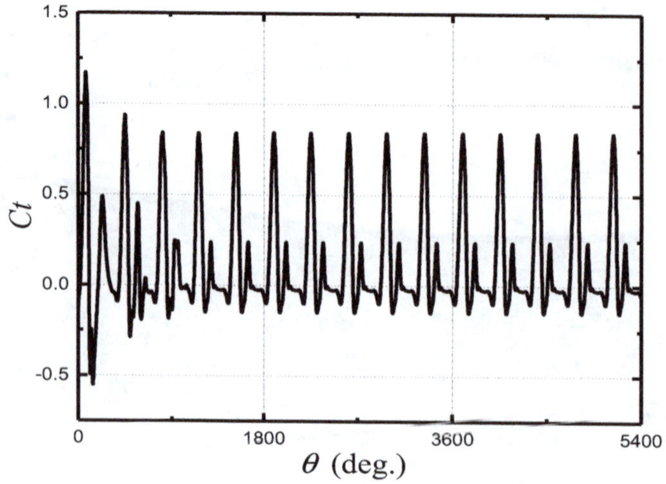

Figure 9. Convergence criterion: the tangential force coefficient with azimuth.

2.5. H-Type VAWT Aerodynamic Parameters

Several major aerodynamic parameters are described below to characterize VAWT dynamic stall.

The tip speed ratio λ is the blade linear velocity to the free stream velocity U_∞, which is used to describe the blades' rotation speed. It is defined as

$$\lambda = R\Omega/U_\infty \tag{3}$$

where R is the radius of the rotor and Ω is the angular velocity of the rotor.

The relative velocity ratio $\hat{U}_{rel}$ is the instantaneous relative velocity of the blade U_{rel} to the free stream velocity. It is defined as

$$\hat{U}_{rel} = \frac{U_{rel}}{U_\infty} \tag{4}$$

The angle of attack based on the instantaneous relative velocity is

$$\alpha_{rel} = \tan^{-1}\left(\frac{\sin\theta}{\lambda + \cos\theta}\right) \tag{5}$$

The α_{rel} varying with azimuth is shown in the Figure 10. It can be seen that the lower the tip speed ratio, the larger the amplitude of α_{rel}.

The tangential force coefficient C_t and the power coefficient C_p represent the ability of the blade and the rotor to generate energy, respectively.

The tangential force coefficient is defined as

$$C_t = \frac{F_t}{1/2\rho U_\infty^2 S} \tag{6}$$

where F_t is tangential force of a blade, and S is the maximum projected area of a blade.

The power coefficient is defined as

$$C_p = \frac{3F_t \cdot \lambda}{1/2\rho U_\infty^2 A} \tag{7}$$

where A is the frontal projected area of wind turbine.

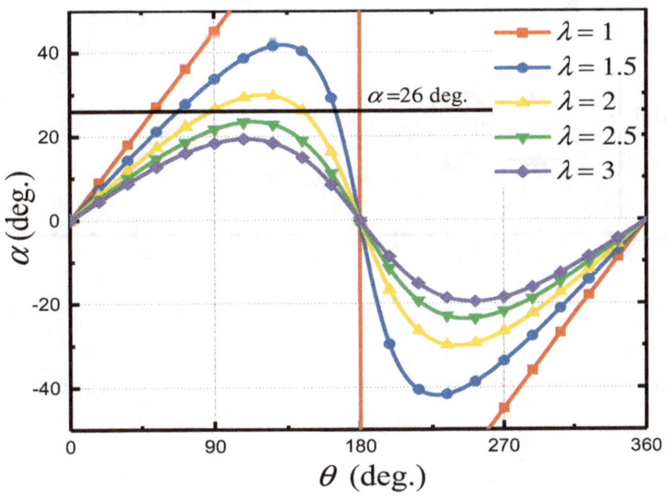

Figure 10. Angle of attack with azimuth.

2.6. Parameter Settings

The research scheme is shown in Table 1. In Scheme 1, the VAWT dynamic stall at five different tip speed ratios is studied. On the basis of Scheme 1, taking $\lambda = 2.15$ as an example, given the actuation duty cycle, and the pulse actuation frequency f, the effects of five plasma actuator positions on VAWT are investigated. In Scheme 3, given the plasma actuator position, the effects of 10 plasma actuation intervals on VAWT dynamic stall are studied.

Table 1. Parameters of the simulations.

Parameter Settings	λ	Actuator Position	*Dtc*	*f*/Hz	Actuation Interval
Scheme 1	1~3	-	-	-	-
Scheme 2	2.15	5 different positions	0.2	383	-
Scheme 3	2.15	0.3*c*	0.2	383	10 different actuation intervals

3. Results and Discussion

3.1. Computational Results Verification

Figure 11 shows the comparison of the experimental data [32] and the two-dimensional numerical simulation data for wind turbine power coefficient. It can be found that the simulation results of both Zuo et al. [22] and this paper can reflect the variation tendency of C_p, but the numerical values are always larger than the experimental values. Because the two-dimensional numerical simulations ignore the end loss of the three-dimensional blades [40] and the influence of the support structure. Reference [40] pointed out that the two-dimensional URANS simulation can capture most of the details for the VAWT flow field. This paper mainly studies the flow separation and control technology on the VAWT, so the two-dimensional URANS simulation is adopted.

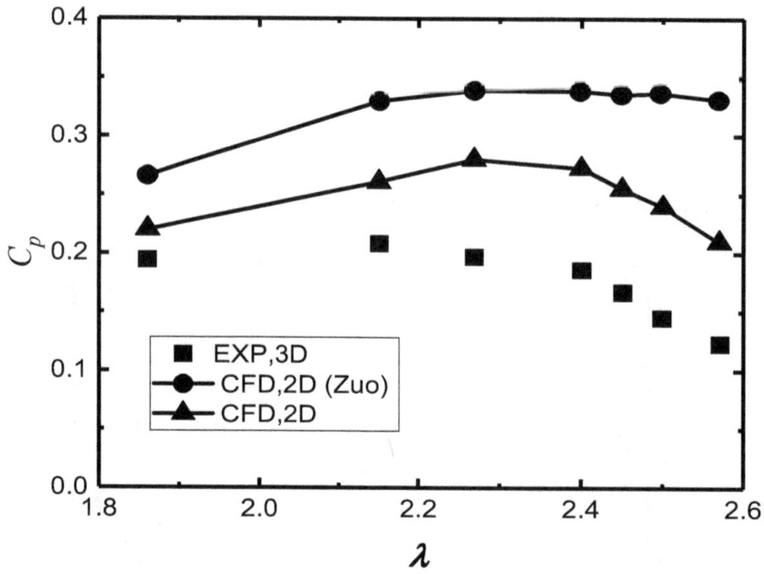

Figure 11. Comparison of experimental data and numerical simulated data for wind turbine power coefficient.

3.2. Influence of Tip Speed Ratio

Five tip speed ratios, $\lambda = 1, 1.5, 2, 2.5$, and 3, were investigated. Figure 12 shows the variation of the blade tangential force coefficient C_t with azimuth at different tip speed ratios. In a rotation cycle, the C_t basically exhibits a double-peak variation. The maximum tangential force coefficient $C_{t\,max}$ and its corresponding azimuth $\theta_{ct.max}$, with increasing tip speed ratio, increase first and then remain unchanged. It can be seen that when the blade moves to the downwind side, the C_t value is around 0 due to the influence of the upstream wakes.

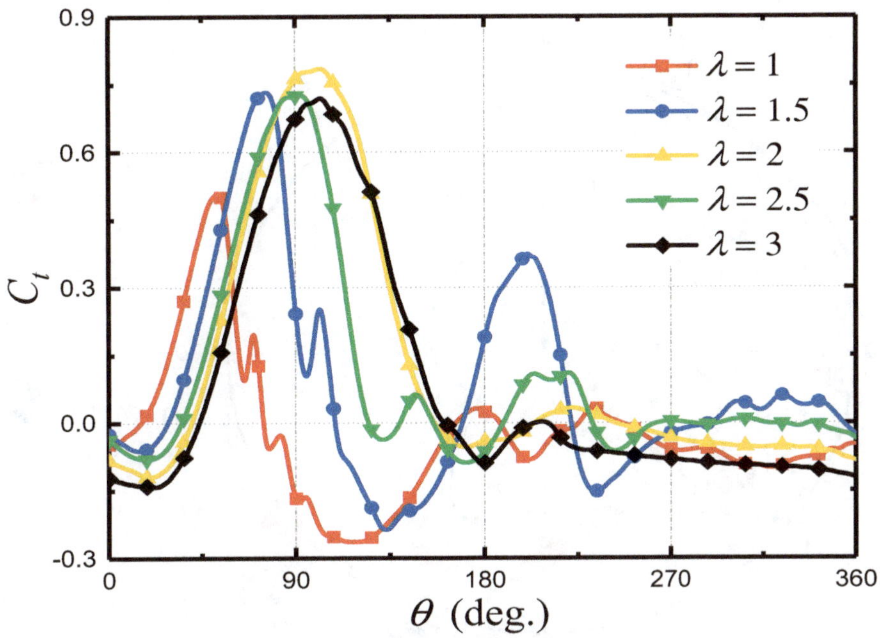

Figure 12. Tangential force coefficient of a blade with azimuth.

In [20], the VAWT consists of NACA0015 airfoil, which is a thin thickness airfoil. The dynamic stall phenomenon of this kind of wind turbine is expressed as the formation, development and shedding of the leading edge vortex on the blade or airfoil. The VAWT selected in this paper consists of the NACA0022 airfoil, which is a medium thickness airfoil. In [41], the aerodynamic performance of the four VAWT airfoils (NACA0012, NACA0022, NACA5522, and LS0421) was compared. The study found that the NACA0022 airfoil performs better at small tip speed ratios. The dynamic stall phenomenon is not exactly the same as the above literature. The main difference is that there is not only the leading edge vortex on the blade, but also a trailing edge vortex with the same direction of rotation.

Figure 13 shows the vorticity (normalized to $c\omega_z/U_\infty$) contour at specific azimuths for blade 1 during a cycle at $\lambda = 1.5$. For ease of understanding, the correspondence between vortex and C_t is indicated by arrows. It should be pointed out that only under this condition, the dynamic stall occurs twice, which is represented by two distinct peaks on the C_t curve, which is the same as the case studied in [11]. The first dynamic stall has gone through four stages a, b, c, and d. The α at point a is small, resulting in a minor C_t. As the angle of attack increases with azimuth, C_t increases significantly. The azimuth of point b corresponds to the maximum of C_t, in which the flow separation from the trailing edge almost rolls up to the leading edge. It can be considered that the movement process of the blade in range a to b is in keeping with the static stall feature. At azimuth c, a leading edge vortex appears, and the vortex adheres to the surface and keep on growing. Under the influence of the leading edge vortex and the non-slip wall condition, a secondary vortex whose rotation direction is opposite to the leading edge vortex is formed. Point d is the minimum of C_t. At this time, the leading-edge vortex extends to one chord length above the suction side, while the boundary layer rolls up and forms a trailing edge vortex close to the surface. Those two vortices interact with each other and obtain circulation in the shear layer to strengthen the vorticity. After that, the leading edge vortex and the trailing edge vortex gradually shed off, and the boundary layer reattaches to the leading edge. The first dynamic stall ends. At the same time, the second dynamic stall occurs on the downwind side. The four typical moments (e, f, g, and h) correspond to the formation, development, and shedding of the leading-edge vortex, and the reattachment of the boundary layer at leading edge. Different from the upwind-side dynamic stall, the trailing edge vortex on the downwind side is not formed on the surface, which is affected by the free flow below the rotor. In this figure, the trajectories of the leading edge vortex and the trailing edge vortex on the upwind and the downwind side can also be observed.

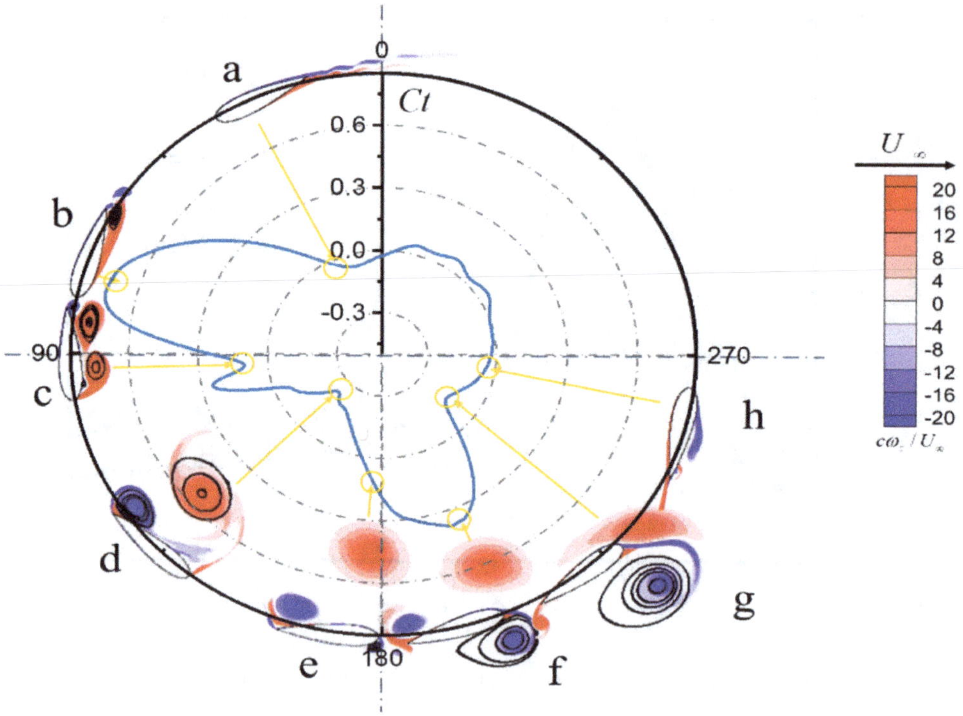

Figure 13. Tangential force coefficient of a blade and vorticity contours with $\lambda = 1.5$.

Figure 14 shows the vorticity (normalized to $c\omega_z/U_\infty$) contours of the blade 1 for selected azimuth at different tip speed ratios. The major vortex structure and its motion state can be clearly observed. The development and their interaction of the leading edge vortex and a pair of trailing edge vortices can be found out. Comparing the vorticity contours, we can see the delay effect of the VAWT dynamic stall at high tip speed ratio. The delay effect means that as the tip speed ratio rises, the VAWT dynamic stall is delayed. The higher the tip speed ratio, the smaller the amplitude of the angle of attack and the relative speed, and the less evidently the dynamic stall occurs.

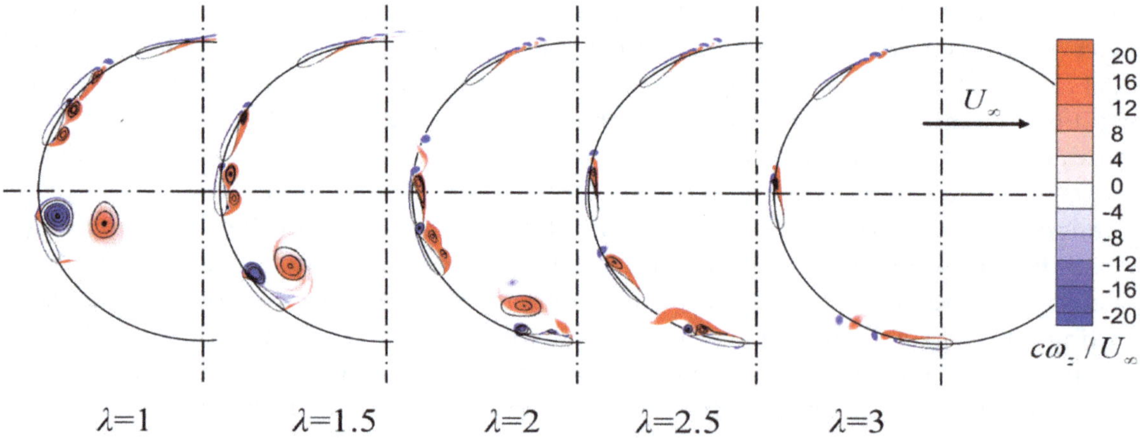

Figure 14. Vorticity contours with different λ for selected azimuth.

Figure 15 shows the vorticity contours for different tip speed ratios at $\theta = 120°$. In order to facilitate the observation and comparison, the coordinates of the blade are transformed. Although the blades rotate to the same azimuth, the vorticity contours are totally different. It can be considered that these five vorticity fields (from $\lambda = 3$ to $\lambda = 1$) can represent the VAWT dynamic stall at different stages. At the same azimuth, the high tip speed ratio corresponds to a small angle of attack, and the dynamic stall delay is more distinct at high tip speed ratios.

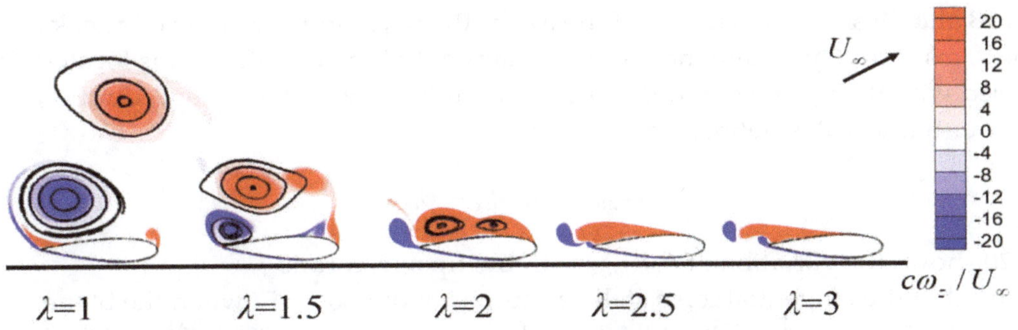

Figure 15. Vorticity contours with different λ at $\theta = 120°$.

Figure 16 shows the azimuth of the maximum tangential force coefficient $\theta_{ct.max}$ as a function of the tip speed ratio, and the horizontal axis is divided by five tip speed ratios. It can be seen from the figure that the $\theta_{ct.max}$ increases first and at a level between 90° and 100° with the increase of λ. Figure 17 shows the angle of attack of the maximum tangential force coefficient $\alpha_{ct.max}$ as a function of the tip speed ratio. It was found that when $\lambda \leq 2$, the $\alpha_{ct.max}$ is approximately above 26°, which is similar to the result observed in [15]. As can be seen from Figure 10, the relative angle of attack of the blade can be higher than 26° only when $\lambda \leq 2$; when $\lambda > 2$, $\theta_{ct.max} = \theta_{\alpha,\, max}$, where $\theta_{\alpha,\, max}$ represents the azimuth of the maximum angle of attack, which is also observed in Figure 17. In summary, it can be considered that when the blade moves to an angle of attack higher than 26°, or when moving to the maximum angle of attack (the maximum angle of attack α_{max} is less than 26°), the tangential force reaches a maximum value and the flow begins to stall. The specific reasons have been analyzed above, as shown in Figures 10 and 15.

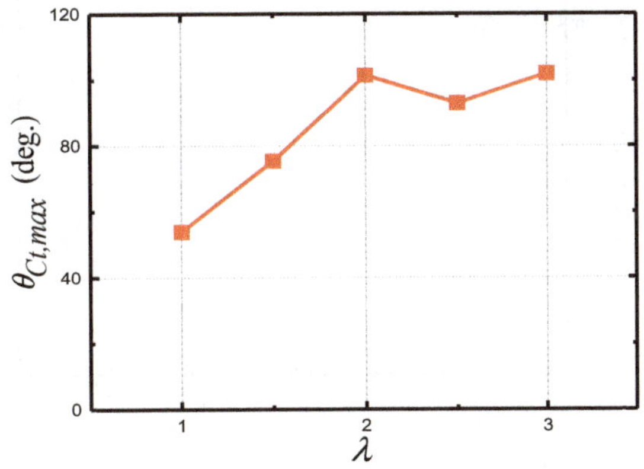

Figure 16. $\theta_{ct.max}$ of a blade with λ.

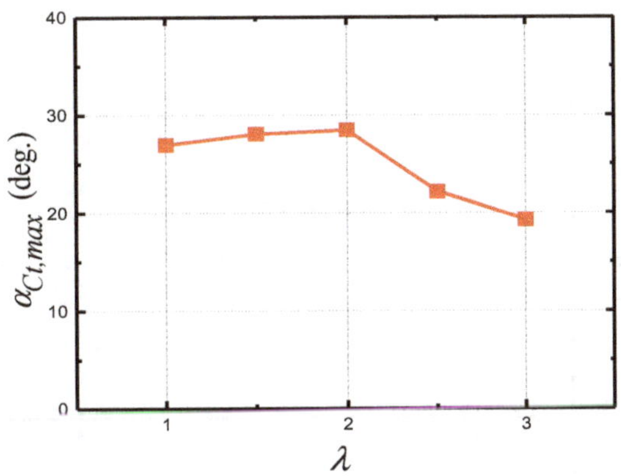

Figure 17. $\alpha_{ct.max}$ of a blade with λ.

Figures 18 and 19 show the change of the azimuth $\theta_{ct.mix}$ and the angle of attack $\alpha_{ct.mix}$ with the tip speed ratio, respectively, where the minimum tangential force coefficient is located. It was found that $\theta_{ct.mix}$ rised with the tip speed ratio, while $\alpha_{ct.mix}$ reduced with the tip speed ratio. The tangential force stall azimuth interval is defined as $\theta_{ct.stall}$,

$$\theta_{ct.stall}, = \theta_{ct.mix} - \theta_{ct.max}. \tag{8}$$

Figure 20 shows the variation of $\theta_{ct.stall}$, with the tip speed ratio.

It is found that the tangential force stall azimuth is about 60° ± 5° when the blade is moving at a low tip speed ratio (i.e., $\alpha_{max} > 26°$); while the stall azimuth is about 80° ± 5° at high tip speed ratio.

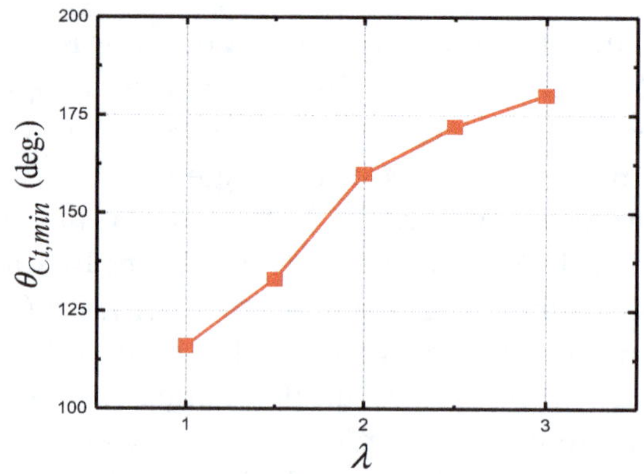

Figure 18. $\theta_{ct.mix}$ of a blade with λ.

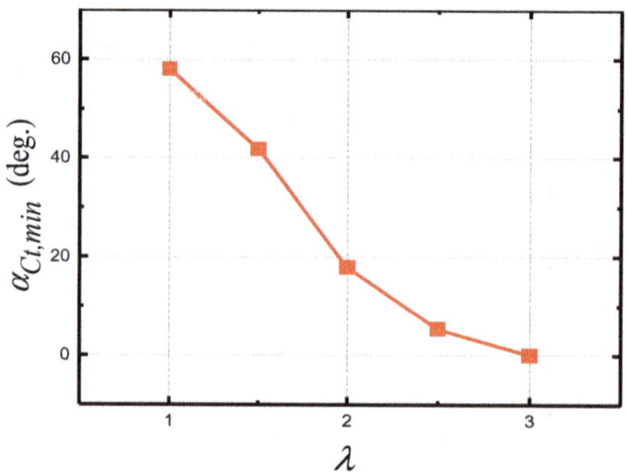

Figure 19. $\alpha_{ct.mix}$ of a blade with λ.

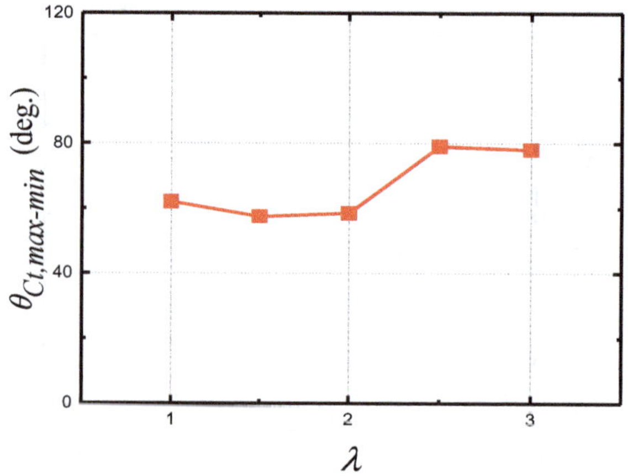

Figure 20. $\theta_{ct.max\text{-}mix}$ of a blade with λ.

3.3. Influence of Plasma Actuation Flow Control Strategy on VAWT

In this part, the wind turbine is selected to operate at a tip speed ratio of 2.15. Because according to Figure 11, when the tip speed ratio is 2.15, the wind turbine has a maximum power coefficient in the experiment.

The effect of the plasma actuator positions on VAWT was studied. The actuator is placed at 5% *c*, 30% *c*, 50% *c*, 70% *c*, and 90% *c* from the leading edge, respectively. C_p of the VAWT with different plasma actuation positions is shown in Table 2. The baseline case is uncontrolled VAWT rotate at a tip speed ratio of 2.15. In contrast, it has been found that using plasma actuator at these five positions can increase the C_p of the VAWT. The optimal position is at 30% *c* and the power is increased by 36%. Figure 21 shows the variation of the tangential force coefficient with azimuth at different plasma actuator positions. In general, the actuation increases the maximum tangential force of the blade. At $15° < \theta < 165°$, plasma actuation can significantly increase the tangential force. However, when $\theta > 165°$, the actuation has less effect.

Table 2. C_p with different plasma actuator positions.

Position of the Plasma Actuator	C_p
Baseline	0.261
5% *c*	0.324
30% *c*	0.358
50% *c*	0.355
70% *c*	0.315
90% *c*	0.262

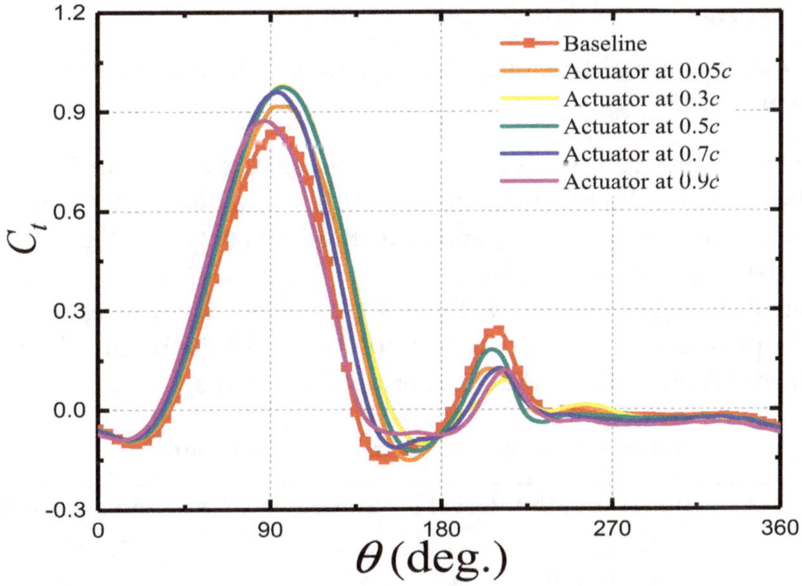

Figure 21. Tangential force coefficient with azimuth at different plasma actuator positions.

Figure 22 shows the relative velocity contour with relative velocity streamlines. The flow fields before and after the controlled (the plasma actuator is at 30% *c*) were compared. When the actuator is at 30% *c*, plasma actuation can suppress flow separation, but does not completely eliminate flow separation. The baseline case forms a secondary vortex at $\theta = 135°$, while the controlled case is formed at $\theta = 155°$. Therefore, arranging the plasma actuator at 30% *c*, plasma actuation will delay the formation of the secondary vortex. It can also be seen from the comparison that the plasma actuation can accelerate the detachment of the vortex on the blade.

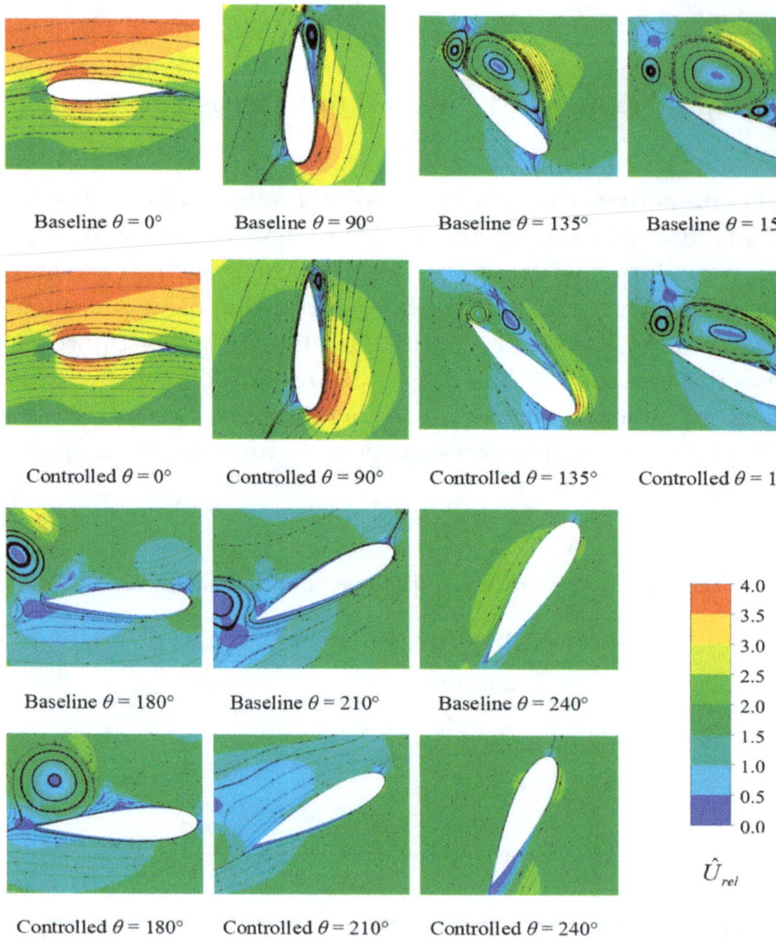

Figure 22. Relative streamlines and relative velocity contours with azimuth for baseline and plasma controlled case (actuator at 30% *c*).

Ten control strategies were studied. The research schemes are shown in Tables 1 and 3. According to that mentioned above, when the tip speed ratio is 2.15, it can be predicted that the H-type VAWT tangential force starts to stall at $\theta = 95° \pm 5°$, and ends the stall at $\theta = 155° \pm 5°$. The plasma actuation control strategies are developed accordingly.

The specific control strategy is shown in Table 3. Case 1 is a baseline and no plasma actuation is applied. In Case 2, the plasma is continuously pulsed on each of the blades during the rotation cycle. In Cases 3–10, the plasma is pulsed in different azimuthal intervals.

Table 3. Strategies of plasma actuation control.

Strategy	Actuation Azimuth Interval	Strategy	Actuation Azimuth Interval
Case 1	No control	Case 6	70° to 150°
Case 2	Global control	Case 7	60° to 150°
Case 3	90° to 160°	Case 8	60° to 130°
Case 4	90° to 150°	Case 9	60° to 120°
Case 5	80° to 150°	Case 10	50° to 110°

Figure 23 shows the variation of the tangential force coefficient with azimuth at the plasma actuation. It can be confirmed from the Case 1 curve that the above prediction is accurate. In general, within $90° < \theta < 180°$, the tangential force coefficient is significantly improved by plasma, and has little effect on other azimuths. Cases 3, 4, and 7–9 show that the end azimuth of the actuation has little impact on the tangential force. Case 4–7 are used to investigate the effect of the starting actuation azimuth on the tangential force. It is found that the early actuation increases the tangential force and the peak value is higher. Cases 9 and 10 are a set of controlled trials. In the case of the same actuation interval size, the influence of the starting and ending position on the tangential force is studied. There is no significant difference between Case 9 and 10, however, when $\theta > 150°$ the tangential force of Case 10 is lower than Case 9.

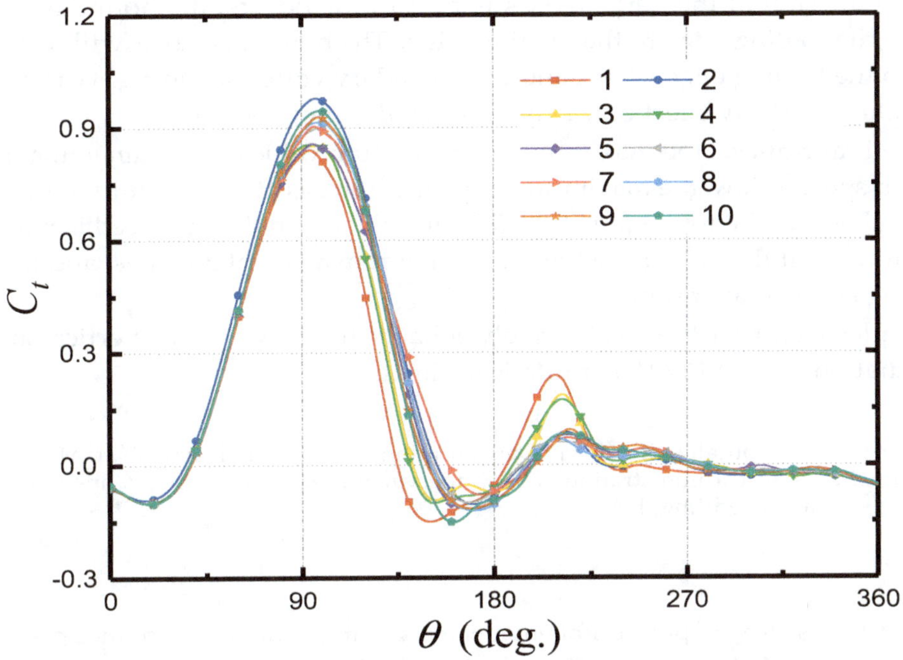

Figure 23. Tangential force coefficient of a blade with azimuth at plasma actuation.

Aerodynamic figure of merit (AFM) [42] is introduced to determine whether the active flow control strategy is cost effective. AFM is defined as

$$\text{AFM} = \frac{P_{\text{controlled}} - P_{\text{actuators}}}{P_{\text{baseline}}} \quad (9)$$

where $P_{\text{controlled}}$ is the power of controlled VAWT, $P_{\text{actuators}}$ is the plasma actuators consumption, and P_{baseline} is the power of uncontrolled VAWT. When AFM > 1, the power consumption of actuators is less than the additional output of VAWT, and the control strategy is feasible.

Table 4 shows the C_p and AFM for different control strategies. The power of VAWT can be directly calculated from the above, and the energy consumption of the plasma exciter is 6.67 W/m [16,17,43]. It can be seen that Case 2 increases the C_p up to 36%, but it is not suitable. Cases 6–10 are feasible, and the maximum C_p can be increased by 34.3%, in which Case 9 is optimal. Considering the consumption of the plasma actuator, Case 9 has the highest AFM value, which means that Case 9 consumes less energy and produces considerable power. Within the scope of the study, it is necessary to apply control in advance to achieve better control results.

Table 4. C_p and AFM at plasma actuation.

Strategy	C_p	AFM	Strategy	C_p	AFM
Case 1	0.261	-	Case 6	0.331	1.033
Case 2	0.358	0.304	Case 7	0.343	1.048
Case 3	0.296	0.928	Case 8	0.331	1.062
Case 4	0.299	0.968	Case 9	0.326	1.072
Case 5	0.310	0.982	Case 10	0.322	1.056

4. Conclusions

In this paper, the phenomenological model and URANS simulation are used to study the VAWT dynamic stall phenomenon with different tip speed ratios. The effects of plasma actuation on the aerodynamic performance of the blade are explored, and the following conclusions are drawn:

1. As the dynamic stall of the vertical-axis wind turbine occurs, the boundary layer separation moves from the trailing edge to the leading edge. The boundary layer will reattach to the blade surface with the leading edge vortex and the secondary vortex shedding. With the increase of the tip speed ratio, the delay effect of the dynamic stall is more obvious.
2. The tip speed ratio affects the maximum and minimum values of the tangential force for a blade. When the tip speed is low, the maximum tangential force is at the azimuth of the $\alpha = 26°$, and the minimum value is at the following $60° \pm 5°$ azimuth. When the tip speed is high, the maximum tangential force is at the azimuth where the maximum angle of attack is, and the minimum is at the following $80° \pm 5°$ azimuth.
3. The pulsed plasma actuation can effectively enhance the power of the vertical-axis wind turbine, and the actuation at 60° to 120° azimuth is optimal.

Author Contributions: Conceptualization, L.M.; Data curation, L.M.; Formal analysis, L.M.; Funding acquisition, X.W.; Investigation, L.M.; Project administration, X.W.; Supervision, J.Z.; Validation, J.Z. and S.K.; Writing—original draft, L.M.; Writing—review & editing, L.M.

References

1. Kantrowitz, A.R. A survey of physical phenomena occurring in flight at extreme speeds. In Proceedings of the Conference on High-Speed Aeronautics, New York, NY, USA, 20–22 January 1955; Volume 1955, pp. 335–339.
2. Roth, J.R.; Sherman, D.M.; Wilkinson, S.P. *Boundary Layer Flow Control with a One Atmosphere Uniform Glow Discharge Surface Plasma*; American Institute of Aeronautics and Astronautics: Reston, VA, USA, 1998.
3. Post, M.L.; Corke, T. Separation control using plasma actuators: Dynamic stall vortex control on oscillating airfoil. *AIAA J.* **2006**, *44*, 3125–3135. [CrossRef]
4. Liu, Y.; Kolbakir, C.; Hu, H.; Meng, X.; Hu, H. An experimental study on the thermal effects of duty-cycled plasma actuation pertinent to aircraft icing mitigation. *Int. J. Heat Mass Transf.* **2019**, *136*, 864–876. [CrossRef]
5. McCroskey, W. *The Phenomenon of Dynamic Stall*; NASA TM-81264; National Aeronuatics and Space Administration Moffett Field Ca Ames Research Center: Mountain View, CA, USA, 1981.
6. Araya, D.B.; Dabiri, J.O. Vertical axis wind turbine in a falling soap film. *Phys. Fluids* **2015**, *27*, 091108. [CrossRef]
7. Liu, Q.; Miao, W.; Li, C. Effects of trailing-edge movable flap on aerodynamic performance and noise characteristics of VAWT. *Energy* **2019**, *192*, 58–79.
8. Liu, Q.; Hao, W.; Li, C.; Miao, W.; Ding, Q. Numerical Simulation on the Forced Oscillation of Rigid Riser with Helical Strakes in Different Section Shapes. *Ocean Eng.* **2019**, *191*, 106439. [CrossRef]
9. Zhu, H.; Hao, W.; Li, C. Numerical investigation on the effects of different wind directions, solidity, airfoils and building configurations on the aerodynamic performance of building augmented vertical axis wind turbines. *Int. J. Green Energy* **2019**, *16*, 1216–1230.
10. Miao, W.; Li, C.; Wang, Y.; Xiang, B.; Liu, Q.; Deng, Y. Study of Adaptive Blades in Extreme Environment using Fluid-Structure Interaction Method. *J. Fluids Struct.* **2019**, *91*, 102734. [CrossRef]
11. Zhu, C.; Wang, T. Comparative Study of Dynamic Stall under Pitch Oscillation and Oscillating Freestream on Wind Turbine Airfoil and Blade. *Appl. Sci.* **2018**, *8*, 1242. [CrossRef]
12. Yanzhao, Y.; Guo, Z.; Zhang, Y.; Jinyama, H.; Li, Q. Numerical Investigation of the Tip Vortex of a Straight-Bladed Vertical Axis Wind Turbine with Double-Blades. *Energies* **2017**, *10*, 1721.
13. Qian, C.; Liu, X.; Ji, H.S.; Kim, K.C.; Yang, B. Aerodynamic Analysis of a Helical Vertical Axis Wind Turbine. *Energies* **2017**, *10*, 575.

14. Sato, S.; Yokoyama, H.; Iida, A. Control of Flow around an Oscillating Plate for Lift Enhancement by Plasma Actuators. *Appl. Sci.* **2019**, *9*, 776. [CrossRef]
15. Phan, M.K.; Shin, J. Numerical investigation of aerodynamic flow actuation produced by surface plasma actuator on 2D oscillating airfoil. *Chin. J. Aeronaut.* **2016**, *29*, 882–892. [CrossRef]
16. Greenblatt, D.; Lautman, R. Inboard/outboard plasma actuation on a vertical-axis wind turbine. *Renew. Energy* **2015**, *83*, 1147–1156. [CrossRef]
17. Benharav, A.; Greenblatt, D. Plasma-based feed-forward dynamic stall control on a vertical axis wind turbine. *Wind Energy* **2016**, *19*, 3–16. [CrossRef]
18. Fujisawa, N.; Shibuya, S. Observations of dynamic stall on Darrieus wind turbine blades. *J. Wind Eng. Ind. Aerodyn.* **2001**, *89*, 201–214. [CrossRef]
19. Wang, X.D.; Ye, Z.; Kang, S.; Hu, H. Investigations on unsteady aerodynamic characteristics of a horizontal-axis wind turbine during dynamic yaw process. *Energies* **2019**, *12*, 3124. [CrossRef]
20. Buchner, A.J.; Soria, J.; Honnery, D.; Smits, A.J. Dynamic stall in vertical axis wind turbines: Scaling and topological considerations. *J. Fluid Mech.* **2018**, *841*, 746–766. [CrossRef]
21. Christian, M.; Christophe, L.; Ion, P. Appropriate Dynamic-Stall Models for Performance Predictions of VAWTs with NLF Blades. *Int. J. Rotating Mach.* **1998**, *4*, 129–139.
22. Zuo, W.; Wang, X.; Kang, S. Numerical simulations on the wake effect of H-type vertical axis wind turbines. *Energy* **2016**, *106*, 691–700. [CrossRef]
23. Buchner, A.J.; Soria, J.; Honnery, D.; Smits, A.J. Dynamic stall in vertical axis wind turbines: Comparing experiments and computations. *J. Wind Eng. Ind. Aerodyn.* **2015**, *146*, 163–171. [CrossRef]
24. Shyy, W.; Jayaraman, B.; Andersson, A. Modeling of glow discharge-induced fluid dynamics. *J. Appl. Phys.* **2002**, *92*, 6434–6443. [CrossRef]
25. Massines, F.; Rabehi, A.; Decomps, P.; Ben Gadri, R.; Ségur, P.; Mayoux, C. Experimental and theoretical study of a glow discharge at atmospheric pressure controlled by dielectric barrier. *J. Appl. Phys.* **1998**, *83*, 2950–2957. [CrossRef]
26. Suzen, Y.B.; Huang, G.; Jacob, J.; Ashpis, D. Numerical simulations of plasma based flow control applications. *AIAA Pap.* **2005**, *4633*, 2005.
27. Abdollahzadeh, M.; Páscoa, J.; Oliveira, P.J.; Abdollahzadehsangroudi, M.; Páscoa, J. Modified split-potential model for modeling the effect of DBD plasma actuators in high altitude flow control. *Curr. Appl. Phys.* **2014**, *14*, 1160–1170. [CrossRef]
28. Abdollahzadeh, M.; Pascoa, J.; Oliveira, P.J.; Abdollahzadehsangroudi, M.; Páscoa, J. Implementation of the classical plasma-fluid model for simulation of dielectric barrier discharge (DBD) actuators in OpenFOAM. *Comput. Fluids* **2016**, *128*, 77–90. [CrossRef]
29. Abdollahzadeh, M.; Pascoa, J.C.; Oliveira, P.J. Comparison of DBD plasma actuators flow control authority in different modes of actuation. *Aerosp. Sci. Technol.* **2018**, *78*, 183–196. [CrossRef]
30. Ma, L.; Wang, X.; Zhu, J.; Kang, S. Effect of DBD plasma actuation characteristics on turbulent separation over a hump model. *Plasma Sci. Technol.* **2018**, *20*, 139–149. [CrossRef]
31. Ebrahimi, A.; Hajipour, M. Flow separation control over an airfoil using dual actuation of DBD plasma actuators. *Aerosp. Sci. Technol.* **2018**, *79*, 658–668. [CrossRef]
32. Howell, R.; Qin, N.; Edwards, J.; Durrani, N. Wind tunnel and numerical study of a small vertical axis wind turbine. *Renew. Energy* **2010**, *35*, 412–422. [CrossRef]
33. Maden, I.; Maduta, R.; Kriegseis, J.; Jakirlić, S.; Schwarz, C.; Grundmann, S.; Tropea, C. Experimental and computational study of the flow induced by a plasma actuator. *Int. J. Heat Fluid Flow* **2013**, *41*, 80–89. [CrossRef]
34. Jayaraman, B.; Thakur, S.; Shyy, W. Modeling of fluid dynamics and heat transfer induced by dielectric barrier plasma actuator. *J. Heat Transf.-Trans. ASME* **2007**, *129*, 517–525. [CrossRef]
35. Jayaraman, B.; Shyy, W. Modeling of dielectric barrier discharge-induced fluid dynamics and heat transfer. *Prog. Aerosp. Sci.* **2008**, *44*, 139–191. [CrossRef]
36. Ostos, I.; Ruiz, I.; Gajic, M.; Gómez, W.; Bonilla, A.; Collazos, C. A modified novel blade configuration proposal for a more efficient VAWT using CFD tools. *Energy Convers. Manag.* **2019**, *180*, 733–746. [CrossRef]
37. Bai, H.; Chan, C.; Zhu, X.; Li, K. A numerical study on the performance of a Savonius-type vertical-axis wind turbine in a confined long channel. *Renew. Energy* **2019**, *139*, 102–109. [CrossRef]

38. Sagharichi, A.; Zamani, M.; Ghasemi, A. Effect of solidity on the performance of variable-pitch vertical axis wind turbine. *Energy* **2018**, *161*, 753–775. [CrossRef]
39. Wong, K.H.; Chong, W.T.; Poh, S.C.; Shiah, Y.-C.; Sukiman, N.L.; Wang, C.-T. 3D CFD simulation and parametric study of a flat plate deflector for vertical axis wind turbine. *Renew. Energy* **2018**, *129*, 32–55. [CrossRef]
40. Qin, N.; Howell, R.; Durrani, N.; Hamada, K.; Smith, T. Unsteady flow simulation and dynamic stall behaviour of vertical axis wind turbine blades. *Wind Eng.* **2011**, *35*, 511–527. [CrossRef]
41. Danao, L.A.; Qin, N.; Howell, R. A numerical study of blade thickness and camber effects on vertical axis wind turbines. *Proc. Inst. Mech. Eng. Part A* **2012**, *226*, 867–881. [CrossRef]
42. Stalnov, O.; Kribus, A.; Seifert, A. Evaluation of active flow control applied to wind turbine blade section. *J. Renew. Sustain. Energy* **2010**, *2*, 063101. [CrossRef]
43. Corke, T.; Post, M.L.; Orlov, D.M. Single-dielectric barrier discharge plasma enhanced aerodynamics: Concepts, optimization, and applications. *J. Propuls. Power* **2008**, *24*, 935–945. [CrossRef]

11

Parametric Study of a Gurney Flap Implementation in a DU91W(2)250 Airfoil

Iñigo Aramendia [1], Unai Fernandez-Gamiz [1,*], Ekaitz Zulueta [2], Aitor Saenz-Aguirre [2] and Daniel Teso-Fz-Betoño [2]

1 Nuclear Engineering and Fluid Mechanics Department, University of the Basque Country UPV/EHU, Nieves Cano 12, 01006 Vitoria-Gasteiz, Spain; inigo.aramendia@ehu.eus

2 Automatic Control and System Engineering Department, University of the Basque Country UPV/EHU, Nieves Cano 12, 01006 Vitoria-Gasteiz, Spain; ekaitz.zulueta@ehu.eus (E.Z.); asaenz012@ikasle.ehu.eus (A.S.-A.); daniel.teso@ehu.eus (D.T.-F.-B.)

* Correspondence: unai.fernandez@ehu.eus

Abstract: The growth in size and weight of wind turbines over the last years has led to the development of flow control devices, such as Gurney flaps (GFs). In the current work, a parametric study is presented to find the optimal GF length to improve the airfoil aerodynamic performance. Therefore, the influence of GF lengths from 0.25% to 3% of the airfoil chord *c* on a widely used DU91W(2)250 airfoil has been investigated by means of RANS based numerical simulations at $Re = 2 \times 10^6$. The numerical results showed that, for positive angles of attack, highest values of the lift-to-drag ratio C_L/C_D are obtained with GF lengths between 0.25% *c* and 0.75% *c*. Particularly, an increase of 21.57 in C_L/C_D ratio has been obtained with a GF length of 0.5% c at 2° of angle of attack AoA. The influence of GFs decreased at AoAs larger than 5°, where only a GF length of 0.25% *c* provides a slight improvement in terms of C_L/C_D ratio enhancement. Additionally, an ANN has been developed to predict the aerodynamic efficiency of the airfoil in terms of C_L/C_D ratio. This tool allows to obtain an accurate prediction model of the aerodynamic behavior of the airfoil with GFs.

Keywords: wind turbine; flow control; Gurney flap; aerodynamics; ANN

1. Introduction

The wind power capacity installed in the last years has been showing an increase and, consequently, the requirement for bigger rotor wind turbines is becoming increasingly more necessary. This growth in size and weight of the wind turbines results in longer rotor blades which cannot be controlled as they were some years ago. Large wind turbines are exposed to severe structural and fatigue loads that must be reduced by means of the use of new flexible-soft materials and with novel load control techniques. Johnson et al. [1] gathered most of the main flow control devices with potential to be applied in wind turbines, assuring a better efficiency and a safe operation under a variety of adverse atmospheric conditions.

Flow control devices were first researched and developed for the aeronautical field with promising results; see the study of Taylor [2]. Then, the aim was to introduce and optimize them for their implementation in wind turbines to improve rotor blades' aerodynamic performance as well as to reduce fatigue loads. Wood [3] developed a scheme to classify the different concepts involved in all flow control devices such as their location, operation principle or working conditions. Aramendia et al. [4] provided an overview of how these devices are classified into passive and active, depending on their operating principle. Active flow control devices need a secondary power source for their activation, while passive flow control devices, as the Gurney flaps (GFs) of the current study, do not require external energy consumption. Other passive devices such as the vortex generators (VGs) have also been studied extensively. Fernandez-Gamiz et al. [5,6] performed numerical simulations to study the characteristics of the primary vortex downstream of a rectangular VG along with a prediction model.

Due to their low cost, simplicity and reliable performance, GFs are being taken into consideration within passive flow control devices, showing promising results to extend the lifetime of future wind turbines [7]. A GF consists of a small tab placed normal to the airfoil surface and close to the trailing edge, either in the upper or in the lower side. The size of these GFs, measured with respect to the airfoil chord length (*c*), usually varies from 0.1% *c* and 3% *c*. They were first used in 1971 by Daniel Gurney, a race car driver who noticed improvements in cornering speeds and in the stability of his vehicle as a consequence of an increase in the downforce. Liebeck [8] discussed first this application and, subsequently, several experiments were carried out by Jeffrey et al. [9] on a NACA 0012 airfoil to investigate the effects of GFs and how they provide a lift improvement and a drag reduction once properly sized. They have been more widely researched for lift enhancement in aeronautics, where their advantages and applications were extensively studied by Wang et al. [10] Additionally, Pastrikakis et al. [11] compared the performance of a helicopter rotor with GFs at low and high forward flight speeds. Tang et al. [12], working with a NACA 0012 airfoil as well, studied a fixed and an oscillating trailing edge GF with the aim of evaluating the aerodynamics loadings by means of an incompressible Navier-Stokes code. Lee et al. [13] also studied the aerodynamic characteristics and the impact of GFs installed in a NACA 0015 along with a trailing edge flap. Different GFs heights and perforations were analyzed using particle image velocimetry (PIV) to measure the development of the tip vortex generated. According to Shukla et al. [14], the implementation of GFs in NACA0012 and NACA0015 symmetrical airfoils results in an improvement in the lift coefficient and lift force. Cole et al. [15] showed with different GF heights the importance of the airfoil shape in the aerodynamic performance of the airfoil. The influence of passive devices such as VGs and GFs was also studied by Fernandez-Gamiz et al. [16]. This work evaluated the improvement of a 5 MW wind turbine in terms of power output. Multiple device configurations and wind speed realizations were studied with the results showing an overall increase on the average power output of the wind turbine. Astolfi et al. [17] presented three test cases to evaluate the wind turbine power curve upgrades under different possible scenarios through operational data. In addition, a multivariate linear method for selecting the most appropriate input for modeling a given output was proposed by Terzi et al. [18] and applied to a multi-megawatt wind turbine with different passive flow control devices installed.

GFs have also been studied as active flow control devices, as shown in the work of Camocardi et al. [19], where the characteristics and structures of the flow pattern downstream the airfoil in the near wake were experimentally investigated. Recently, Han et al. [20] analyzed the influence of fixed and retractable GFs in the performance of variable speed helicopter rotors. Their results showed that both fixed and retractable GFs enhanced the performance of the rotor and expanded the flight envelope. However, the retractable one led to a better behavior in rotor power savings. In the wind turbine research field, the goal of this type of devices is to increase the lift on the rotor blades and, therefore, to increase the mechanical torque applied by the wind in the rotor. A lift enhancement of nearly 15% was observed in the study of Storms et al. [21] with a GF length of 0.5% *c* on a NACA 4412. Similarly, in the work of Mohammadi et al. [22], lift improvements were achieved with different GF shapes when compared to the clean DU91W(2)250 airfoil. Numerical methods have been developed and improved over the last years to study, as shown in the work of Gebhardt et al. [23], to study the behavior of large horizontal-axis wind turbines. Computational Fluid Dynamic (CFD) techniques are also frequently used to study the advantages and limitations of different flow control devices [24]. Recently, Woodgate et al. [25] used an in-house CFD solver to evaluate the implementation of GFs on wings and rotors. Different methods of modeling a GF were discussed and 2D cases were simulated to compare thick and infinitely thin GFs. Furthermore, they tested the solver also with 3D cases including rotors in hover and forward flight. In the study of Fernandez-Gamiz et al. [26], CFD simulations were made to find the optimal position of a microtab to improve the power output of a 5 MW wind turbine.

Additionally, Fernandez-Gamiz et al. [27] carried out a parametric study to find the most favorable dimension of a GF on a S810 airfoil by means of Proper Orthogonal Decomposition (POD) methods.

A parametric study is presented in the present work to analyze the effect of the GF length on the aerodynamic performance in a widely used DU91W(2)250 airfoil in multi-megawatt Horizontal Axis Wind Turbines (HAWT). In order to achieve this purpose, numerical RANS based simulations have been made and validated with the wind tunnel experimental data provided by Timmer et al. [28]. In addition, an Artificial Neural Network (ANN)-based model is presented to predict the effects of GFs on the aerodynamic efficiency in the DU91W(2)250 airfoil.

2. Numerical Setup

The behavior of the implementation of GFs was studied by means of Computational Fluid Dynamic (CFD) tools. In the current work, the commercial code STAR CCM+ v.11.02 [29] has been chosen to simulate the behavior of different GF lengths on a DU91W(2)250 airfoil.

The UpWind algorithm was employed for the pressure-velocity coupling and a linear upwind second order scheme was used to discretize the mesh. The numerical simulations were performed in steady state and ran fully turbulent using Reynolds Averaged Navier-Stokes (RANS) equations. Specifically, the Menter [30] k-ω SST shear stress turbulence model was used since it leads to a significant improvement in handling non-equilibrium boundary layer regions such as those close to separation, as addressed, as addressed by Kral [31] and Gatski [32]. This turbulent model combines both the standard k-ε model and k-ω model, retaining the properties of k-ω close to the wall and gradually blending into the standard k-ε model away from the wall. Mayda et al. [33] presented different tab configurations applying RANS calculations with the SST turbulence model. An O-mesh domain was defined for the numerical simulations, where the size of the computational domain was set to be 42 times the airfoil chord length c, R = 42 c, according to the recommendation of Sørensen et al. [34].

All the simulations were performed with a Reynolds number of Re = 2×10^6. The generation of an optimized grid represents the most important step before running the numerical solution in order to achieve reliable results. The grid domain consists of 65348 structured elements, where the height of the first cell normalized by the airfoil chord length was determined to be $\Delta z/c$ of 1.35×10^{-6}. The stretching in both normal and chord-wise direction is achieved by *tanh* functions based on Vinokur [35]. With the aim of resolving the viscous sublayer inside the turbulent boundary layer, the mesh was designed to achieve a dimensionless wall distance $y+$ at the first node adjacent to the airfoil wall less than 1 ($y+ < 1$). Figure 1 shows the cell distribution close to the trailing edge of the airfoil and around the GF, where the mesh refinement plays a major role due to the high velocity gradients expected in this region.

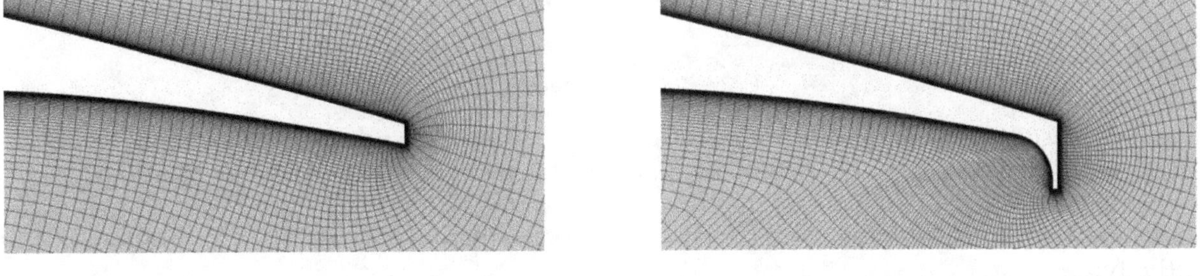

Figure 1. Mesh distribution on the trailing edge; with the clean airfoil and with a GF of 1.5% c implemented.

The walls of the airfoil and the GFs were set as non-slip boundary type. The validation of the numerical simulations without the implementation of the GF on the DU91W(2)250 airfoil was carried out with the wind tunnel results obtained by Timmer [28].

A mesh dependency study was made by means of the Richardson's extrapolation to verify that the solution obtained numerically is not dependent on the mesh size, following the procedure carried out by Fernandez-Gamiz et al. [27]. Three different grids were created with a mesh refinement ratio of 2. A fine, medium and coarse mesh of 65348, 32674 and 16337 cells have been designed, respectively. The results are summarized in Table 1, where RE indicates Richardson's extrapolation solution, p defines the order of accuracy and R the ratio of error. R values less than 1 were obtained, indicating that we are within the asymptotic range of convergence for all angles of attack. Figure 2 illustrates the lift-to-drag ratios achieved with each mesh level. The fine mesh provided the best results compared with the experimental wind tunnel results of Timmer et al. [28]. Thus, the fine mesh was used in the numerical simulations presented in the current work.

Table 1. Mesh dependency study results.

AoA	Mesh			Richardson Extrapolation		
	Coarse	Medium	Fine	RE	p	R
−6	−35.09	−49.94	−53.98	−52.47	1.87	0.27
−5	−25.97	−37.36	−39.95	−39.19	2.13	0.23
−4	−15.62	−22.35	−24.04	−23.47	2.00	0.25
−3	−4.86	−6.72	−7.15	−7.02	2.12	0.23
−2	6.24	8.98	9.61	9.42	2.13	0.23
−1	16.90	24.18	26.00	25.39	2.00	0.25
0	28.53	39.43	41.95	41.19	2.12	0.23
1	37.37	53.47	57.50	55.14	1.44	0.37
2	47.24	66.87	72.68	70.23	1.75	0.30
3	56.90	78.79	87.54	81.70	1.32	0.40
4	66.38	95.49	102.13	100.16	2.13	0.23
5	74.38	105.86	114.44	111.22	1.87	0.27

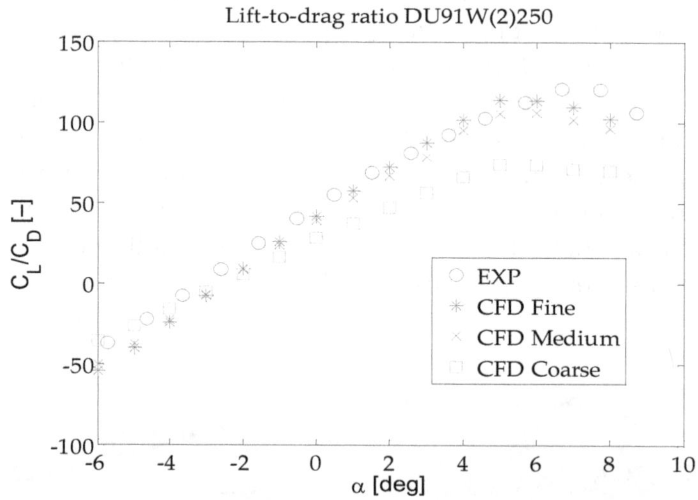

Figure 2. Results of the lift-to-drag ratio with each mesh level of the numerical simulations (CFD) vs. the experimental data of the DU91W(2)250 clean airfoil without the implementation of any GF.

Both lift and drag dimensionless coefficients were calculated with the Equations (1) and (2), respectively:

$$C_L = \frac{L}{\frac{1}{2}\rho U_\infty^2 c} \tag{1}$$

$$C_D = \frac{D}{\frac{1}{2}\rho U_\infty^2 c} \tag{2}$$

An air density value of $\rho = 1.204\ \mathrm{kg/m^3}$ was set, the dynamic viscosity was defined by $\mu = 1.855 \times 10^{-5}$ Pa·s and the free stream velocity corresponds to $U_\infty = 30$ m/s. The airfoil chord length is $c = 1$ m. Since the numerical simulations have been performed in two-dimensions, the parameters L and D represent the lift and drag forces per unit of area.

Gurney Flap Layout

The position and size of the GF is displayed in Figure 3. The dimension c represents the airfoil chord length and the dimension y the GF length. Twelve cases have been considered depending on the GF length, which is expressed as a percentage of the airfoil chord length, as shown in Table 2. Each case has been studied for fifteen different angles of attack, from $-6°$ to $8°$. The combination of all these GF positions for each angle of attack gives 195 different cases to study. All these cases have been designed following the procedure of previous studies carried out by Fernandez-Gamiz et al. [26,27]. The airfoil without any flow control device was also taken into account and simulated in order to compare the influence of the GFs on the airfoil aerodynamic performance.

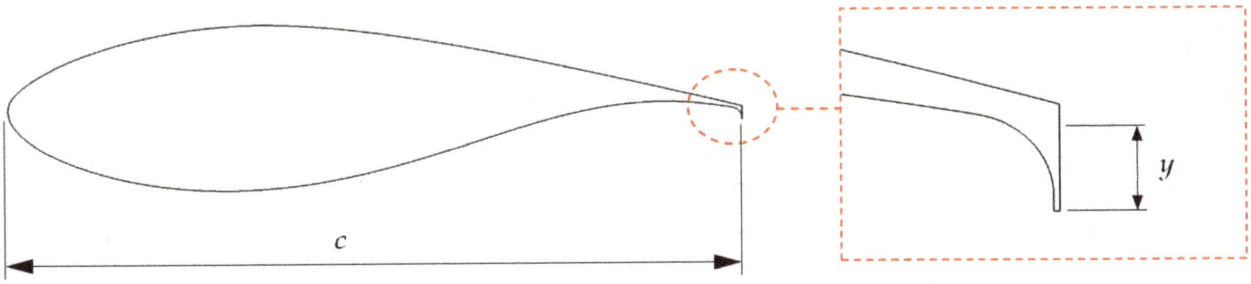

Figure 3. Detailed view of GF on a DU91W(2)250.

Table 2. Test cases considered in the study.

Test ID	TEST CASE	y (% c)
0	DU91W250	no GF
1	DU91W(2)250GF025	0.25
2	DU91W(2)250GF05	0.50
3	DU91W(2)250GF075	0.75
4	DU91W(2)250GF1	1.00
5	DU91W(2)250GF125	1.25
6	DU91W(2)250GF15	1.50
7	DU91W(2)250GF175	1.75
8	DU91W(2)250GF2	2.00
9	DU91W(2)250GF225	2.25
10	DU91W(2)250GF25	2.50
11	DU91W(2)250GF275	2.75
12	DU91W(2)250GF3	3.00

3. Results

The influence of the GF length was evaluated with the lift-to-drag ratio C_L/C_D for every angle of attack α studied, as illustrated in Figure 4. For each AoA, the evolution of the lift-to-drag ratio C_L/C_D with the GF length was compared with the clean airfoil case, i.e., no GF implementation. With regard to negative AoAs, from $-6°$ to $-4°$, as the GF length increases the lift-to-drag ratio increases as well. In all these cases the effect of the GF enhances the aerodynamic efficiency of the airfoil, except for the case of $-6°$ with a GF length of 0.25% c. However, in the range of AoAs from $-3°$ to $-1°$, a peak value of C_L/C_D is achieved before the maximum GF length of 3% c. For $-3°$ of AoA, a C_L/C_D ratio of 18.99 was obtained with a GF length of 2% c. Similarly, the peak values of C_L/C_D for $-2°$ and $-1°$ of AoA correspond to GF lengths of 1.5% c and 1% c, respectively.

On the other hand, a different behavior is observed for positive angles of attack. From 0° to 3°, the influence of the GF length on the C_L/C_D ratio follows the same pattern. With 0° of AoA, a maximum C_L/C_D value of 61.75 is obtained corresponding to a GF length of 1% *c*. Furthermore, the optimal GF length for 1° of AoA corresponds to 0.75% *c* with a C_L/C_D ratio of 78.91. As the angle of attack increases larger C_L/C_D maximum values are obtained, 92.37 and 102.65 for 2° and 3° of AoA, respectively. As can be seen from the plots at 2° and 3° of AoA, the effect of the GF is not favorable for all lengths. For 2° of AoA, GFs with lengths of 2.75% *c* and 3% *c* do not provide an increase in the C_L/C_D ratio compared with the clean airfoil. Similarly, in the case of 3° of AoA, GF lengths larger than 2% *c* do not present beneficial effects in the aerodynamic performance of the airfoil.

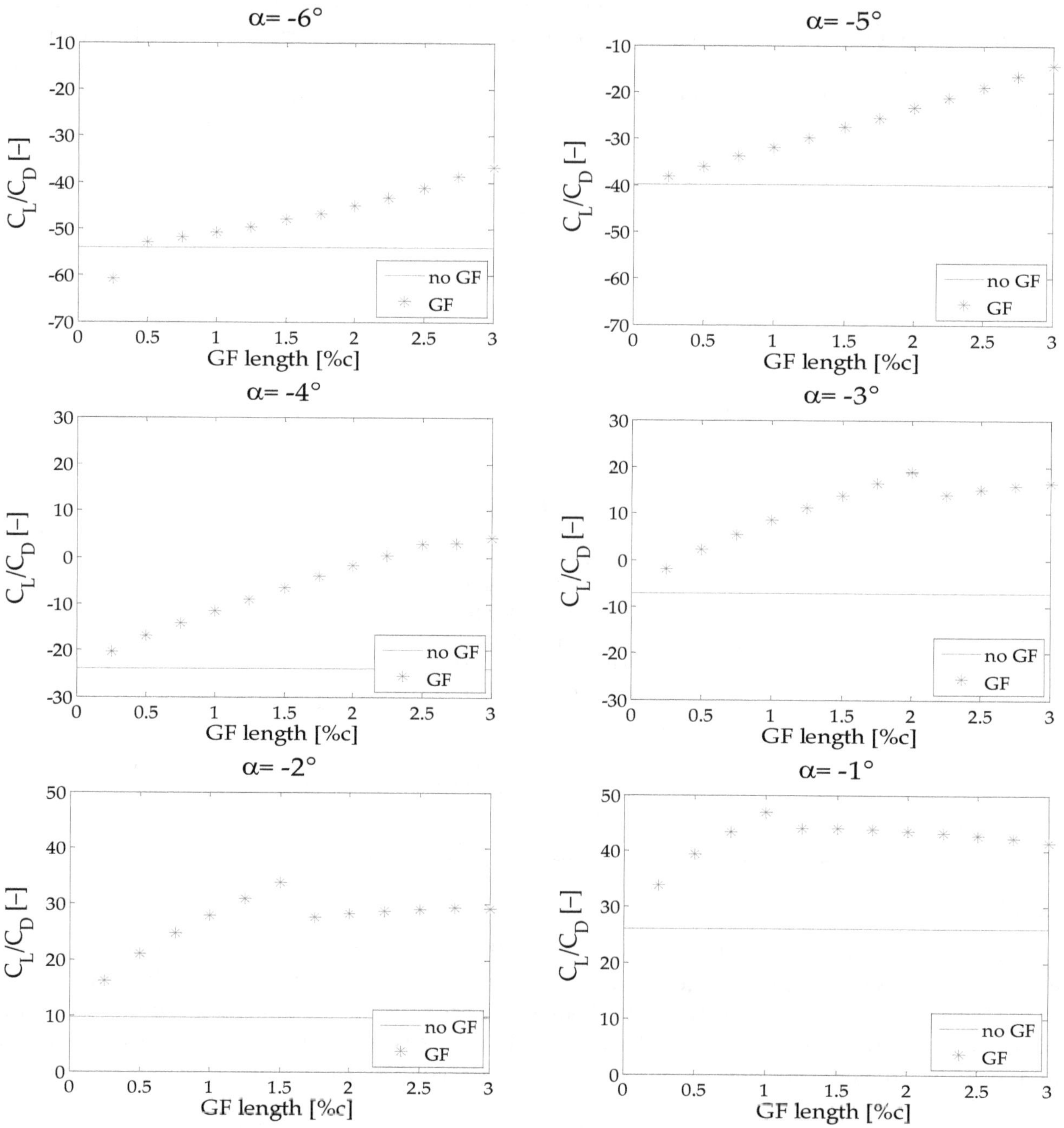

Figure 4. *Cont.*

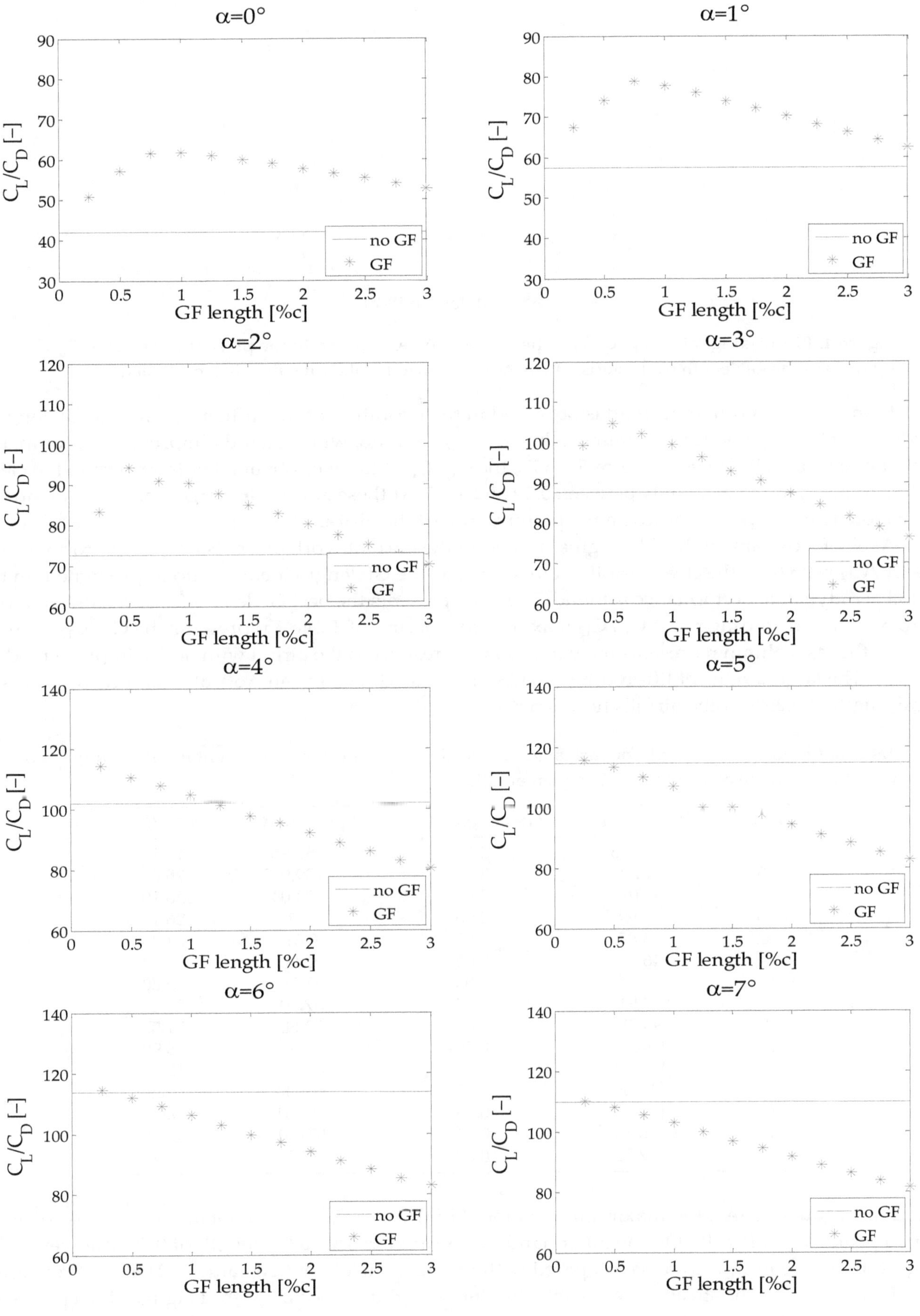

Figure 4. *Cont.*

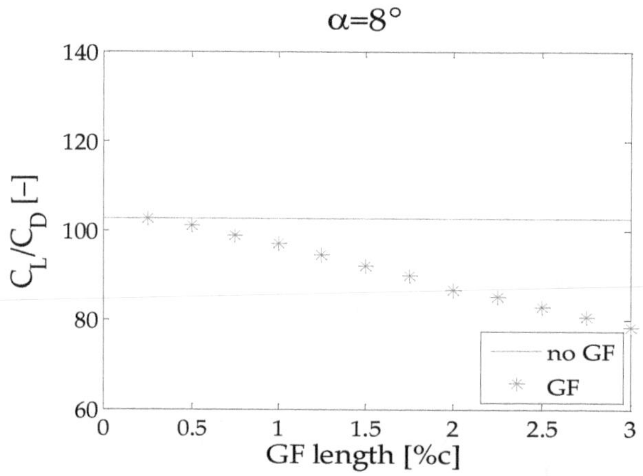

Figure 4. Lift-to-drag ratio (C_L/C_D) for the range of AoAs studied in the present work. The C_L/C_D ratio along the GF length (red asterisks) vs. the clean airfoil (solid blue line) is represented.

From 4° to 7° a different trend is observed in the evolution of the influence with the GF lengths studied. A GF length of 0.25% *c* presents the best C_L/C_D ratio, with a slightly improvement compared with the clean airfoil. However, from 5° to 7°, the C_L/C_D difference obtained with respect to the clean airfoil at these AoAs varies only between 0.29 and 1.25. At these AoAs, GFs larger than 0.25% *c* do not contribute to enhance the aerodynamic performance of the airfoil.

At 8° of AoA, any of the GF lengths studied in the current work presents a better performance in the C_L/C_D ratio than the clean airfoil. At this AoA, as the GF length increases no improvement in the airfoil aerodynamic performance is found. Table 3 represents the best GF length for each AoA in terms of maximum lift-to-drag ratio: C_L/C_D max. Fourth column of Table 3 represents the C_L/C_D with no GF and the last column represents the variation with respect to the airfoil with no GF implemented: Δ C_L/C_D. The largest value of lift-to-drag C_L/C_D 115.68 is achieved at an AoA of 5° and corresponds to a GF length of 0.25% of the airfoil chord length *c*.

Table 3. Optimum GF length for each angle of attack studied and the increment in lift-to-drag ratio with respect to the case with no GF implementation.

AoA (°)	C_L/C_D Max.	GF Length (% *c*)	C_L/C_D no GF	Δ C_L/C_D
−6	−36.86	3.00	−53.98	17.13
−5	−14.27	3.00	−39.95	25.68
−4	4.16	3.00	−24.04	28.19
−3	18.99	2.00	−7.15	26.14
−2	33.87	1.50	9.61	24.26
−1	46.93	1.00	26.00	20.93
0	61.75	1.00	41.95	19.80
1	78.91	0.75	57.50	21.41
2	94.25	0.50	72.68	21.57
3	104.52	0.50	87.54	16.98
4	114.48	0.25	102.13	12.36
5	115.68	0.25	114.44	1.25
6	114.41	0.25	113.81	0.60
7	110.10	0.25	109.81	0.29
8	102.55	0.00	102.73	0.00

For negative AoAs, the maximum increment in the C_L/C_D ratio is obtained with −4° of AoA with a peak value of 28.19. On the other hand, for positive AoAs, a GF length of 0.5% *c* achieves the largest difference in C_L/C_D ratio compared with the clean airfoil, with a value of 21.41 at 2° of AoA.

Figure 5 shows the streamwise velocity distribution around the GF for the lengths which provided the best cases in terms of C_L/C_D ratio. The presence of the GF changes the trailing edge flow pattern and, therefore, modifies the effective camber of the airfoil resulting in a lift enhance. As the GF size increases, the presence of the two vortices behind the GF is more visible. The GF jets the flow in the boundary layer away from the airfoil surface and, thus, the circulation in the region behind the GF is enhanced.

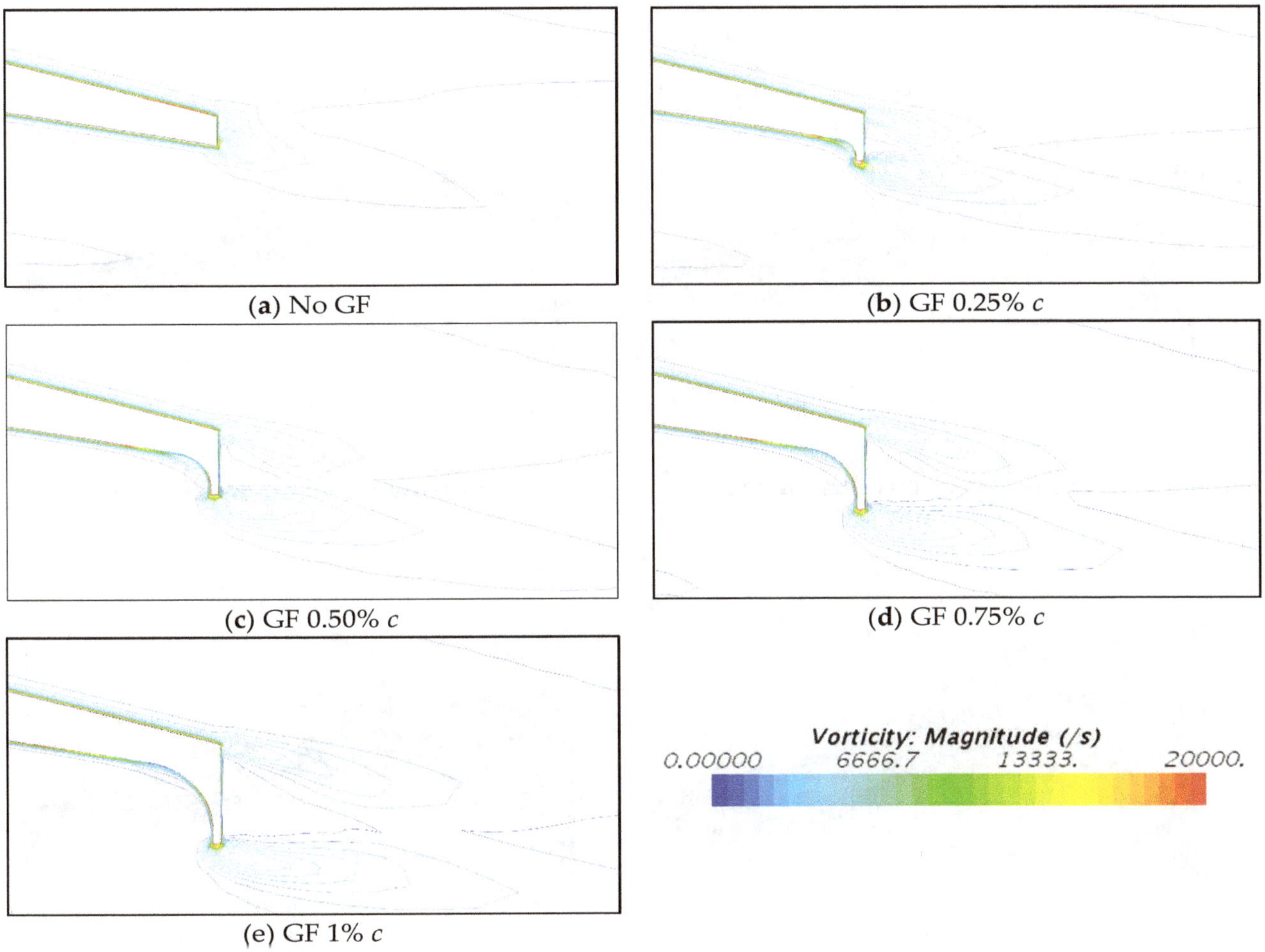

Figure 5. Streamwise vorticity fields for the case without GF and GF lengths of 0.25% *c*, 0.50% *c*, 0.75% *c* and 1% *c* at an angle of attack $\alpha = 0°$.

Figure 6 shows the pressure distributions on the surface of the clean airfoil DU91W(2)250 and the airfoil with GFs of 0.25% *c*, 0.50% *c*, 0.75% *c* and 1% *c* which have been considered that provide the best aerodynamic performance in terms to C_L/C_D ratio. The airfoil geometry, with a continuous black line, is illustrated as well. In the cases with the GF implemented, the aft loading of the airfoil increases and a positive gap with respect to the clean airfoil is visible at all AoAs.

ANN-Based Predicition Model

Artificial Neural Networks (ANNs) represent an exceptional tool in the modeling of different type of systems due to their several advantages and outstanding properties. One of them is related with the capability of learning, as they can learn complex non-linear black box models with an appropriate selection of the training algorithm and the input/output values. If neural networks are properly trained, i.e., the training patterns are carefully chosen, their response in new situations (new inputs) will be suitable with high probability, which means that will have the so-called generalization property. Lastly, as neural networks have an inherent parallel internal structure, their response after they have been trained is very fast, providing them real time capabilities to develop prediction models with large datasets, see Lopez-Guede et al. [36]. All these properties make ANNs a method of major importance in a wide number of applications.

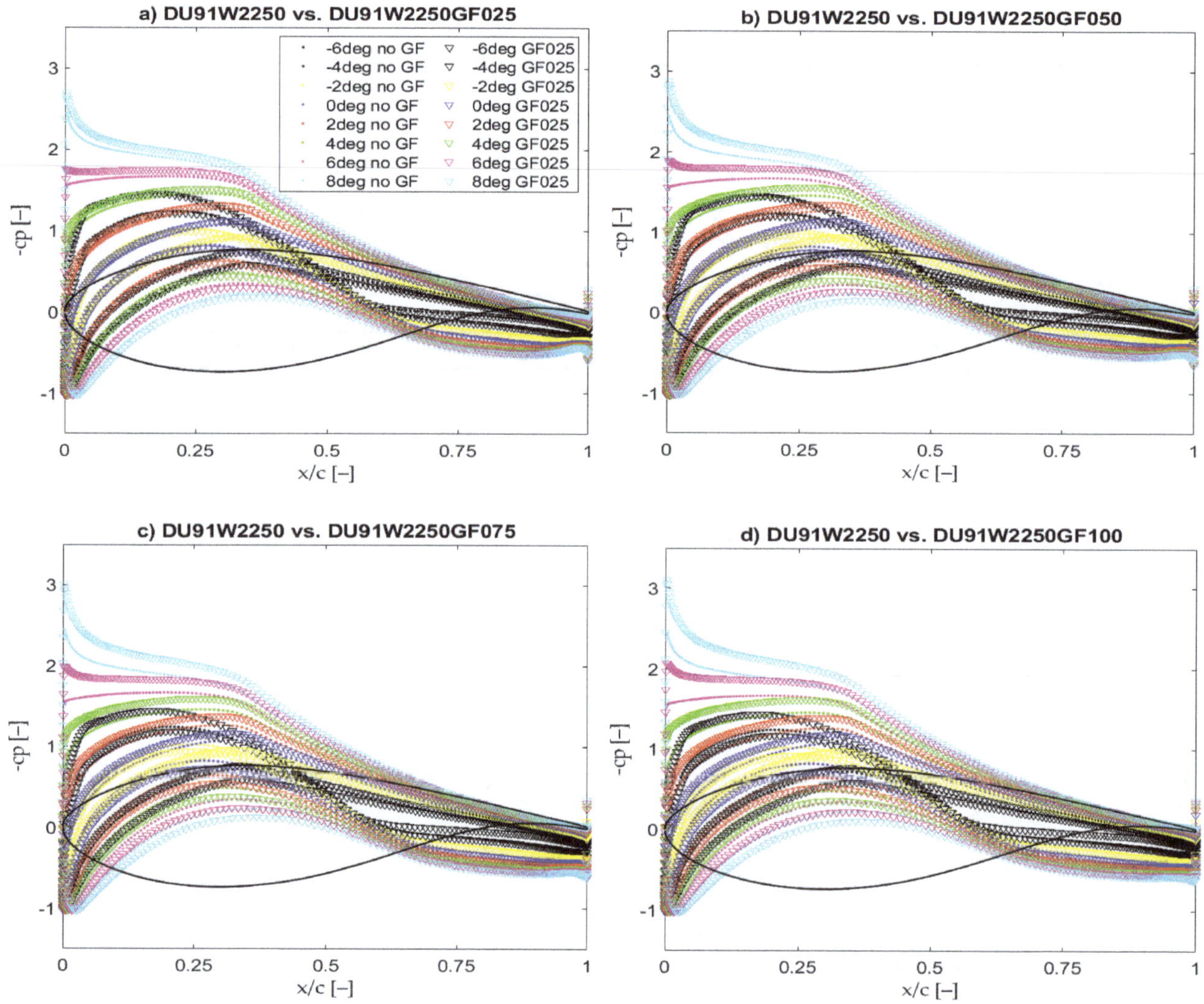

Figure 6. Pressure coefficient cp comparison between the clean airfoil DU91W2250 (no GF) and the best cases: (**a**) DU91W(2)250GF025; (**b**) DU91W(2)250GF050; (**c**) DU91W(2)250GF075; (**d**) DU91W(2) 250GF100.

Once studied the influence of different GF lengths and angles of attack in the airfoil aerodynamic performance by means of CFD techniques, a prediction model has been developed with an artificial neural network (ANN). The aim of this model is to predict the aerodynamic efficiency (C_L/C_D ratio) of the airfoil following the strategy based on the study by Lopez-Guede et al. [37]. In the current study, a multi-layer model with two hidden layers has been used. The C_L/C_D ratio is obtained by Equation (3), whereas the outputs of each hidden neuron follow a sigmoid function as defined in Equation (4). All these parameters represent a typical configuration of Multilayer Perceptron with Backpropagation (BP-MLP) neural networks. The postsinaptical h_i of each i neuron is calculated using a linear combination by means of Equation (5), where ω_i denotes the output layer weights and ω_{ij} the input (hidden) layer weights:

$$\frac{C_L}{C_D} = \sum_{i=1}^{i=Nhidden} \omega_i \cdot g_i\left(\vec{x}\right) + \theta \tag{3}$$

$$g_i(\vec{x}) = \frac{1}{1+e^{-h_i}} \tag{4}$$

$$h_i(\vec{x}) = \sum_{j=1}^{j=Ninputs} \omega'_{i,j} \cdot x_j + \theta'_i \quad (5)$$

Matrices 6–9 contain the synaptic weights obtained by the neural network for the prediction of the C_L/C_D ratio:

$$Input\ hidden\ layer\ weights\ (\omega_{ij}) = \begin{bmatrix} -2.4229 & -0.9914 \\ 1.7982 & 0.2139 \end{bmatrix} \quad (6)$$

$$Output\ layer\ weights\ (\omega_i) = \begin{bmatrix} 1.2220 & 3.6533 \end{bmatrix} \quad (7)$$

$$Hidden\ layer\ threshold\ parameters\ (\sigma) = \begin{bmatrix} 1.4471 \\ 0.7852 \end{bmatrix} \quad (8)$$

$$Output\ layer\ threshold\ parameters\ (\sigma') = [-2.9805] \quad (9)$$

The ANN has two input neurons corresponding to the two real valued inputs (AoA and GF length) and one unique output neuron for the valued target (CL/CD ratio), as shown in Figure 7. The datasheet has been partitioned, i.e, 70% for training the ANN, 20% for validation and 10% for testing the quality of learned model.

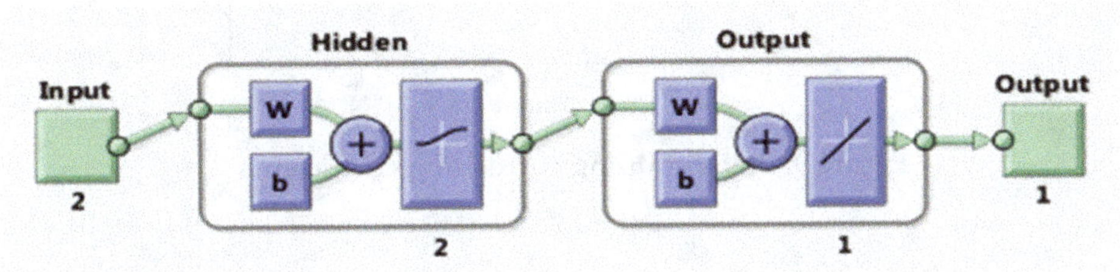

Figure 7. Diagram of the internal structure of the ANN designed.

Figure 8 shows the ANN results for all GF lengths and all AoAs studied in the present work. The colored surface illustrates the C_L/C_D values calculated by the proposed ANN and the CFD results are represented by black asterisks. The neural network predicts accurately the lift-to-drag ratios obtained by the numerical computations.

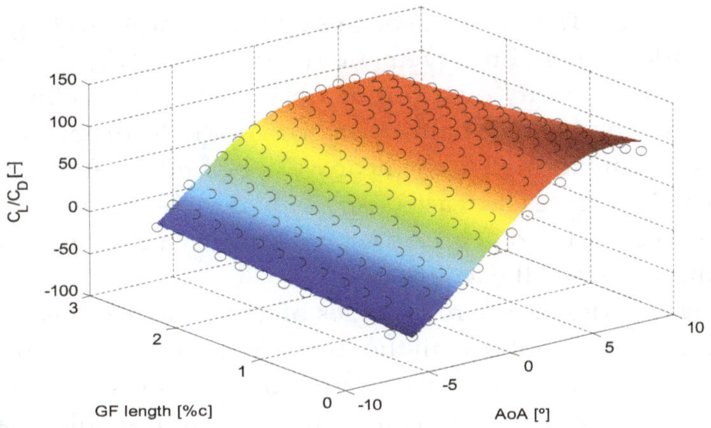

Figure 8. Neural network prediction surface of the CFD results of the airfoil aerodynamic performance C_L/C_D as a function of the GF length and the angle of attack (AoA). Black circles represent the CFD results.

The correlation coefficients (R-value) are shown in Figure 9. The regression line (highlighted in red) is drawn over expected C_L/C_D ratio and its neural prediction. The horizontal axis shows the expected C_L/C_D ratio and the vertical axis the C_L/C_D ratio prediction given by the neural network. The correlation coefficient value was determined as 0.99671 which indicates a good correlation between the results given by the ANN and the followed target. This result for the simulated GF lengths allows the algorithm to calculate and predict the C_L/C_D ratio for any GF length in the conditions studied of Reynolds number Re = 2×10^6.

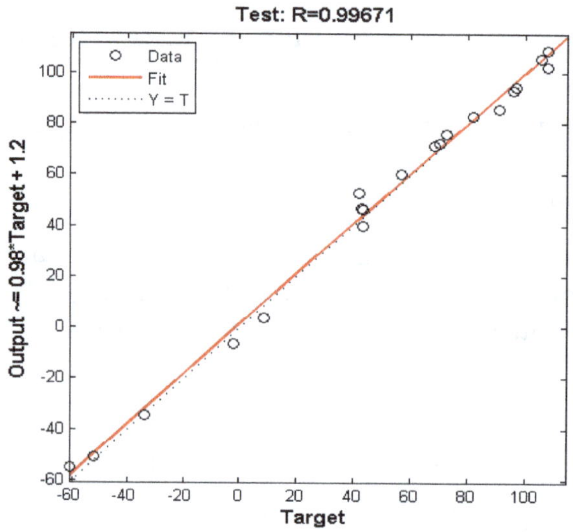

Figure 9. ANN training correlation coefficients.

4. Conclusions

The current work presents a parametric study to evaluate the influence of different GF lengths in the aerodynamic performance of a DU91W(2)250 airfoil. Two-dimensional CFD simulations have been carried out, using RANS equations at Reynolds number of Re = 2×10^6. Firstly, a clean airfoil without any flow control device implemented has been simulated and the results in terms of lift-to-drag ratio have been validated with experimental data. Then, simulations have been performed with GFs from 0.25% *c* to 3% *c* at angles of attack from $-6°$ to $8°$ to investigate the airfoil aerodynamic performance. The results showed that, for positive angles of attack, peak values are obtained with GF lengths between 0.25% *c* and 0.75% *c*. Specifically, an increase of 21.57 in C_L/C_D ratio has been obtained with a GF length of 0.5% *c* at 2° of AoA. The influence of GFs decreased at AoAs larger than 5°, where only a GF length of 0.25% *c* provides a slight improvement in terms of C_L/C_D ratio enhancement.

Secondly, the streamwise velocity distribution around the GF has been addressed with those GF lengths that provided the best aerodynamic performance in terms of C_L/C_D ratio. As expected, two counter-rotating vortices are formed in the near wake behind the flow control device. As the GF length increases the size of the vortices increases, especially the one on the lower region. Moreover, the pressure distribution has been studied at all AoAs, resulting in a clearly visible increment in the cp due to the GF implementation for all the length cases.

The optimal GF length varies at different angles of attack, which means that a fixed GF length would not reach the optimal C_L/C_D ratio at all the AoAs. Therefore, this result suggests that a careful analysis of the GF length is needed to yield an efficient flow control system and to consider the study of GFs as active flow control devices. In that case, the GF length would change along the AoA to achieve the best C_L/C_D ratio depending on the working conditions.

To that end, an ANN has been developed and trained to predict the aerodynamic efficiency of the airfoil in terms of C_L/C_D ratio. This tool allows to obtain an accurate prediction model of the aerodynamic behavior of the airfoil, which can be a suitable method to optimize the GF lengths for different wind and /or rotor blade airfoil geometry scenarios and its implications in the wind turbine control system.

Author Contributions: I.A., U.F.-G. and E.Z. conceived and performed the CFD simulations; A.S.-A. and D.T.-F.-B. analyzed the results and provided constructive instructions in the process of preparing the paper.

Acknowledgments: The authors are grateful to the Government of the Basque Country and the University of the Basque Country UPV/EHU through the SAIOTEK (S-PE11UN112) and EHU12/26 research programs, respectively. The funding of Fundation Vital Fundazioa is also acknowledged.

References

1. Johnson, S.J.; van Dam, C.P.; Berg, D.E. *Active Load Control Techniques for Wind Turbines*; Sandia Report SAND 2008-4809; Sandia National Laboratories: Albuquerque, NM, USA; Livermore, CA, USA, 2008.
2. Taylor, H.D. *The Elimination of Diffuser Separation by Vortex Generators*; R-15064-5; United Aircraft Corporation: East Hartford, CT, USA, 1947.
3. Wood, R.M. A Discussion of Aerodynamic Control Effectors Concepts (ACEs) for Future Unmanned Air Vehicles (UAVs). In Proceedings of the AIAA 1st Technical Conference and Workshop on Unmanned Aerospace Vehicle, Systems, Technologies and Operations, Portsmouth, VA, USA, 20–23 May 2002.
4. Aramendia, I.; Fernandez-Gamiz, U.; Antonio Ramos-Hernanz, J.; Sancho, J.; Manuel Lopez-Guede, J.; Zulueta, E. Flow Control Devices for Wind Turbines. In *Energy Harvesting and Energy Efficiency: Technology, Methods, and Applications*; Springer: Cham, Switzerland, 2017; Volume 37, pp. 629–655. [CrossRef]
5. Fernandez-Gamiz, U.; Velte, C.M.; Rethore, P.; Sorensen, N.N.; Egusquiza, E. Testing of self-similarity and helical symmetry in vortex generator flow simulations. *Wind Energy* **2016**, *19*, 1043–1052. [CrossRef]
6. Fernandez-Gamiz, U.; Errasti, I.; Gutierrez-Amo, R.; Boyano, A.; Barambones, O. Computational Modelling of Rectangular Sub-Boundary Layer Vortex Generators. *Appl. Sci.* **2018**, *8*, 138. [CrossRef]
7. Holst, D.; Bach, A.B.; Nayeri, C.N.; Paschereit, C.O.; Pechlivanoglou, G. Wake Analysis of a Finite Width Gurney Flap. *J. Eng. Gas Turbines Power Trans. ASME* **2016**, *138*, 062602. [CrossRef]
8. Liebeck, R. Design of Subsonic Airfoils for High Lift. *J. Aircr.* **1978**, *15*, 547–561. [CrossRef]
9. Jeffrey, D.; Zhang, X.; Hurst, D. Aerodynamics of Gurney flaps on a single-element high-lift wing. *J. Aircr.* **2000**, *37*, 295–301. [CrossRef]
10. Wang, J.J.; Li, Y.C.; Choi, K. Gurney flap-lift enhancement, mechanisms and applications. *Prog. Aerosp. Sci.* **2008**, *44*, 22–47. [CrossRef]
11. Pastrikakis, V.A.; Steijl, R.; Barakos, G.N. Effect of active Gurney flaps on overall helicopter flight envelope. *Aeronaut. J.* **2016**, *120*, 1230–1261. [CrossRef]
12. Tang, D.; Dowell, E.H. Aerodynamic loading for an airfoil with an oscillating gurney flap. *J. Aircr.* **2007**, *44*, 1245–1257. [CrossRef]
13. Lee, T.; Su, Y.Y. Lift enhancement and flow structure of airfoil with joint trailing-edge flap and Gurney flap. *Exp. Fluids* **2011**, *50*, 1671–1684. [CrossRef]
14. Shukla, V.; Kaviti, A.K. Performance evaluation of profile modifications on straight-bladed vertical axis wind turbine by energy and Spalart Allmaras models. *Energy* **2017**, *126*, 766–795. [CrossRef]
15. Cole, J.A.; Vieira, B.A.O.; Coder, J.G.; Premi, A.; Maughmer, M.D. Experimental Investigation into the Effect of Gurney Flaps on Various Airfoils. *J. Aircr.* **2013**, *50*, 1287–1294. [CrossRef]
16. Fernandez-Gamiz, U.; Zulueta, E.; Boyano, A.; Ansoategui, I.; Uriarte, I. Five Megawatt Wind Turbine Power Output Improvements by Passive Flow Control Devices. *Energies* **2017**, *10*, 742. [CrossRef]
17. Astolfi, D.; Castellani, F.; Terzi, L. Wind Turbine Power Curve Upgrades. *Energies* **2018**, *11*, 1300. [CrossRef]
18. Terzi, L.; Lombardi, A.; Castellani, F.; Astolfi, D. Innovative methods for wind turbine power curve upgrade assessment. In Proceedings of the WindEurope Conference 2018, Hamburg, Germany, 25–28 September 2018.
19. Camocardi, M.E.; Maranon Di Leo, J.; Delnero, J.S.; Lerner, J.L.C. Experimental Study Of A Naca 4412 Airfoil With Movable Gurney Flap. In Proceedings of the 49th AIAA Aerospace Sciences Meeting including the New Horizons Forum and Aerospace Exposition, Orlando, FL, USA, 4–7 January 2011.
20. Han, D.; Dong, C.; Barakos, G.N. Performance improvement of variable speed rotors by Gurney flaps. *Aerosp. Sci. Technol.* **2018**, *81*, 118–127. [CrossRef]

21. Storms, B.L.; Jang, C.S. Lift Enhancement of an Airfoil using a Gurney Flap and Vortex Generators. *J. Aircr.* **1994**, *31*, 542–547. [CrossRef]
22. Mohammadi, M.; Doosttalab, A.; Doosttalab, M. The effect of various gurney flaps shapes on the performace of wind turbine airfoils. In Proceedings of the ASME Early Career Technical Conference, Atlanta, GA, USA, 2–3 November 2012.
23. Gebhardt, C.G.; Preidikman, S.; Massa, J.C. Numerical simulations of the aerodynamic behavior of large horizontal-axis wind turbines. *Int. J. Hydrogen Energy* **2010**, *35*, 6005–6011. [CrossRef]
24. Pastrikakis, V.; Woodgate, M.; Barakos, G. CFD Method for Modelling Gurney Flaps. In *Recent Progress in Flow Control for Practical Flows*; Doerffer, P., Barakos, G., Luczak, M., Eds.; Springer: Berlin, Germany, 2017; pp. 23–49.
25. Woodgate, M.A.; Pastrikakis, V.A.; Barakos, G.N. Rotor Computations with Active Gurney Flaps. *Adv. Fluid-Struct. Interact.* **2016**, *133*, 133–166. [CrossRef]
26. Fernandez-Gamiz, U.; Zulueta, E.; Boyano, A.; Ramos-Hernanz, J.A.; Manuel Lopez-Guede, J. Microtab Design and Implementation on a 5 MW Wind Turbine. *Appl. Sci.* **2017**, *7*, 536. [CrossRef]
27. Fernandez-Gamiz, U.; Gomez-Marmol, M.; Chacon-Rebollo, T. Computational Modeling of Gurney Flaps and Microtabs by POD Method. *Energies* **2018**, *11*, 2091. [CrossRef]
28. Timmer, W.; van Rooij, R. Summary of the Delft University wind turbine dedicated airfoils. *J. Sol. Energy Eng. Trans. ASME* **2003**, *125*, 488–496. [CrossRef]
29. Siemens STAR CCM+ Version 11.06.011. Available online: http://mdx.plm.automation.siemens.com/ (accessed on 10 June 2018).
30. Menter, F.R. 2-Equation Eddy-Viscosity Turbulence Models for Engineering Applications. *AIAA J.* **1994**, *32*, 1598–1605. [CrossRef]
31. Kral, L. Recent experience with different turbulence models applied to the calculation of flow over aircraft components. *Prog. Aerosp. Sci.* **1998**, *34*, 481–541. [CrossRef]
32. Gatski, T.B. *Turbulence Modeling for Aeronautical Flows*; von Karman Institute for Fluid DynamicsVKI Lecture Series: CFD-Based Aircraft Drag Prediction and Reduction; Von Karman Institute for Fluid Dynamics: Sint-Genesius-Rode, Belgium, 2003.
33. Mayda, E.A.; van Dam, C.P.; Nakafuji, D. Computational Investigation of Finite Width Microtabs for Aerodynamic Load Control. In Proceedings of the 43rd AIAA Aerospace Sciences Meeting and Exhibit, Reno, NV, USA, 10–13 January 2005.
34. Sorensen, N.N.; Mendez, B.; Munoz, A.; Sieros, G.; Jost, E.; Lutz, T.; Papadakis, G.; Voutsinas, S.; Barakos, G.N.; Colonia, S.; et al. CFD code comparison for 2D airfoil flows. *J. Phys. Conf. Ser.* **2016**, *753*, 082019. [CrossRef]
35. Vinokur, M. On One-Dimensional Stretching Functions for Finite-Difference Calculations. *J. Comput. Phys.* **1983**, *50*, 215–234. [CrossRef]
36. Lopez-Guede, J.M.; Ramos-Hernanz, J.A.; Zulueta, E.; Fernandez-Gamiz, U.; Oterino, F. Systematic modeling of photovoltaic modules based on artificial neural networks. *Int. J. Hydrogen Energy* **2016**, *41*, 12672–12687. [CrossRef]
37. Lopez-Guede, J.M.; Ramos-Hernanz, J.A.; Zulueta, E.; Fernandez-Gamiz, U.; Azkune, G. Dual model oriented modeling of monocrystalline PV modules based on artificial neuronal networks. *Int. J. Hydrogen Energy* **2017**, *42*, 18103–18120. [CrossRef]

12

Multiple Wind Turbine Wakes Modeling Considering the Faster Wake Recovery in Overlapped Wakes

Zhenzhou Shao, Ying Wu, Li Li, Shuang Han and Yongqian Liu *

State Key Laboratory of Alternate Electrical Power System with Renewable Energy Sources, North China Electric Power University, Changping District, Beijing 102206, China; ncepushao@163.com (Z.S.); wu_ying@ncepu.edu.cn (Y.W.); lili@ncepu.edu.cn (L.L.); hanshuang1008@sina.com (S.H.)

* Correspondence: yqliu@ncepu.edu.cn

Abstract: In a wind farm some wind turbines may be affected by multiple upwind wakes. The commonly used approach in engineering to simulate the interaction effect of different wakes is to combine the single analytical wake model and the interaction model. The higher turbulence level and shear stress profile generated by upwind turbines in the superposed area leads to faster wake recovery. The existing interaction models are all analytical models based on some simple assumptions of superposition, which cannot characterize this phenomenon. Therefore, in this study, a mixing coefficient is introduced into the classical energy balance interaction model with the aim of reflecting the effect of turbulence intensity on velocity recovery in multiple wakes. An empirical expression is also given to calculate this parameter. The performance of the new model is evaluated using data from the Lillgrund and the Horns Rev I offshore wind farms, and the simulations agree reasonably with the observations. The comparison of different interaction model simulation results with measured data show that the calculation accuracy of this new interaction model is high, and the mean absolute percentage error of wind farm efficiency is reduced by 5.3% and 1.58%, respectively, compared to the most commonly used sum of squares interaction model.

Keywords: wind farm; analytical model; wake interaction model; turbulence intensity; mixing coefficient; wind farm efficiency

1. Introduction

The wake from upwind wind turbines leads to decreased wind power output and increased fatigue load of downwind turbines. In a large wind farm, the power loss caused by the wake effect normally accounts for about 10% to 20% of all output produced in an entire year [1]. The accurate prediction of wake deficits is of vital importance for calculating wind farm power output [2], optimizing operations [3], the micro-siting of wind farms [4], etc. Wake models, the most frequently used tool for the prediction of wake effects, are now generally classified into three families: the roughness length-based model, the field model, and the kinematic model. The roughness length-based model treats wind turbines as obstacle elements or roughness elements that impact the atmospheric wind profile, which is specifically used in predicting the wake loss over a large wind farm or the wake interaction between wind farms [5]. The field model, or the computational fluid dynamics (CFD) model solves the Navier–Stokes equation to obtain detailed flow field information [6,7], and has high precision but is time consuming and computationally expensive, which hinders applications in engineering fields that require multiple computations of the wind farm power, such as wind farm layout optimization. The kinematic model (or empirical model) analytically obtains the wake velocity distribution behind a single turbine based on some ideal hypotheses. With the merits of a simple formulation, acceptable precision, and quick calculation speed, it has been widely used in engineering.

Many previous studies have introduced, analyzed, and compared the commonly used empirical wake models in detail [8–11].

One wind turbine may lie within the superimposed area of wakes from upwind turbines, and the interaction mechanism of multiple wakes is not well understood due to the extreme complexity of the turbulent structures within them. The commonly used approach in engineering to simulate the interaction effect of different wakes is to combine the single analytical wake models with some simple assumptions about superposition. When many turbines are aligned with the wind direction, it has been experimentally observed that the second turbine suffers the maximum power loss, while the subsequent turbines have relatively smaller further losses (e.g., turbines in a row under the west wind in the Horns Rev offshore wind farm [12]). Based on this observation, Leuven [13] assumed the given wind turbine was influenced only by the wake from the nearest upwind turbine in his wind farm model (WINDPARK) to calculate power. The result showed that this method obtained a good agreement between its prediction and the measurements in the Zeebrugge wind farm. A more frequently used method is using superposition models to analytically describe the effect of multiple wakes. There are four available superposition models: namely the geometric superposition (GS) model, linear superposition (LS) model, sum of squares (SS) model, and energy balance (EB) model. Among them, the latter three are the most common. They are based on a similar principle that the flow characteristics in the superposed area are caused by the accumulation of all wakes from upwind turbines, while the only distinction is in the different mathematical expressions of wake deficits from the upwind turbines.

To be specific, the LS model was proposed by Lissaman [14] and predicts the velocity deficit of superposed wakes by summing the velocity deficit of all upwind turbines. Crespo [15] argued that this assumption would lead to an overestimation of wake deficit in the superposed area and could even obtain a negative speed for large perturbations. However, when the wake velocity in the interference region is relatively small, such as the wake interaction from upwind turbines abreast of each other, the LS model performs well in predicting the overlapped wake deficit [16]. The SS model was presented by Katic [17] in his classical literature about the Park model and assumes the velocity deficit in the superposed area equals the square root of the sum of squares of the velocity deficit from each upwind turbine. The prediction of this model was verified to be in better agreement with experiments compared to the LS model when many wakes are combined [15]. Voutsinas [18] applied an explicit energy equation to develop the EB model, assuming that the total energy loss in the superposition wake equals the sum of the energy losses for each individual upwind turbine.

In order to evaluate the precision and effect of the different superposition models, Erik [19] compared the model predictions with wind tunnel measurements and found that the SS model could obtain more accurate results for almost all cases, followed by the EB model. Tian [20] integrated the two-dimensional Jensen model into different superposition models to calculate the power loss of the Horns Rev wind farm and reached the same conclusion. In fact, associating the SS model with the classical Jensen model [17,21], also known as the Park model, is the most popular approach to predict the wake deficit of wind farms [22–25] and is also the standard implementation in the wind resource assessment of many commercial software such as WAsP [26] and WindPRO [27], despite being proposed for more than 30 years.

In recent years, the SS model has been found to be unsuitable for simulating deep array effects, as the power outputs calculated for downwind turbines reach a constant too quickly. However, this does not preclude the widespread use of the SS model as before. The GS, LS, and SS models are all experience-based analytical models without definite physical basis, which makes it difficult to improve these engineering interaction models by theoretical analysis and numerical simulation. The only one that employs physics is the EB model, based on a simplified energy equation without considering the energy exchange between the wake region and ambient atmosphere. Kuo [28] took such energy into account and introduced the kinetic energy correction factor to revise the EB model. It is worth noting that the correction factor should be determined experimentally. The author suggested the value of this factor to be 1 if no measurements or CFD simulation data are available, which corresponds to the conventional EB model. An experimental study by Smith and Taylor [29] found that for a configuration of two turbines in a row, the wake velocity of the downwind turbine recovers more rapidly than the one upwind; i.e., at the same relative position, the velocity deficit of the downwind turbine wake is smaller. This might be because of the higher turbulence level and shear stress profile generated by upwind turbines in the superposed wake area, which enhances momentum diffusion and thus leads to faster wake recovery. In order to reflect this character in the engineering interaction model aiming to improve its precision, this study introduces a mixing coefficient to revise the classical EB model based on this phenomenon and proposes a new interaction wake model with higher accuracy.

This paper is organized as follows. In Section 2, a complete simulation method for wake flow in the wind farm is presented. First, a brief introduction of the single wake model used in this study is given in Section 2.1. Then, a new interaction model that considers the effect of increased turbulence intensity in the superposed area is proposed in Section 2.2. In addition, when predicting the wake flow in a wind farm, the relative location between wind turbines under an arbitrary wind direction is needed, and so a simple and convenient approach for this challenge is provided in Section 2.3. The newly proposed model and method are implemented for the offshore Lillgrund and Horns Rev I wind farms, and the results are compared with measurements in Section 3. Finally, a summary and conclusions are given in Section 4.

2. The New Analytical Wind Farm Wake Model

2.1. Wind Turbine Wake Model

The widely used 1D Jensen model is adopted to quantify the reduction of the downwind wind speed behind a wind turbine in this study. It is a linear wake model based on mass conservation, first proposed by N.O. Jensen [21] of the Denmark Risø Laboratory in 1982. Assuming a circular cylinder wake zone, the width of the wind turbine wake is expanded linearly with increasing downwind distances, and the wake velocity on the cross-section is equal at all spots; i.e., a top-hat distribution. The wind speed in the wake u_w, is a function of the thrust coefficient C_t, and wake decay coefficient k, which can be formulated as:

$$u_w = u_0\left[1 - \left(1 - \sqrt{1 - C_t}\right)\left(\frac{r_0}{r_0 + kx}\right)^2\right] \tag{1}$$

where u_0 is the inflow velocity, x is the distance behind the wind turbine, r_0 is the rotor radius, and $r_0 + kx$ is the wake radius at the downwind position x. For a given thrust coefficient and downwind distance, the only adjustable parameter in the Jensen wake model is the wake decay coefficient k, which is sensitive to many factors including the roughness, hub height, turbine-induced turbulence, ambient turbulence, or atmospheric stability. Based on long-term project practice, the value of k recommended in WAsP [26] is 0.075 and 0.05 for onshore and offshore wind farms, respectively. Several analytical models have been developed to estimate the wake decay coefficient, such as in References [9,30,31]. The focus of this paper is on developing the interaction model, and so the recommended values in WAsP that are most widely used are employed in our calculations.

2.2. Wake Superposition

The four frequently used wake interaction models in engineering are listed as follows:

$$\text{Geometric Sum (GS)} \quad \frac{u_i}{u_0} = \prod_j^N \frac{u_{ji}}{u_j}$$

$$\text{Linear Sum (LS)} \quad \left(1 - \frac{u_i}{u_0}\right) = \sum_j^N \left(1 - \frac{u_{ji}}{u_j}\right)$$

$$\text{Energy Balance (EB)} \quad u_0^2 - u_i^2 = \sum_j^N \left(u_j^2 - u_{ji}^2\right) ,$$

$$\text{sum of Squares (SS)} \quad \left(1 - \frac{u_i}{u_0}\right)^2 = \sum_j^N \left(1 - \frac{u_{ji}}{u_j}\right)^2$$

where N is the total number of upwind turbines which affect the target wind turbine i by their wakes; u_i is the inflow speed of target turbine i; u_j is the inflow speed of the upwind turbine j; and u_{ji} denotes the wind speed at turbine i due to the single wake from turbine j that can be obtained by using empirical wake models such as the Jensen wake model (Equation (1)).

Except for the GS model, all models first calculate the wake losses of each upwind turbine at the position of the target turbine and then overlay them as the total loss due to upwind turbines. Hence, the only distinction between the superposition models is that the wake deficit is expressed in different ways. Figure 1 gives a schematic for the typical wake superposition along a row of four turbines. As shown, wind turbine 3 is located in the superimposed zone of upwind turbines 1 and 2. When calculating the velocity distribution before turbine 3 in scenario I, first, the superposition should be divided into scenario II and scenario III. Then, the wake deficits caused by turbines 1 and 2 are computed separately and finally combined as the total loss caused by upwind turbine.

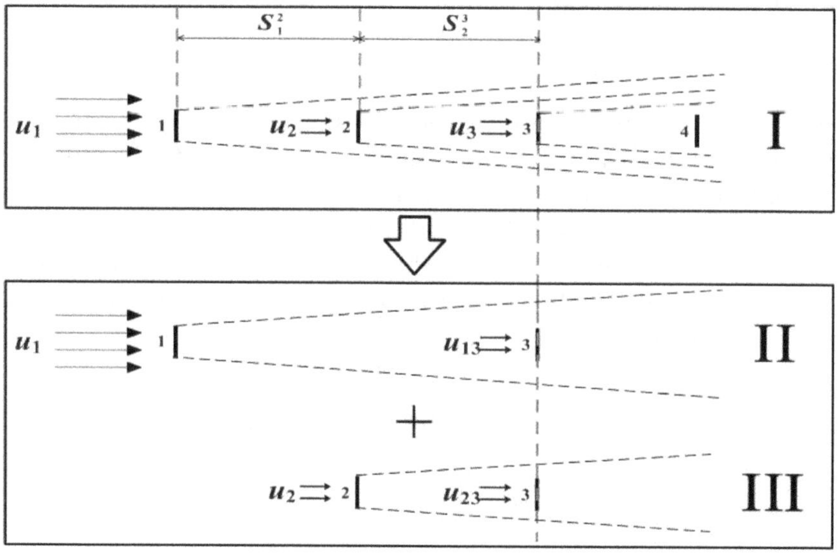

Figure 1. Schematic diagram of three overlapping turbine wakes. The simulation of the inflow velocity at turbine 3 in scenario I is decomposed into scenarios II and III, and then superimposed.

In the far wake region, turbulence plays a leading role in the flow development. The turbulent environment re-energises the low-momentum wake regions within wind farms [32]. The wind speed gradient between the wake and the free flow outside the wake results in additional shear-generated turbulence, which assists the transfer of momentum into the wake from the surrounding flow. Thus the wake and the surrounding flow start to mix. The rate of decay of the velocity deficit is strongly dependent on the ambient turbulence levels [33,34]. As a general rule of thumb [35,36], the stronger the ambient turbulent intensity, the faster the wake recovery. Some wind tunnel experiments and CFD simulations [37,38] have shown that the turbulence intensity in the wake overlay area is higher than that of the undisturbed wind turbine at the same location. Meanwhile, as mentioned above, for a particular experiment of two turbines in a row, the wake velocity of the downwind turbine recovers more rapidly than that of the upwind one. So this can be explained by the fact that the turbulence level generated by the upwind turbines is higher, which creates more momentum diffusion, leading to a faster recovery in the superposed area. Traditional interaction models are merely the mathematical expressions of the wake losses and simple superimposition. They cannot take account of the effect induced by the increased turbulence intensity. To overcome this limitation, this study multiplies the superimposed total loss—i.e., terms at the right of equal-sign in the traditional models—by a mixing coefficient less than 1 to characterize the faster velocity recovery caused by an increased turbulence intensity in the superposed wake area. The mixing coefficient can be described by an empirical formula as follows:

$$\begin{aligned} S_{\text{ave}} &= \frac{1}{N-1}\sum_{j}^{N-1} S_j^{j+1} \\ \alpha &= 1 - \frac{d_T}{S_{\text{ave}}} \end{aligned} \tag{2}$$

where S_j^{j+1} refers to the spacing between every two upwind adjacent turbines along the wind direction as shown in Figure 1; d_T is the rotor diameter; and S_{ave} is their average value.

The mixing coefficient can be calculated fast using Equations (2) which is only based on the spacing between upwind turbines. This expression meets two requirements: First, the coefficient is always less than 1 characterizing less total loss in the wake overlay area due to the higher turbulence level. Second, according to some experimental results [39,40], smaller upwind turbine spacing could lead to larger turbulence intensity in the superimposed area. So the mixing coefficient tends to be smaller with denser upwind turbines, which means faster velocity recovery.

With the exception of the EB model, all traditional interaction models are empirical descriptions without a clear physical basis. With this in mind, the mixing coefficient is combined with the EB model, and then the modified energy balance model (MEB for short) is proposed as follows:

$$u_0^2 - u_i^2 = \alpha \sum_{j}^{N} \left(u_j^2 - u_{ji}^2 \right) \tag{3}$$

Having a similar form to the EB model, the MEB model can be used with ease and no additional complexity is added to the flow field calculation. However, it should be noted that this modification is appropriate only if there are large perturbations in the wake superposition, such as an array of wind turbines along the wind direction. When considering the superposition of two wakes abreast, the relative spacing along the wind direction between the two turbines is 0, so the mixing coefficient which is based on this spacing cannot be solved and the MEB model does not apply. For this case the traditional EB model is recommended to predict the velocity deficit in the interference region.

2.3. Relative Position of Wind Turbines under Arbitrary Directions

When the wind direction changes, the wind turbine conducts yaw control to ensure that the rotor swept plane is always perpendicular to the inflow direction, which changes the relative positions of the turbines and the wake effect. Therefore, determining the relative position of wind turbines efficiently under arbitrary wind direction is a required process for predicting wake deficits [41]. We use a simple method to solve this problem in this research. As shown in Figure 2, north is regarded as 0° and has the established coordinate frame XOY. The coordinate origin can be chosen from any point in the wind farm, and the Y axis points to the 0° wind direction. T1 represents an arbitrary wind turbine with position (x, y). When the wind direction changes to β, the coordinates are rotated β degrees clockwise and a new coordinate system X′OY′ is obtained. The coordinates of the wind turbine in frame X′OY′ can be obtained by:

$$\begin{pmatrix} x' \\ y' \end{pmatrix} = \begin{pmatrix} \cos\beta & -\sin\beta \\ \sin\beta & \cos\beta \end{pmatrix} \begin{pmatrix} x \\ y \end{pmatrix} \tag{4}$$

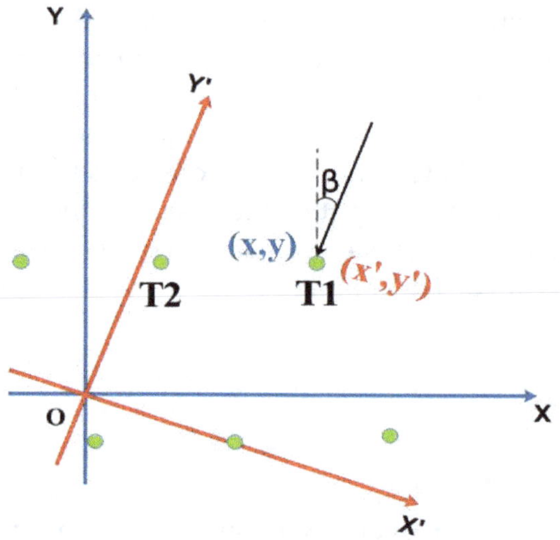

Figure 2. Schematic representation of the coordinate transformation. Green points represent wind turbines.

After the coordinate transformation, the relative positions of the wind turbines can be determined. Combining with the single wake model and the above-mentioned interaction wake model, the wake velocity distribution and power output can be obtained for varying wind directions in wind farm.

3. Benchmarking Study

3.1. Lillgrund Wind Farm

3.1.1. Introduction

The Lillgrund wind farm consists of 48 SWT-2.3-93 Simens wind turbines with a wind rotor diameter of 92.6 m and hub height of 65 m. Since the typical characteristic of this park is that the internal spacing between adjacent turbines is relatively small, the power loss due to the wake effect is bound to be very large. Figure 3 gives the layout of the Lillgrund offshore wind farm. As shown, the turbine spacing is 3.3D and 4.3D rotor diameters under the wind direction of 120° and 222°, respectively.

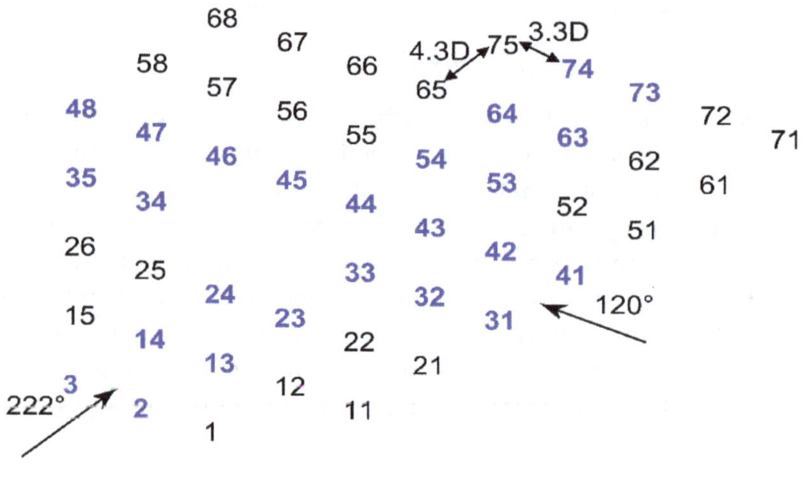

Figure 3. Layout of the Lillgrund offshore wind farm.

3.1.2. Results and Discussion

The co-axial arrangement of wind turbines can be observed in the above two directions. In order to test the accuracy of different models, the Jensen model is combined with three previous interaction wake models (LS, EB, SS) and with the proposed MEB model to calculate the power deficit along rows with and without missing turbines (in blue in Figure 3). The inflow mean velocity at hub height is 9 m/s in all cases. The wake expansion rate of the Jensen model is set to be 0.05 according to the recommended value for offshore conditions. The calculations are performed for various wind directions, consistent with those of the inflow sector of processed SCADA (Supervisory Control and Data Acquisition System) data [−2.5°,2.5°] [11], and are performed in 0.5° steps for both arrays in this research. The results for these cases are given in Figures 4 and 5. The power deficit of the wind turbine can be defined as $1 - P_i/P_0$, where P_i is the actual power of a specific turbine i and P_0 is the power of a turbine not disturbed by a wake.

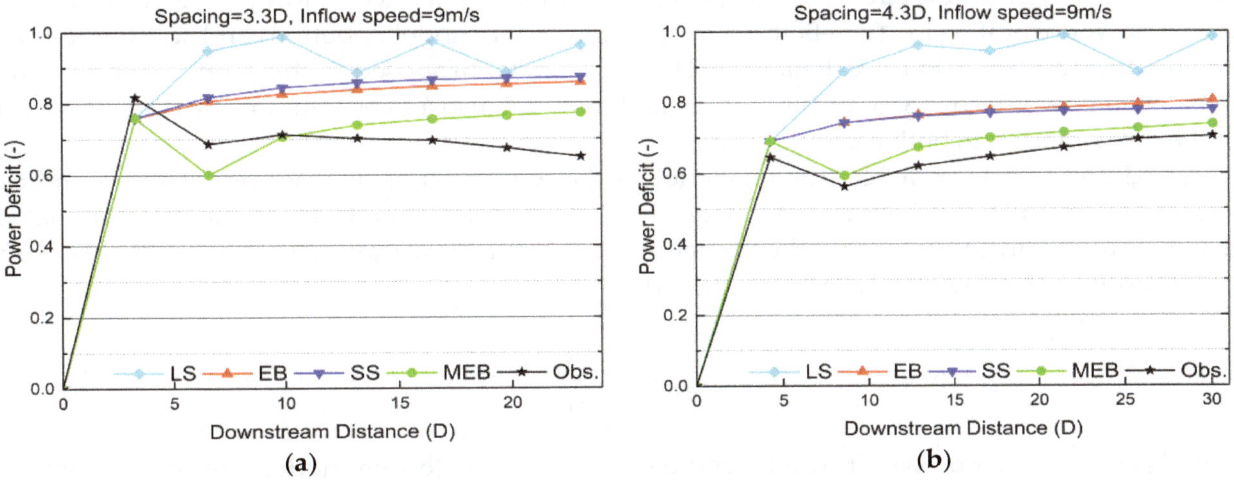

Figure 4. Lillgrund power deficit in a complete row with (**a**) 3.3D spacing and (**b**) 4.3D spacing.

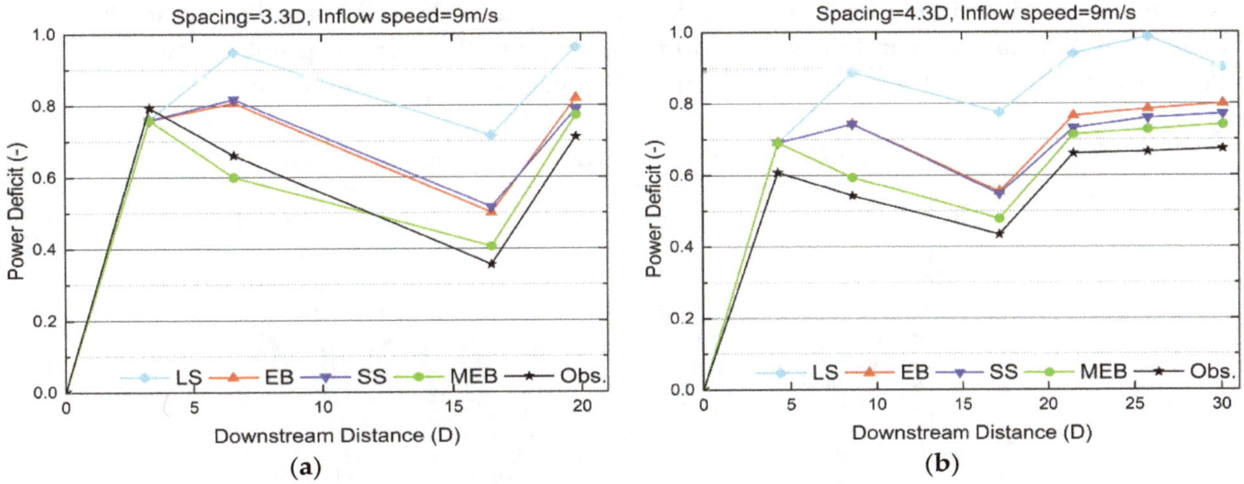

Figure 5. Lillgrund power deficit in a row with (**a**) 3.3D spacing with two missing turbines and (**b**) 4.3D spacing with one missing turbine.

Figure 4 shows the variation of power deficit for different turbine separation distances in the complete arrays along the two directions. In these two cases, the results from the EB model and SS model are very similar, while the LS model seriously over-predicts the deficit and presents a zig-zag pattern. This is because when the calculated inflow is less than the rated speed of the wind turbine, the turbine will stop operating, and no wake is produced in this study. Specifically, without a superposition effect, the power deficits of the second turbine predicted by different models are the same and reach up to about 82% and 65%, respectively, because of the small spacing. As for the third turbine, the power deficit significantly decreases compared with the former one. Neither of the three typical models could capture this trend, but show rather a continuous decrease of power. Only the proposed MEB model perfectly predicts the power recovery of the third turbine. For the case with 4.3D spacing, the MEB model simulation results of the fourth and subsequent turbines agree very well with observations. However, what should be noted is that none of the superposition models predict the trend of power recovery after the third turbine for the case of 3.3D spacing. Perhaps this is because of the error in the single wake model due to the overly close spacing of 3.3D. Katic [17] also admitted that the calculation of the near-wake zone will involve large errors using the Jensen wake model.

The input data and the characteristics of the cases shown in Figure 5 are more or less the same as in the previous ones, except that some turbines are missing in the studied rows. The absence of turbines causes wakes to further recover to some extent from the third turbine to the fourth. The traditional model results also show this pattern, simply because a smaller predicted inflow before the third turbine leads to a lower overall velocity deficit (the right hand of the model equation) and so a higher inflow before the fourth turbine in the calculation. The MEB model proposed in this study not only catches the sustained power recovery of turbines three and four, but also agrees with the measurements better than the other models in both cases.

In addition to power deficit, the efficiency of the wind farm is another performance indicator measuring the precision of the interaction wake model, which is defined as the ratio between the wind farm actual total output power and the power of the wind farm without considering the wake effect for each turbine. The efficiency can be expressed as:

$$eff = \frac{\sum_{i} P_i}{MP_0} \tag{5}$$

where M is the number of wind turbines in the wind farm. The Jensen model is integrated with four superposition models to calculate the park efficiency under all the wind directions with a span of 3°. In every direction, the inflow mean velocity is 9 m/s. Comparisons of the simulation results with field measurements are given in Figure 6. In general, all the four models could broadly predict the variation of efficiency with changing wind direction and the location of the extreme points. However, there still exist some biases between the model results and observations, especially for the LS model.

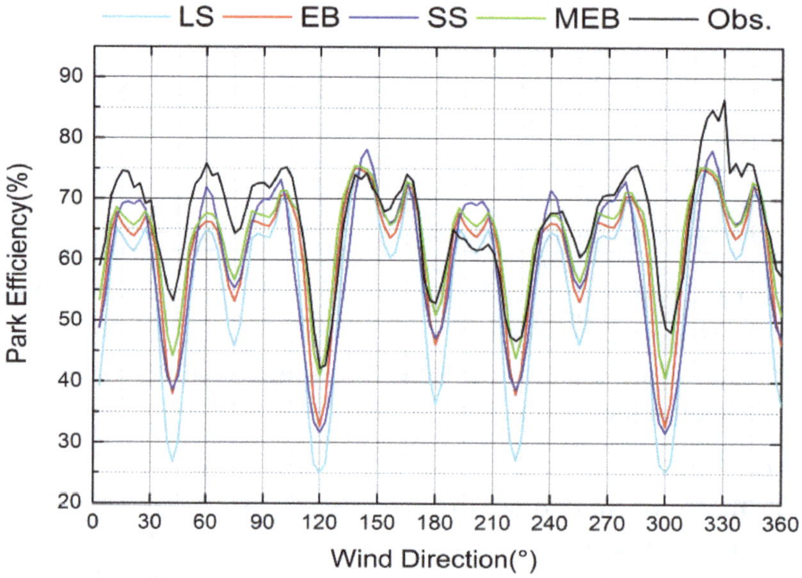

Figure 6. Park efficiency of the Lillgrund wind farm for the inflow sector 0–360° with 3° increments.

To conduct quantitative analysis, the root mean square error (RMSE) and the mean absolute percentage error (MAPE) are adopted to measure the difference between predictions and observations and to evaluate the model performance. Their definitions are as follows:

$$\text{RMSE} = \sqrt{\frac{1}{K}\sum_{i}^{K}(eff_{model,i} - eff_{obs,i})^2} \quad (6)$$

$$\text{MAPE} = \frac{1}{K}\sum_{i}^{K}\left|\frac{eff_{model,i} - eff_{obs,i}}{eff_{obs,i}}\right| \times 100\% \quad (7)$$

where K is the total number of simulated cases in different wind directions; $eff_{\text{model},i}$ and $eff_{\text{obs},i}$ refer to the efficiency from the model prediction and field measurement of the ith given wind direction. The RMSE and MAPE of different model simulation results are listed in Table 1, and the error of models under prevailing wind direction sectors (240°–270°) is also shown. It can be seen that under the inflow sector (0–360°), the park efficiency calculated by the MEB model reduces the RMSE by 3.79% and decreases the MAPE by 5.3% in comparison to the most commonly used SS model. Moreover, these values are reduced by 1.96% and 2.48%, respectively, in the dominant wind direction range.

Table 1. Simulation error of the park efficiency using different models.

Error	0°–360°				Prevailing Wind Directions			
	LS	EB	SS	MEB	LS	EB	SS	MEB
Root mean square error (RMSE) (%)	12.55	7.26	8.99	**5.20**	8.99	4.60	5.07	**3.11**
Mean absolute percentage error (MAPE) (%)	17.06	9.24	11.78	**6.48**	12.84	6.20	6.22	**3.74**

3.2. Horns Rev I Wind Farm

3.2.1. Introduction

The Horns Rev I offshore wind farm is located about 14 km off the west coast of Denmark. It has been extensively studied in evaluating its wake effect and power production, with plenty of meteorological and SCADA data available for comparison. The wind farm consists of 80 Vestas V80 2 MW turbines with a rotor diameter of 80 m and hub height of 70 m. The wind farm's 10 columns and 8 rows are aligned in a parallelogram shape, as seen in Figure 7. The spacing distance between turbines in both lines and columns is 7D, and those under the wind direction of 221° and 312° are 9.4D and 10.4D, respectively.

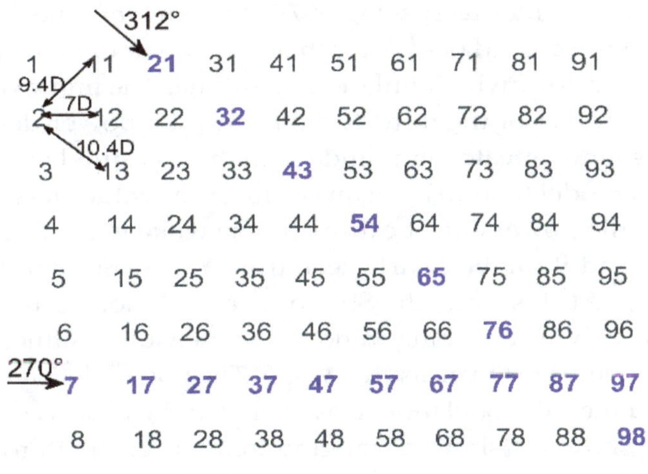

Figure 7. Layout of the Horns Rev I offshore wind farm.

3.2.2. Results and Discussion

The Jensen model is also chosen in this benchmark with $k = 0.05$. Four different interaction models are applied to predict the power deficit of each turbine along rows with an internal spacing of 7D and 10.4D, respectively (in blue in Figure 7). The comparisons between simulation results obtained from different models and extracted SCADA data [22] with an inflow velocity of 7.5 m/s are given in Figures 8 and 9. The cases in Figure 8 correspond to narrow wind direction sizes [−2°,2°], and the cases in Figure 9 correspond to a broader direction interval [−15°,15°], which is considered to decrease the error of wind direction uncertainty as much as possible. All of the runs were performed at a 1° step.

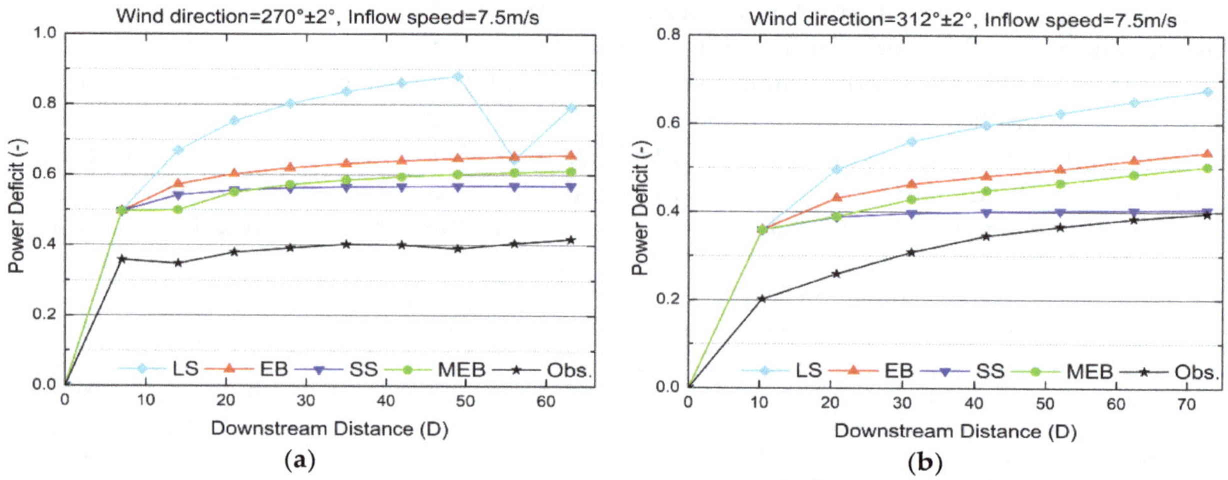

Figure 8. Horns Rev I power deficit in (**a**) a row with 7D spacing and 270 ± 2°, and (**b**) a row with 10.4D spacing and 312 ± 2°.

When the wind directions are 270° and 312°, the spacing between two turbines undergoes a minimum value of 7D and a maximum value of 10.4D, respectively. Observing the power deficit of the second turbine in Figure 8, there is a large discrepancy when comparing numerical simulations using the Jensen model with wind farm production data. In addition to the reasons for the calculation accuracy of the wake model itself, another important reason is the large wind direction uncertainty included in the datasets while the numerical simulations are carried out for narrow wind direction sectors. Thus, when comparing the performance of different wake interaction models, the primary concern is whether the simulation results can match the changing trend of the measured data. It can be seen from Figure 8a that, with an internal spacing of 7D, the power of the third turbine has recovered more or less compared with the second one. After that, the power deficit tends to gradually increase with a very small range further downwind until the last turbine. The improved MEB model simulation catches the recovery of the power of the third turbine, and the power deficit of the latter turbines increases gradually with a slightly greater amplitude than the measured value. On the other hand, the simulation results of the SS model tend to be in an equilibrium value starting from the third turbine. When the spacing is 10.4D, the power deficit caused by the wake is relatively small due to the larger interval between turbines, and it can be clearly seen from Figure 8b that the deficit along the wind direction increases in steps. At this time, the SS model calculation result, which reach a constant, are obviously inconsistent with the changing trend of the measured values, although the calculated downwind turbine power is closer to the measured value. The power deficit variation curves simulated by the new MEB model and the EB model more closely match the observed values. In fact, the results of the MEB and EB model show the similar changing trend after the third turbine. This is because the MEB model just multiplies a coefficient to the original EB model. In essence, the mathematical structure of these two models is consistent.

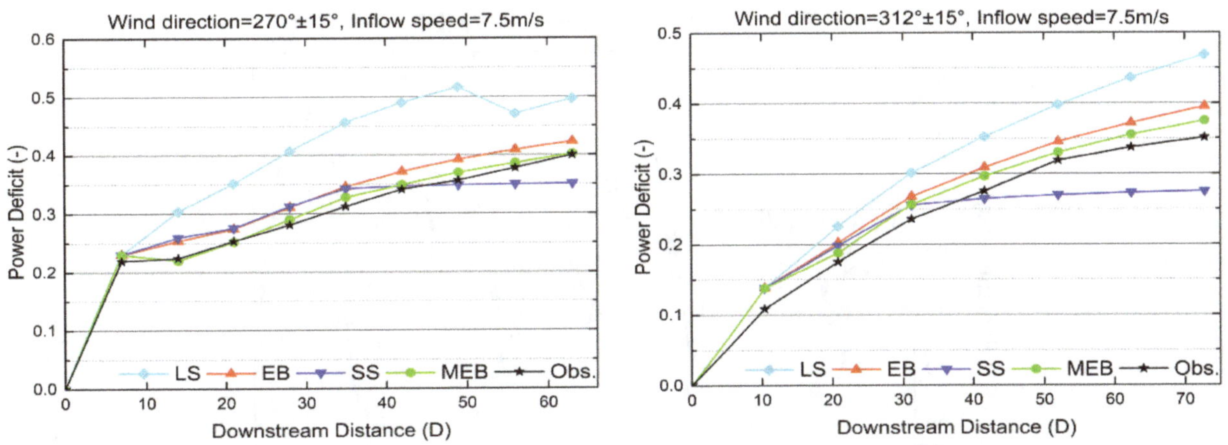

(a) (b)

Figure 9. Horns Rev I power deficit in (**a**) a row with 7D spacing and 270 ± 15°, and (**b**) a row with 10.4D spacing and 312 ± 15°.

With a broader sector [−15°,15°], the comparisons in Figure 9 show that the Jensen model coupled with interaction models matches much better with all the measured points due to the smaller direction uncertainty error, especially for the new MEB and EB models. The LS model still seriously overestimates the wake loss in the superposition region, while the SS model fails to model the deep array effect as it quickly reaches an equilibrium state. Note that, with the internal spacing of 7D, as seen in Figure 9a, the power of the third turbine restores compared to the second one simulated by the MEB model; nevertheless, the measurements show a slight decrease. This departure may be due to the different frequency of wind directions within the sector [−15°,15°]. The authors in Reference [24] pointed out that a normal distribution fits well with the measured wind direction variations within an averaging period at Horns Rev I around the wind direction of 270°. That is to say, there are a greater number of available measurements for data analysis near 270°, and the number of measurements which are close to the wind direction interval boundary are relatively small. Thus, data for directions near 270°, of which the wake losses of downwind turbines are much heavier, contribute more to the final processing results. However, simulations in this study are performed for a wide range of wind directions (±15°) with a resolution of 1°. These single simulated powers are then averaged within the range. Therefore, the effect of the cases under directions further away from 270° is overestimated, which results in less deficit in the turbines. In addition, since the turbines are far apart from each other, the upwind wake has restored to a certain degree when it develops to the downwind turbine. The turbulence intensity in the far wake superimposed area tends to be similar to the single wake, which leads to the calculation results of the MEB model and EB being relatively close, as shown in Figure 9b. The farther the spacing, the closer the mixing coefficient is to 1, and so the closer the calculation result of the new model is to the EB model.

Similar models as in the previous benchmark are considered to calculate the park efficiency, and the inflow sector is taken as 0–360° with a span of 5°. In every direction, the inflow mean velocity is 8 m/s. The comparison of the simulation results with measurements is shown in Figure 10. There are significant differences in some wind directions where the wake has a large impact, such as for 0° and 270°. The RMSE and MAPE of different model simulation results are listed in Table 2. It can be seen that under the inflow sector (0–360°), compared with the SS model, the MEB model reduces the wind farm efficiency RMSE by 1.39% and MAPE by 1.58%. Under the dominant wind direction, the wind farm efficiency RMSE decreased by 1.79%, and the MAPE decreased by 2.52%. The results show that for wind farms such as Horns Rev I, which conducts reasonable layout optimization and has a relatively large spacing distance between turbines, the performance of the MEB model has a certain degree of improvement compared with the EB and SS model. However, this improvement is not as good for the Lillgrund wind farm with smaller intervals between turbines.

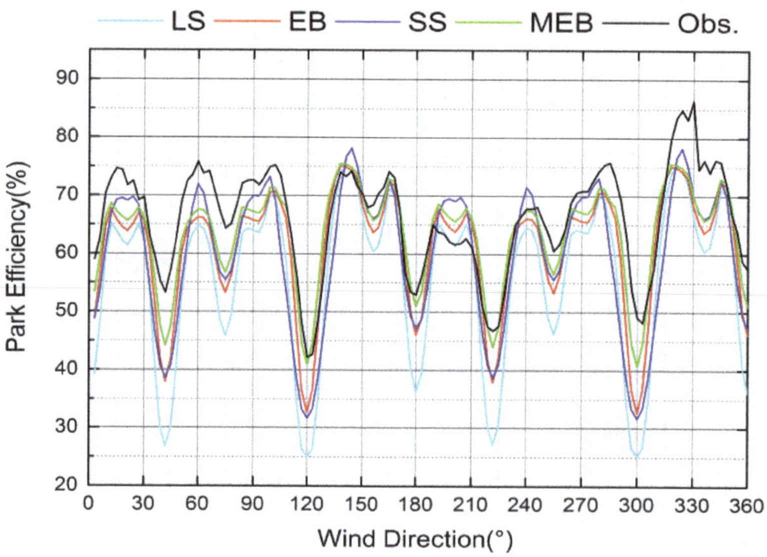

Figure 10. Park efficiency of the Horns Rev I wind farm for inflow sector 0–360° with a 5° increment.

Table 2. Simulation error of park efficiency using different models.

Error	0°–360°				Prevailing Wind Directions			
	LS	EB	SS	MEB	LS	EB	SS	MEB
RMSE (%)	10.04	6.08	6.59	**5.20**	11.65	6.76	7.52	**5.73**
MAPE (%)	8.94	5.92	6.78	**5.20**	10.73	6.70	8.33	**5.81**

4. Conclusions

The conventional interaction wake models separately compute the velocity deficit from each upwind turbine and then linearly superpose them without considering the faster wake recovery caused by a higher turbulence intensity in the superposed wake region, when compared to a single wake area. This omission might contribute to lower model precision. To compensate for this, a new interaction wake model based on the improvement of the energy balance model is proposed to predict the velocity distribution and wind farm power in this paper. A mixing coefficient is introduced to characterize the above phenomenon in the energy balance model, and an empirical expression is also given to calculate this parameter. In order efficiently to provide support for simulating the wake speed distribution under varying wind directions, a simple method to determine the relative position between wind turbines is also given.

A validation of the proposed interaction model is performed using the observations in the Lillgrund and Horns Rev I offshore wind farms, mainly focusing on the power loss of every single turbine and the efficiency of the wind farms. Compared to other commonly used superposition models in engineering, the presented model can better capture the power changing tendency of wind turbines arranged in an array with different spacing, regardless of whether there are missing turbines or not. Especially for turbines in a row in the Lillgrund offshore wind farm with smaller turbine spacing, an obvious power recovery appears at the third turbine, which is only predicted by the presented model. As for the wind farm efficiency, all the interaction models can obtain the changing trend of wind farm output with varying wind direction and determine the location of limit points, while the proposed model in particular is observed to be in better agreement with the measurements.

The proposed interaction model has similar simplicity but higher precision compared with the commonly used interaction models, without increasing the complexity and computational cost of the simulation because the former only multiplies the traditional one by a mixing coefficient. This coefficient is used to characterize the phenomenon that the higher turbulence level and shear stress profile generated by upwind turbines in the superposed wake area leads to a faster wake recovery, which is obtained by an empirical expression in this paper. Although the modification yields reasonable predictions, a more physical and more accurate model describing the effect of turbulence intensity on the velocity recovery of multiple wakes will be the focus of future research.

Author Contributions: The paper was a collaborative effort among the authors. Z.S. performed the modeling, analyzed the data, and wrote the paper. Y.L. supervised the related research work. L.L. and S.H. contributed analysis tools. Y.W. polished the language.

Acknowledgments: Thanks to the support by the National Key Research and Development Program of China (No. 2017YFE0109000), and Open Fund of Operation and Control of Renewable Energy & Storage Systems (No. 1810-00895).

References

1. Barthelmie, R.J.; Rathmann, O.; Frandsen, S.T.; Hansen, K.; Politis, E.; Prospathopoulos, J.; Rados, K.; Cabezón, D.; Schlez, W.; Phillips, J.; et al. Modelling and measuring flow and wind turbine wakes in large wind farms offshore. *Wind Energy* **2009**, *12*, 431–444. [CrossRef]
2. Wang, Y.; Liu, Y.; Li, L.; Infield, D.; Han, S. Short-term wind power forecasting based on clustering pre-calculated CFD method. *Energies* **2018**, *11*, 854. [CrossRef]
3. Tian, J.; Zhou, D.; Su, C.; Soltani, M.; Chen, Z.; Blaabjerg, F. Wind turbine power curve design for optimal power generation in wind farms considering wake effect. *Energies* **2017**, *10*, 395. [CrossRef]
4. Herbert-Acero, J.F.; Probst, O.; Réthoré, P.E.; Larsen, G.C.; Castillo-Villar, K.K. A review of methodological approaches for the design and optimization of wind farms. *Energies* **2014**, *7*, 6930–7016. [CrossRef]
5. Peña, A.; Rathmann, O. Atmospheric stability-dependent infinite wind-farm models and the wake-decay coefficient. *Wind Energy* **2014**, *17*, 1269–1285. [CrossRef]
6. Ferrer, E.; Browne, O.M.F.; Valero, E. Sensitivity Analysis to Control the Far-Wake Unsteadiness behind Turbines. *Energies* **2017**, *10*, 1599. [CrossRef]
7. Munters, W.; Meyers, J. Dynamic strategies for yaw and induction control of wind farms based on large-eddy simulation and optimization. *Energies* **2018**, *11*, 177. [CrossRef]
8. Renkema, D.J. Validation of Wind Turbine Wake Models. Master's Thesis, Delft University of Technology, Delft, Holland, 2007.
9. Niayifar, A.; Porté-Agel, F. Analytical modeling of wind farms: A new approach for power prediction. *Energies* **2016**, *9*, 741. [CrossRef]
10. Jeon, S.; Kim, B.; Huh, J. Comparison and verification of wake models in an onshore wind farm considering single wake condition of the 2MW wind turbine. *Energy* **2015**, *93*, 1769–1777. [CrossRef]
11. Göçmen, T.; Laan, P.V.D.; Réthoré, P.E.; Diaz, A.P.; Larsen, G.C.; Ott, S. Wind turbine wake models developed at the technical university of Denmark: A review. *Renew. Sust. Energy Rev.* **2016**, *60*, 752–769. [CrossRef]
12. Hasager, C.B.; Rasmussen, L.; Peña, A.; Jensen, L.E.; Réthoré, P.-E. Wind farm wake: The Horns Rev photo case. *Energies* **2013**, *6*, 696–716. [CrossRef]
13. Van Leuven, J. The Energetic Effectiveness of a Cluster of Wind Turbines. Master's Thesis, Universite Catholique de Louvain, Louvain-la-Neuve, Belgium, 1992.
14. Lissaman, P.B.S. Energy effectiveness of arbitrary arrays of wind turbines. *J. Energy* **1979**, *3*, 323–328. [CrossRef]
15. Crespo, A.; Hernández, J.; Frandsen, S. Survey of modelling methods for wind turbine wakes and wind farms. *Wind Energy* **1999**, *2*, 1–24. [CrossRef]
16. Crespo, A.; Manuel, F.; Hernández, J. Numerical modelling of wind turbine wakes. In Proceedings of the 1990 European Community Wind Energy Conf., Madrid, Spain, 8–12 March 1990; pp. 66–170.
17. Katic, I.; Højstrup, J.; Jensen, N.O. A simple model for cluster efficiency. In Proceedings of the European Wind Energy Association Conference & Exhibition, Rome, Italy, 7–9 October 1986; pp. 407–410.
18. Rados, K.G.; Voutsinas, S.G.; Zervos, A. Wake effects in wind parks. A new modelling approach. In *Scientific Proceedings*; EWEC: Travemunde, Germany, 1993; pp. 444–447.
19. Erik, D. Evaluation of the Software Program Windfarm and Comparisons with Measured Data from Alsvik. Master's Thesis, Royal Institute of Technology, Stockholm, Sweden, 2000.

20. Tian, L.L.; Zhu, W.J.; Shen, W.Z.; Yong, Y.L.; Zhao, N. Prediction of multi-wake problems using an improved Jensen wake model. *Renew. Energy* **2017**, *102*, 457–469. [CrossRef]
21. Jensen, N.O. *A Note on Wind Generator Interaction*; Technical Report Risoe-M-2411(EN); Risø National Laboratory: Roskilde, Denmark, 1983.
22. VanLuvanee, D.R. Investigation of Observed and Modeled Wake Effects at Horns Rev Using WindPRO. Master's Thesis, Technical University of Denmark, Copenhagen, Denmark, 2006.
23. Peña, A.; Réthoré, P.E.; Hasager, C.B.; Hansen, K.S. *Results of Wake Simulations at the Horns Rev I and Lillgrund Wind Farms Using the Modified Park Model*; DTU Wind Energy-E-Report-0026(EN); Risø Campus: Roskilde, Denmark, 2013.
24. Gaumond, M.; Rethoré, P.E.; Ott, S.; Peña, A.; Bechmann, A.; Hansen, K.S. Evaluation of the wind direction uncertainty and its impact on wake modelling at the Horns Rev offshore wind farm. *Wind Energy* **2014**, *17*, 1169–1178. [CrossRef]
25. Gao, X.X.; Yang, H.X.; Lu, L. Optimization of wind turbine layout position in a wind farm using a newly-developed two-dimensional wake model. *Appl. Energy* **2016**, *174*, 192–200. [CrossRef]
26. Mortensen, N.G.; Heathfield, D.N.; Myllerup, L.; Landberg, L.; Rathmann, O. *Getting Started with WAsP 9*; Technical Report Risø-I-2571(EN); Risø National Laboratory: Roskilde, Denmark, 2007.
27. Thørgersen, M.; Sørensen, T.; Nielsen, P.; Grötzner, A.; Chun, S. *WindPRO/PARK: Introduction to Wind Turbine Wake Modelling and Wake Generated Turbulence*; EMD International A/S: Aalborg, Denmark, 2005.
28. Kuo, J.Y.J.; Romero, D.A.; Amon, C.H. A mechanistic semi-empirical wake interaction model for wind farm layout optimization. *Energy* **2015**, *93*, 2157–2165. [CrossRef]
29. Smith, D.; Taylor, G.J. Further analysis of turbine wake development and interaction data. In Proceedings of the 13th BWEA Wind Energy Conf., Swansea, Wales, 20–26 October 1991; pp. 325–331.
30. Frandsen, S. On the wind speed reduction in the center of large clusters of wind turbines. *J Wind Eng. Ind. Aerod.* **1992**, *39*, 251–265. [CrossRef]
31. Peña, A.; Réthoré, P.-E.; van der Laan, M.P. On the application of the Jensen wake model using a turbulence-dependent wake decay coefficient: The Sexbierum case. *Wind Energy* **2016**, *19*, 763–776. [CrossRef]
32. Tobin, N.; Chamorro, L.P. Modulation of turbulence scales passing through the rotor of a wind turbine. *J. Turbul.* **2018**, 1–11. [CrossRef]
33. Burton, T.; Sharpe, D.; Jenkins, N.; Bossanyi, E. *Wind Energy Handbook*; John Wiley & Sons Ltd.: Chichester, UK, 2001; pp. 35–37. ISBN 0-471-48997-2.
34. Makridis, A. Modelling of Wind Turbine Wakes in Complex Terrain Using Computational Fluid Dynamics. Master's Thesis, the University of Edinburgh, Edinburgh, UK, 2012.
35. Crespo, A.; Herna´ndez, J. Turbulence characteristics in wind-turbine wakes. *J. Wind Eng. Ind. Aerod.* **1996**, *61*, 71–85. [CrossRef]
36. Duckworth, A.; Barthelmie, R.J. Investigation and validation of wind turbine wake models. *Wind Eng.* **2008**, *32*, 459–475. [CrossRef]
37. Veisi, A.A.; Shafiei Mayam, M.H. Effects of blade rotation direction in the wake region of two in-line turbines using Large Eddy Simulation. *Appl. Energy* **2017**, *197*, 375–392. [CrossRef]
38. Lee, S.; Vorobieff, P.; Poroseva, S. Interaction of Wind Turbine Wakes under Various Atmospheric Conditions. *Energies* **2018**, *11*, 1442. [CrossRef]
39. Liu, H.; Hayat, I.; Jin, Y.; Chamorro, L.P. On the evolution of the integral time scale within wind farms. *Energies* **2018**, *11*, 93. [CrossRef]
40. Chamorro, L.P.; Porté-Agel, F. Turbulent flow inside and above a wind farm: A wind-tunnel study. *Energies* **2011**, *4*, 1916–1936. [CrossRef]
41. Gu, B.; Liu, Y.; Yan, J.; Li, L.; Kang, S. A wind farm optimal control algorithm based on wake fast-calculation model. *J. Sol. Energy Eng.* **2016**, *138*, 024501:1–024501:6. [CrossRef]

13

Study of Floating Wind Turbine with Modified Tension Leg Platform Placed in Regular Waves

Juhun Song and Hee-Chang Lim *

School of Mechanical Engineering, Pusan National University, 2, Busandaehak-ro 63beon-gil, Geumjeong-gu, Busan 46241, Korea; ralph366@nate.com

* Correspondence: hclim@pusan.ac.kr

Abstract: In this study, the typical ocean environment was simulated with the aim to investigate the dynamic response under various environmental conditions of a Tension Leg Platform (TLP) type floating offshore wind turbine system. By applying Froude scaling, a scale model with a scale of 1:200 was designed and model experiments were carried out in a lab-scale wave flume that generated regular periodic waves by means of a piston-type wave generator while a wave absorber dissipated wave energy on the other side of the channel. The model was designed and manufactured based on the standard prototype of the National Renewable Energy Laboratory (NREL) 5 MW offshore wind turbine. In the first half of the study, the motion and structural responses for operational wave conditions of the North Sea near Scotland were considered to investigate the performance of a traditional TLP floating wind turbine compared with that of a newly designed TLP with added mooring lines. The new mooring lines were attached with the objective of increasing the horizontal stiffness of the system and thereby reducing the dominant motion of the TLP platform (i.e., the surge motion). The results of surge translational motions were obtained both in the frequency domain, using the response amplitude operator (RAO), and in the time domain, using the omega arithmetic method for the relative velocity. The results obtained show that our suggested concept improves the stability of the platform and reduces the overall motion of the system in all degrees-of-freedom. Moreover, the modified design was verified to enable operation in extreme wave conditions based on real data for a 100-year return period of the Northern Sea of California. The loads applied by the waves on the structure were also measured experimentally using modified Morison equation—the formula most frequently used to estimate wave-induced forces on offshore floating structures. The corresponding results obtained show that the wave loads applied on the new design TLP had less amplitude than the initial model and confirmed the significant contribution of the mooring lines in improving the performance of the system.

Keywords: floating offshore wind turbine; modified Morison equation; omega arithmetic method; tension leg platform; hydrodynamic motion response

---

1. Introduction

Offshore wind turbines generate more electricity than their onshore counterparts owing to higher wind speeds occasioned by the low surface roughness of the ocean. Due to increased interest in the design of offshore wind farms, the foundation of traditional fixed-bottom offshore wind turbines has been changing to the prolific floating type, which may be more feasible and less expensive for deep water. The diversity in the specifications of each floating platform type and the fact that wind turbines encounter a variety of wind, current, and wave loading at sea have resulted in active research into the dynamic response of offshore floating wind turbine support systems under various conditions with a view to ensuring security and reliability.

Jonkman et al. [1] provided the characteristics and specifications of the National Renewable Energy Laboratory (NREL) 5 MW offshore wind turbine, which is now considered the baseline wind turbine. This model has been mounted on various types of floating platforms and studies conducted on each structure. Wayman and Scalavonous [2] conducted comparative static and dynamic analyses of a Tension Leg Platform (TLP) FOWT and a shallow drafted barge FOWT to explore the effects of water depth, wind speed, and combined wind–wave conditions on system performance. In their study, they evaluated the response amplitude operator (RAO) standard deviation of the system in order to define the natural frequency at each degree of freedom. Lee [3] investigated the dynamic response of a spar buoy wind turbine by defining the nacelle acceleration, dynamic tensions on catenaries, and maximum tension acting on the anchors based on an extreme sea state condition for a 100-year return period. Lee showed that as water depth decreases the dynamic performance of the system becomes more challenging. Naqvi [4] performed experiments using a 1:100 scale TLP and spar buoy derived from the NREL prototype to study the motion and tension in the cables in response to regular waves and operational wind speed.

Hong et al. [5] utilized a 1:100 spar buoy to analyze the effects of center of gravity, mooring line tension, and fairlead location on the translational and rotational deviations of the system in response to regular waves. Most of the previous studies in this area were concentrated on analyzing the behavior of the platform under various environmental conditions. However, several studies were conducted in which either the best features of the platforms were combined or a new design was developed in order to increase the performance of the structure and minimize the drawbacks of each platform in a manner that enhanced the structural stability of the system to satisfactorily withstand all the environmental loads. Withee [6] studied the coupled dynamic analysis of a floating wind turbine system that combined the best features of both the TLP and the spar buoy with the objective of dampening the pitch and roll and reducing the loads on the structure. In the study, all modes of motion and the tether forces were investigated experimentally and numerically for a range of operational as well as severe wave and wind conditions. Murray et al. [7] proposed a 1:92 model of an extended TLP for deployment at a water depth of 1500 m and investigated its corresponding extreme responses, tether tensions, and platform deviations.

Karimirad and Moan [8] subsequently analyzed a spar-type offshore wind turbine with two distinct mooring systems under operating and survival conditions and focused on their stochastic dynamic motion and structural responses. The first concept was a catenary-moored spar wind turbine that used a taut system instead of the catenary mooring lines normally used for the spar buoy, while the second design was a spar with a tension leg mooring. The latter design consisted of a spar platform mounted in a single pre-tensioned mooring line. Experiments were carried out on the effect of hydrodynamic damping and the extreme structural responses under intense environmental conditions to examine the wind–wave-induced and wave-induced responses on the two mooring model concepts. Nianxin et al. [9] carried out experiments on a 1:60 TLP model to investigate the performance of the floating wind turbine in a coupled wind–wave environment. The model was derived from another prototype of TLP and additional catenaries were attached to it. The platform motion in surge, sway, and heave, as well as the tether forces, were investigated. Martinez et al. [10] studied the feasibility of a semi-floating spar buoy wind turbine attached to mooring lines strongly anchored to the bed with a spherical joint, with the objective of reducing the oscillations of the system and thereby reducing the fatigue and maximum loads compared to the spar buoy turbine. Regarding the data modeling of sea height, there have been a series of measurements taken at the Northern Sea near Scotland (see Hadjihosseini et al. [11–13]). In addition, the approach to analyze the evolution of stochastic properties associated to the operation of wind turbines has been made, particularly in what concerns loads on the wind turbine (see Lind et al. [14,15]). Regarding the aeroelastic behavior, such as the case when the turbine is in power production the dynamic response, the behavior of wind turbine is completely different due to a significant increase of aerodynamic damping (see Marino et al. [16,17]). Regarding the improvements of Morison equation for steep waves and for diffraction regimes, there have been some reference works (see Nielsen et al. [18]; Paulsen et al. [19]; Bredmose et al. [20]; Patel and Witz [21]).

In this study, the design of the traditional TLP floating platform derived from the NREL prototype and a conceptual TLP design with additional mooring lines were experimentally investigated. The additional mooring lines reinforce the horizontal stiffness of the system and thereby diminish the dominant motion, which is obviously the surge motion, to enable safe performance in deep water. In the study, a series of experiments were conducted using a 1:200 scale model to determine the impact of waves on the performance of the model. The range of operational and extreme waves used in the experiments was based on real data for a 100-year return period of the Northern Sea of California to ensure the feasibility of installing the new system in the chosen site. Another objective of this study was determination of the wave loads on the platform; this was achieved experimentally using the Morison equation.

Previous studies predominantly focused on measurement of the wave loads acting on a fixed-bottom offshore wind turbine. Peeringa [22] showed that an acceptable prediction of the wave loads for non-breaking waves is achievable by using the Morison equation and the stream function wave model. However, the hydrodynamic forces are underestimated by the applied methods in the case of breaking and post-breaking waves. Journee and Massie [23] and Sarpkay and Isaacson [24] proposed various methods to define the drag and inertia coefficients for better estimation of the solution to the Morison equation. Burrows et al. [25] also defined several techniques, such as least square, cross-spectral, and the methods of moments, to identify these coefficients, and subsequently measured the wave loads using the Morison equation to prove the accuracy of predicting random forces using the equation. The measured data were later shown to sufficiently match the predicted values, which confirmed the suitability of the equation for measuring wave loads. Henderson [26] subsequently enumerated several methods to determine wave load models and structural dynamics models depending not only on the turbine selected but also the support structure. The studies conducted revealed that the selection of a relevant approach for defining the resulting loads plays a fundamental role in the cost of the structure and its ability to resist environmental loads. Dean and Dalrymple [27] presented several methods to define appropriate wave kinematics theoretically or experimentally.

Considering the results of the above studies, in this study, the Morison equation was applied to measure the wave loading on the offshore wind turbine's TLP platform to verify its accuracy in predicting the wave loading on an offshore floating structure. The remainder of this paper is organized as follows: Section 2 describes the experimental setup used to perform the experiments on the scale model. An enumeration of the main properties of the model along with the laws that permit scaling it down and a sketch of the new design are also given, and the difficulties and errors dealt with while performing the experiments outlined. Section 3 provides a theoretical outline of all the governing equations used in our study in order to facilitate a thorough understanding of the physics underlying the phenomena under investigation. Section 4 analyzes, discusses, and interprets the data collected in the experiments. The data are associated with both the dynamic performance of the two concepts in response to regular and extreme waves and also the wave loading on the structure. Finally, Section 5 concludes by summarizing all the results obtained and outlines plans for further work.

2. Design of the Laboratory Experiment

Considering the dimensional restrictions imposed by the wave flume, the scale of the model was set at $\lambda_m/\lambda_p = 1{:}200$. The subscript "m" denotes the scale of the model and "p" the scale of the prototype (see Figure 1). Froude scaling was applied in order to establish scaling factors between the model and the full-scale prototype (Jonkman et al. [1]). The latter was derived from the NREL 5 MW wind turbine. In the flume experiment, the piston-type wave maker was used to generate a propagating wave more effectively because the horizontal water particle velocities of the wave and paddle are nearly constant over the water column. In the flume, a passive wave absorbers comprised a hard filling material was installed in the channel. Based on the wave theory and reflectance, two wave gauges (KENEK, Wave height Meter CH-606) were used to take proper samples for a period, approximately 30 s (see more details in the papers of Lim [28] and Zhu and Lim [29]). To make reliable result, the experiment considered the effect of evanescent waves in the wave flume, which could contaminate test areas leading unexpected data. To avoid this effect, the model was placed around three times away from the wave maker. However, due to the size limitation of the wave flume, the influence of evanescent waves was not severe, but still remained (see Lim [28] and Zhu and Lim [29]).

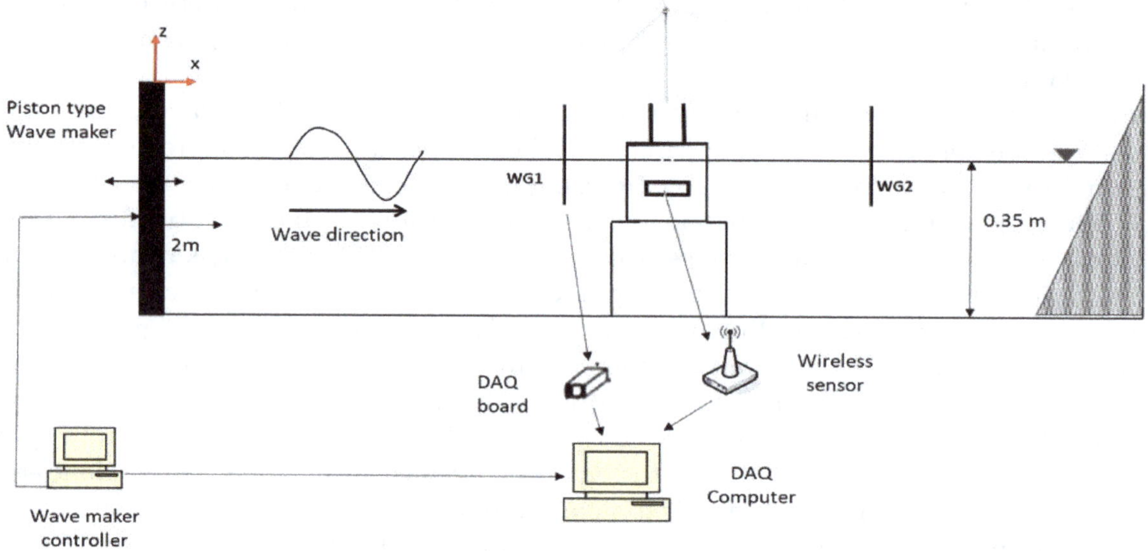

Figure 1. Measurement setup sketch.

The model was made from acrylic plastic; the design parameters for all the various dimensions, which were scaled according to Froude law, are listed in Table 1. The shape was identical to that of the prototype and the model surface was smooth in order to prevent deterioration of the surface over time and model deformations. However, scaling all the parameters precisely was difficult to achieve because of the large scale chosen for our study and the fact that we were primarily interested in the hydrodynamic motion of the platform. For these reasons, the aerodynamic properties of the wind turbine were not scaled, which means that the blades, rotor, and nacelle were not taken into consideration and only the height of the tower along with the overall mass of both the tower and the rotor nacelle assembly were matched with the real wind turbine parameters for the 5 MW wind turbine specifications (see Jonkman et al. [1]). In addition, for the TLP model, as proven previously, the larger is the length scale, the more dominant will be the scale effects, and the more complicated it will be to obtain identical ratios between the model and its prototype. Thus, the model was scaled more to the perfect weight than the perfect length in such a manner as not to impair the scaling of the moment of inertia of the system, which is a parameter that is directly proportional to both the length and the weight.

The modified concept developed in this study was actualized by connecting additional mooring catenaries to the system (see Figure 2). The catenaries were directly attached to the platform in the bottom surface in order to increase the horizontal stiffness and give more stability to the system. As shown in Figure 2, the rigid body, comprising the wind turbine and the floating support, undergoes oscillatory translational and rotational motions as a result of wind and wave loadings. Surge and sway (ξ_1 and ξ_2) describe horizontal motions along the x and y axes, respectively. The horizontal axes of the system were aligned with the four spokes to which the floater tethers were connected. Heave (ξ_3) represents the vertical motion of the floater. The oscillatory angular motions are referred to as the roll, pitch, and yaw. The surge, sway, and yaw describe the three rigid-body motions. The tether lengths were assumed to remain constant during rigid-body motion.

Table 1. Main design parameters of the Tension-Leg Platform (TLP) model.

$\lambda = \frac{\text{fullscale}}{\text{model}} = 200$			
TLP Properties	**Full Scale**	**Scale Ratio**	**Model**
Diameter	22 m	λ	11 cm
Cylinder height	21.5 m	λ	10.75 cm
Concrete ballast height	4.5 m	λ	2.25 cm
Installed draft	20.01 m	λ	10 cm
Deck clearance	1.49 m	λ	1.5 cm
Steel thickness	0.01 m	$\sqrt{\lambda}$	0.2 cm
Concrete mass	4.345×10^6 kg	λ^3	455 g
Turbine mass	0.698×10^6 kg	λ^3	110 g
Buoyant mass	7.797×10^6 kg	λ^3	780 g
Total mass	5.249×10^6 kg	λ^3	715 g
Center of gravity	−9.4 m	λ	−4.5 cm
Center of buoyancy	−10.01 m	λ	−5 cm
Surge resonant period	15 s	$\sqrt{\lambda}$	1.1 s
Pitch resonant period	0.5 s	$\sqrt{\lambda}$	0.2 s
No. of tethers	4	-	4

Figure 3 illustrates the configuration of the modified TLP with the added inclined lines. Four small holes were drilled in the bottom surface of the cylinder and filled with glue. The steel wire ropes, containing a soft spring at each line, were then introduced into the holes and glue applied to make it harden. The combination of spring and steel wire resulted in a non-homogeneous model for modeling the axial stiffness and neglecting the bending stiffness (Bartrop [30]).

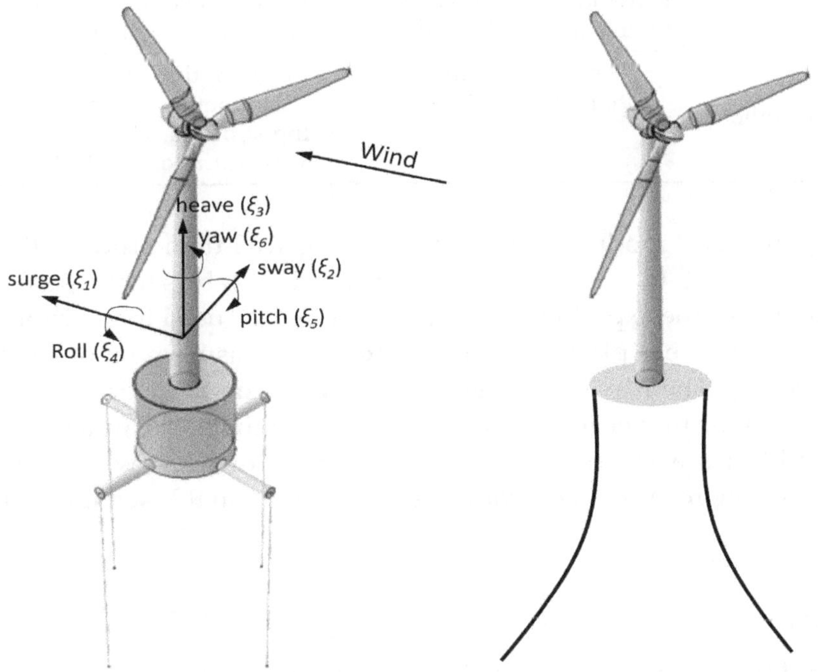

Figure 2. Sketch of TLP system (**left**); and catenary mooring system (**right**).

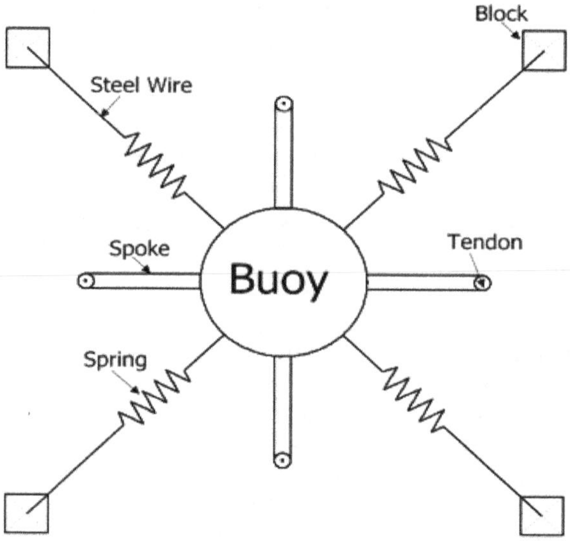

Figure 3. Modified tension leg configuration.

The scale model tests were then carried out in the wave flume of our laboratory. (see Table 2) On one side, the wave tank was fitted with a piston-type wave maker that generated regular monochromatic waves with different heights and periods. On the downstream section of the flume, a wave absorber was installed to dissipate wave energy and to avoid reflected waves inside the tank.

Table 2. Spoke and mooring lines design parameters.

Parameter	Full Scale	Model
Spoke	Length = 14 m Diameter = 3 m	Length = 17 cm Diameter = 1.5 cm
Tension leg	Length = 46 m Diameter = 0.48 m Fairlead location = 24 m Pretension = 2.3 × 10^4 N	Length = 23 cm Diameter = 0.2 m Fairlead location = 12 cm Pretension = 2.3015 N
New tethers	Length = 60 m Diameter = 0.48 m - -	Length = 30 cm Diameter = 0.2 mm Spring stiffness = 1.471 N/mm Maximum load = 23.2 N

The response of the TLP model to the wave loading was then studied through a sensor via a wireless data acquisition system, which eliminated the need for cables that would otherwise affect the platform motions and hence the experimental results. The sensor used in the experiments was placed inside the cylinder of the model's platform; it facilitated the measurement of six degrees-of-freedom deviations simultaneously. The waves generated by the wave maker were recorded using the wave gauge placed upstream of the model, and the depth of water in the flume was kept constant. The order of the test wave amplitudes was varied from 0 to 40 mm with a test wave period from 0.4 to 2 s. The experimental data were recorded and controlled using National Instruments LabVIEW software.

3. Theoretical Approach

3.1. Equation of Motion

An offshore wind turbine mounted on a TLP undergoes six degrees-of-freedom motion forming three rigid body translational motions and three rotational motions. When it is excited by waves, the assumption based on the linear hydrodynamic analysis and a linear damping allows the motion equations for surge, heave, pitch, and yaw to be solved in the frequency domain. Sway and roll are identical to surge and pitch, thus are not treated here. It was also assumed that no waves are generated by the floater, since the wavelengths of the ambient waves were assumed to be much greater than the floater diameter. The wave–floater interaction problem can therefore be solved in a simplified manner. The governing equation describing the balancing of the exciting forces for the surge behavior is derived from Newton's second law (see Withee [6]), and is stated as follows:

$$\begin{bmatrix} M_{11}+A_{11} & A_{15} \\ M_{51} & M_{55}+A_{55} \end{bmatrix}\begin{bmatrix} \ddot{\xi}_1 \\ \ddot{\xi}_5 \end{bmatrix}+\begin{bmatrix} B_{11} & B_{15} \\ B_{51} & B_{55} \end{bmatrix}\begin{bmatrix} \dot{\xi}_1 \\ \dot{\xi}_5 \end{bmatrix}+\begin{bmatrix} C_{11} & C_{15} \\ C_{51} & C_{55} \end{bmatrix}\begin{bmatrix} \xi_1 \\ \xi_5 \end{bmatrix}=\begin{bmatrix} F_1(t) \\ F_5(t) \end{bmatrix} \quad (1)$$

and the uncoupled heave and yaw equations of motion are:

$$\begin{array}{c} (M_{33}+A_{33})\,\ddot{\xi}_3+(B_{33})\,\dot{\xi}_3+C_{33}\xi_3=F_3(t) \\ (M_{66}+A_{66})\,\ddot{\xi}_6+(B_{66})\,\dot{\xi}_6+C_{66}\xi_6=F_6(t) \end{array} \quad (2)$$

where M_{jk} are the components of the mass matrix of the system; B_{jk} and A_{jk} are the frequency-dependent damping and added mass coefficients, respectively; F_j are the external exciting forces due to waves; and C_{jk} are the hydrostatic restoring coefficients.

The term $\ddot{\xi}$ represents the dimensional acceleration of the system in each mode of motion at each frequency, $\dot{\xi}$ is the system's velocity in each mode, and ξ is the system's displacement. The solution to these coefficients is fundamental for solving the equation of motion; a solution procedure was described in detail by Withee [6], with the assumption that the restoring is achieved by a stiff mooring tether that limits any significant motion in pitch and drives the restoring in pitch toward infinity (Wayman and Sclavonous [2]). The eigenvalues of this equation give the natural frequencies for the modes of motion in this study.

In order for the system to be analyzed using linear frequency domain analysis, all of the linear quantities must be analyzed relative to the wave elevation:

$$\eta(t)=Re(Ae^{-ik(xcos\beta+ysin\beta)}e^{i\omega t}) \quad (3)$$

where A is the wave amplitude. The linear wave excitation forces f_i acting on the platform are defined relative to the local coordinate system (x, y, z) and by virtue of linearity adopt the complex representation:

$$f_i=Re(A\cdot X_i e^{i\omega t}),\quad i=1,...,6 \quad (4)$$

where X_i is the complex amplitude of the exciting force or moment in the i-direction. The platform motions ξ_i in the six degrees of freedom also adopt the complex representation:

$$\xi_i=Re(A\cdot \Xi_i e^{i\omega t}),\quad i=1,...,6 \quad (5)$$

where Ξ_i is the complex response amplitude for the platform in the j-direction (i.e., RAO).

The water particle components are essential in defining the wave-induced forces on floating structures. Therefore, the Airy wave theory, often called the linear wave theory, was used to predict the horizontal water velocity and acceleration where the cylinder is located (Dean and Dalrymple [27]).

3.2. Modified Morison Equation

In deep water, the waves are assumed to have a low steepness (small wave height compared to wavelength). This enabled us to use the Airy linear wave theory. This latter assumption and the assumption that the waves are unidirectional are considered basic assumptions for the use of the Morison equation for calculating the wave loading on offshore structures.

A simple equation was developed by Morison for calculating the wave loads on floating submerged structures in water. The equation is stated as follows:

$$dF = \frac{\rho\pi D^2}{4}\dot{U} + \frac{\rho\pi D^2}{4}C_a(\dot{U} - \dot{V}) + \frac{\rho D}{2}C_D\,|U - V|\,(U - V) \tag{6}$$

where dF is total wave force; U and $\dot{U}$ are fluid velocity and acceleration, respectively; V and $\dot{V}$ are the relative velocity and acceleration of the floating structure; C_a and C_D are the added-mass and drag coefficient; ρ is density; and D is the cylinder diameter. The first two terms on the right side of the equation represent the inertia forces, while the last term determines the drag forces on the platform. The inertia forces comprise the Froude–Krylov force and the hydrodynamic mass force.

4. Results and Analysis

The scale floating wind turbine model was experimentally tested against a series of regular waves generated by the wave maker to obtain the dynamic response of the system in the six degrees of freedom. Regular wave parameters were chosen based on historical wave data for the North Sea near Scotland, where it was assumed that the offshore TLP wind turbine will operate.

4.1. Surge Dynamic Analysis of the TLP in Response to Regular Wave

All the physical structures including the floaters displayed natural frequencies in their motion. It is well known that a structure tends to oscillate at this frequency when stimulated by external forces. Under resonance conditions, a small force applied at this frequency produces a large oscillatory response. Thus, a dynamic force applied to a physical object causes it to vibrate. It is therefore important to know, on the one hand, the natural frequencies within a system and, on the other hand, the frequencies at which excitation is likely to occur to ensure that they do not coincide. It is generally not possible to control excitation frequencies of excitation, but the natural frequency of a system—which depends on physical characteristics such as mass and stiffness—can be altered.

When the TLP model is subjected to regular waves, the mooring lines may help to reduce the rotational motion around the three axes, even though they were mainly designed to reduce the surge acceleration and displacement. Incoming regular waves were assumed to be uniform and incident on the platform in the surge direction and a wave with an amplitude of approximately 12 mm and a wave period of T = 0.8 s was chosen for this experiment. Figure 4 shows the temporal variation of surge acceleration for the initial TLP and modified TLP concepts, while the corresponding surge amplitude spectrum comparison of the initial and modified concepts after applying Fast Fourier Transform (FFT) to the signal's temporal variation is given in Figure 5.

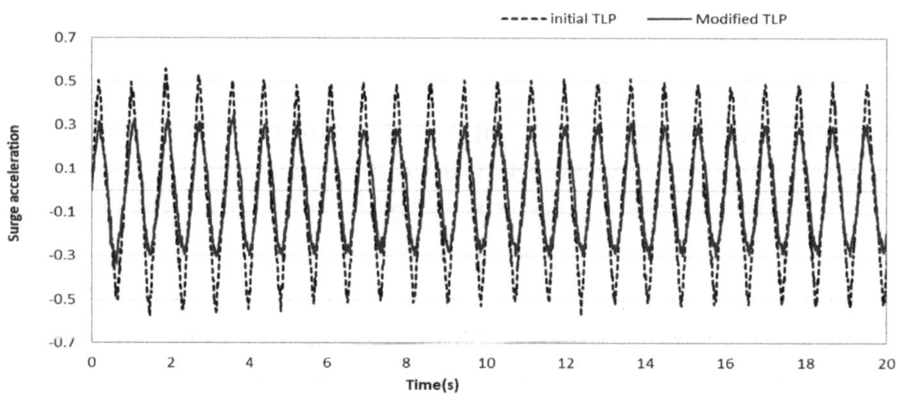

Figure 4. Time domain surge acceleration comparison for new and initial TLP concept.

As shown in Figures 4 and 5, the amplitude of the surge acceleration decreased as a result of addition of the mooring lines. This indicates that the new configuration contributes significantly to enable effective damping of the surge motion compared to the vertical tendons only, and hence improves the performance of the platform under wave attack. In addition, the spectra display well-defined peaks because of the periodic harmonic motion displayed in the time domain, even though the spectrum of the initial model appears to be noisier than that of the modified TLP for the same wave frequency. There are a couple of reasons to explain the improvement. One would be larger amplitude motion in the initial model, and the other an additional mooring line in modified TLP configuration; nevertheless, the peaks remain well defined in both cases. These findings confirm that the modified concept helps to stabilize the system's motion, which is essential to improving the performance of the wind turbine, because reducing the motion positively affects power production and turbine structural loading. In other words, platform motion causes larger inertial loads on the entire structure, which potentially leads to reduced efficiency and increased fatigue loading. Furthermore, the tether pretension also plays a fundamental role in the stability of the system. This also explains why the modified TLP with its higher initial pretension is more stable than the initial four-legged TLP. The new tethers were inclined, instead of being straight, in order that they not exceed the operational range of the tether's pretension.

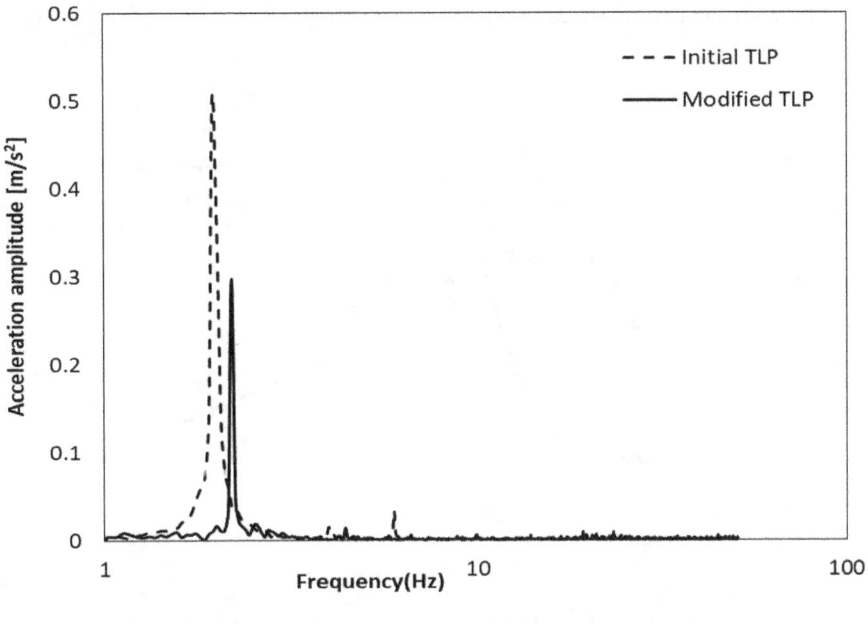

Figure 5. Surge spectrum for initial and modified TLP concept.

4.2. RAO Analysis for Regular Wave Tests

In this study, the water depth was kept constant at 35 cm in the wave flume, and a series of regular wave tests was carried out on both the initial model and the new design. This was done to validate previous results and to study the natural period and natural frequency of the new model compared to the initial one using the RAO method. The RAO is defined as the non-dimensional ratio of the surge to the surface elevation, which is used to analyze the dynamic motion of floating structures.

Figure 6 shows the RAOs for the NREL TLP, the modified TLP, and a numerical simulation of the same prototype (i.e., modified TLP) developed by Nematbakhsh [31]. For the translational accelerations (surge, sway, and heave), the RAO is determined by

$$RAO_{acc}(w) = \left| \frac{\sqrt{S_R(f)}}{\sqrt{S_I(f)} * (2\pi f)^2} \right| \quad (7)$$

where $S_R(f)$ represents the dimensional response of the structure in each mode of motion in the frequency domain and $S_I(f)$ is the wave power spectral density. As shown in the figure, both the initial TLP and the new design have the same trend with a slight decrease in the surge RAO response. However, the natural frequency, defined graphically as the frequency at which the peak occurs, was slightly shifted to a higher frequency for the new design but remained smaller than 0.4 rad/s. In addition, it was considered as the operational natural wave condition for FOWT, which is defined by ITTC spectrum for different sea states and provides a significant wave height of 5.0 m and a modal period of 12.4 s (see Kim and Hong [32]). On top of this condition, the water flume was only operated by a wave generator so that it was purely affected by the wave loading because wind tunnel was not turned on. On the other hand, the reference data of numerical simulation were also based on the nonlinear hydrodynamic forces on the platform, while wind speed was assumed constant. Therefore, the natural frequency can be inferred from the surge RAO response of the test model itself exerted by the wave loading. Consequently, both concepts can operate safely in the chosen site. Furthermore, to validate the results, the experimental surge RAO data were compared to the numerical representation. This latter was performed for a range of wave periods, from 0.6 to 1.2 s. The results obtained show that, for the same range of wave frequencies, a good similarity was found between the numerical and experimental data as the RAO motion increases while moving close to the natural frequency. It was expected that the peak in the numerical model would occur within the same range of wave frequency since the natural frequency in surge was defined to be 15 s.

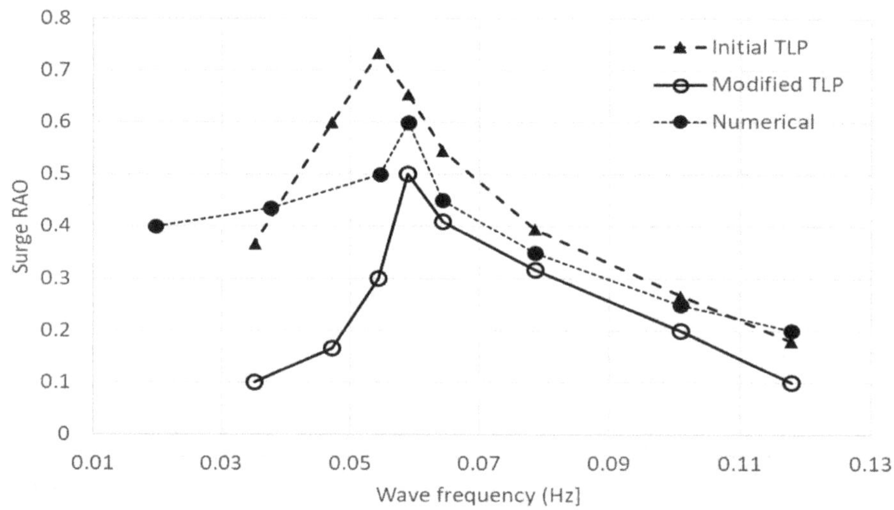

Figure 6. The surge RAO comparison for the two concepts.

4.3. Water Depth Effect

Considering that offshore wind turbines are suited to operate in rated power case as well as in deep water, the effect of water depth was investigated and experiments conducted to ensure the capability of the modified concept TLP to operate in deep water. Figure 7 compares the surge RAO's response in two different water depths for the TLP with the additional cables.

As shown in Figure 7, the RAO shows that the motion of the platform decreases with increasing water depth, which confirms that the TLP is a suitable structure for deep water. On the other hand, the natural frequency of the system was shifted to lower frequency in the case of the TLP with additional mooring cables in greater water depth. This was achieved as a result of the increase in the length of the tendons and the added mooring lines when operating in deeper water. The increase in the length of the tethers and the added mooring lines influenced the amplitude motion as well as the natural frequency, and they enabled an acceptable restoration in the surge of the system reducing the natural frequency. The same results were obtained by Wayman and Sclavonous [2], who investigated the water depth effect for an NREL TLP. Those results and the results of this actual study are in good agreement as both studies indicate that the water depth not only contributes to reduce the motion but also lowers the natural frequency. Thus, it is possible to tune the tether stiffness of a system to match a certain experimental natural frequency by means of restoring.

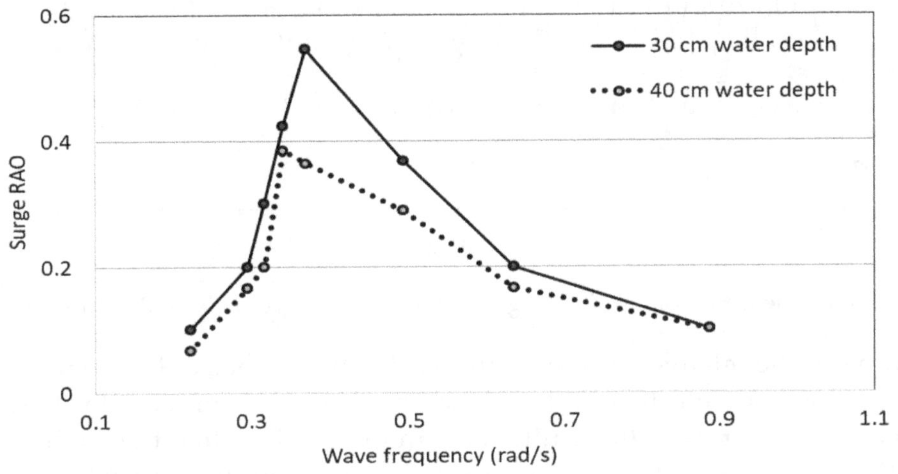

Figure 7. The water depth effect on the motion of the modified TLP with cables.

4.4. Wave Loading Using Morison Equation for Regular Waves

The wave loads were calculated using the Morison equation. The experimental values obtained for the initial TLP were then compared to those for the TLP with the additional cables. The experimental conditions were similar for both cases and the same wave height and wave period were applied in the case of a regular wave test.

4.4.1. Omega Arithmetic Method for Relative Velocity

One of the requirements of the Morison equation for predicting wave loading in floating structures is the definition of the relative velocity of the system while being subjected to the action of the incoming waves. The relative velocity was integrated from the acceleration data using the omega arithmetic method, which is considered to be the best method for the integration to convert acceleration into velocity (see Appendix A for omega arithmetic code).

As shown in Figure 8, the relative velocity is represented after being integrated from the acceleration data using the omega arithmetic method outlined above. The method determines the corresponding velocity from an acceleration dataset without going through the formal integration in time. Meanwhile, this method is obviously influenced by the frequency behavior of the signal, as the conversion process was carried out in the frequency domain rather the time domain signal. In other words, the time series were converted to the frequency domain using FFT, integrated by dividing by ω, and then converted back to the time domain using the inverse FFT. Thus, this procedure is the most accurate for the conversion process because neither the time domain integration nor the application of a high pass filter before performing the integration was a good solution for integrating the relative velocity in this case.

Because the acceleration motion response is a harmonic function, the velocity has a sinusoidal variation, but it is not in phase. The phase relationship between acceleration and velocity is such that the velocity is 90° out of phase with acceleration. This is noticeable from the figures; when the structure passed through the equilibrium position, the velocity was at its maximum and the acceleration was zero.

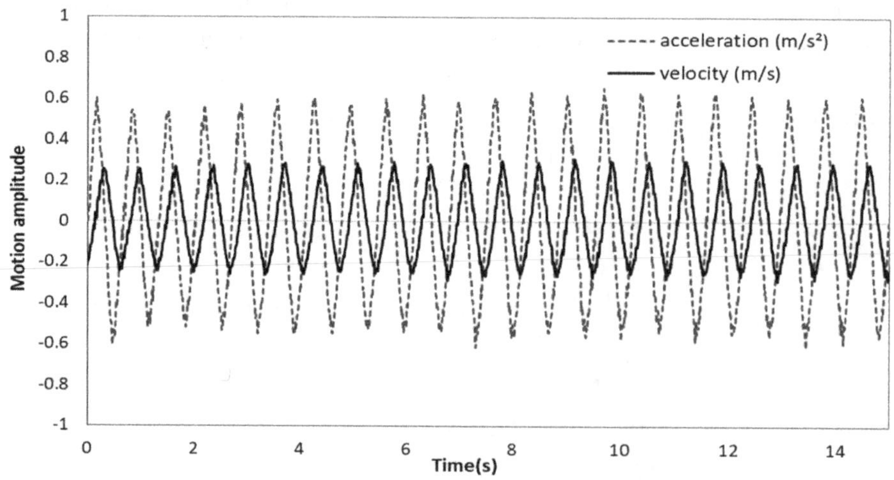

Figure 8. The time domain for surge acceleration and velocity of the initial TLP.

Figure 9 compares the relative velocity of the TLP with tendons only and the modified concept with additional mooring lines attached. It shows that the magnitude of the velocity was smaller in the new model than in the prototype. The results confirm the findings that the mooring lines helped to reduce not only the surge acceleration but also the velocity of the structure when it was subjected to monochromatic regular waves.

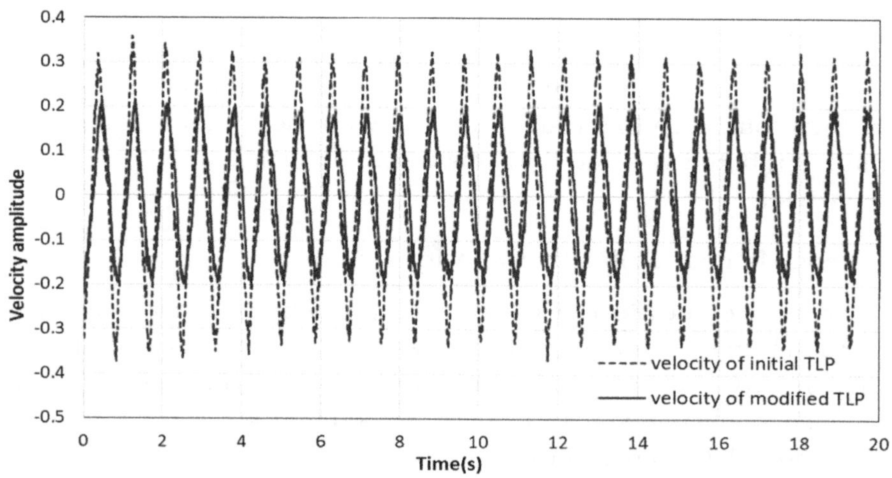

Figure 9. Time domain surge velocity comparison for new and initial TLP concept.

4.4.2. Wave Loading Results

The water particle kinematics in the case of a regular wave can be defined on the basis of the extrapolated Airy wave theory. The Airy theory was used because of its suitability for a wide range of waves at any site, including extreme and fatigue cases. However, it has a drawback in that it does not consider the peaks and troughs of waves. Conversely, the extrapolated Airy theory, known as the wheeler stretching method, makes the original theory more flexible by calculating the kinematics both at the mean water level and at the water elevation surface, which results in a kinematics prediction that is closer to reality.

Numerous methods also exist for defining the inertia and drag coefficients. The results published by the DNV (Det Norske Veritas) are widely accepted (Journee and Massie [23]). The DNV has calculated coefficients for different types of cylinders based on their roughness with respect to the KC

(Keulegan–Carpenter) number because of the primary dependence of the coefficients on the number. The KC number is expressed by the following equation (which is valid for the deep-water studies):

$$KC = UT/D = \pi H/D \tag{8}$$

where U is the water particle velocity, T is the wave period, D is the cylinder diameter, and H is the wave height. Based on the DNV report, the inertia coefficient of a smooth floating cylindrical platform is related to the added mass coefficient by the following relation:

$$C_M = 1 + C_a \tag{9}$$

where $C_a = 1$ in the case of a circular cylinder. This leads to an inertia coefficient with an approximate value of two and a drag coefficient close to one for low KC numbers, as is the case in this study where the inertia forces are dominant and the drag forces are not important. In this study, to get both coefficients, the Morison's method was applied (see Journee and Massie [23]). This method is based on the measured force (N), velocity (m/s) and acceleration (m/s^2) from the flume experiment. Given that the fundamental data were provided, the Morison equation can be decomposed into two parts: inertia and drag terms. Therefore, based on the Morison's method, the drag coefficient can be obtained while making the relative acceleration becomes zero, whereas the inertia coefficients can also be calculated while having relative velocity zero. More details are also well described in the paper (see Wolfram and Naghipour [33]).

From the results for operational range, it is clear that the forces caused by the interaction between the cylinder and the waves on the new design are less than those on the initial model (see Figure 10). This confirms that the new mooring lines not only contribute to a stable system but also results in lower load on the platform. Moreover, the horizontal component of the total force per unit length has a harmonic shape because of the accelerating and decelerating movements in the body motion, which imply a change in the kinetic energy of the fluid within the control volume. Furthermore, the unstable drag and inertia coefficients also have an effect on the forces as they depend primarily on the KC number, which in turn depends on the velocity of the fluid. As a result, the better estimation of the water kinematics enabled the Morison equation to be more accurate in predicting the wave loading on the TLP's cylinder. It is also notable that the loads have a nonzero mean value. This offset may be explained by the excess buoyancy and the pretension on the tether, which requires that the cylinder support some of the load in order to keep the system sufficiently balanced. This is because the pretension of the tethers is related to the buoyancy force and the overall mass of the system times the gravitational acceleration.

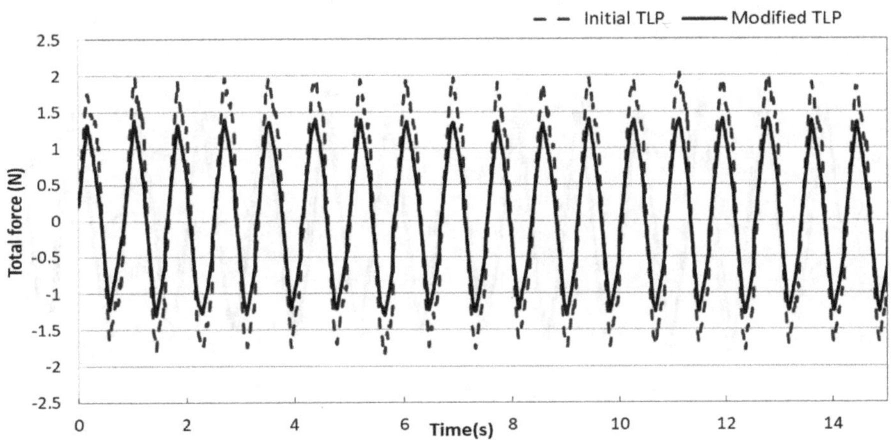

Figure 10. Experimental wave loads comparison in a regular wave.

4.5. Extreme Design Sea Condition Analysis

The ability of the new system with added mooring lines to withstand extreme sea conditions in Northern sea of California was verified. This verification was performed with the objective of ensuring that the new design can operate in a severe sea state, which makes it suitable for both regular and extreme sea states. The extreme sea conditions were issued by the Center for Offshore Safety, USA, and the parameters of the waves were chosen based on their report (Berg [34]). The chosen wave amplitude was 5 cm with a wave period of 1.2 s based on the scale ratio of 1:200 used in this study.

As shown in Figures 11 and 12, the surge acceleration remained the most obvious of the three translational motions with a larger amplitude than regular waves because of the severe sea state. However, the extreme design case caused the platform to have significant motion in the sway direction, which cannot be neglected as in the case of the regular wave, because of the substantial contribution of the mooring lines in effectively damping all the other rigid body and rotational motions. In addition, the ability of the system to respond was demonstrated and it was also shown that severe wave conditions tend to increase the motion amplitude of the modified TLP above the regular and operational range. However, this amplitude remained in the acceptable range where a structure can ensure its operability without affecting the performance of the installed wind turbine.

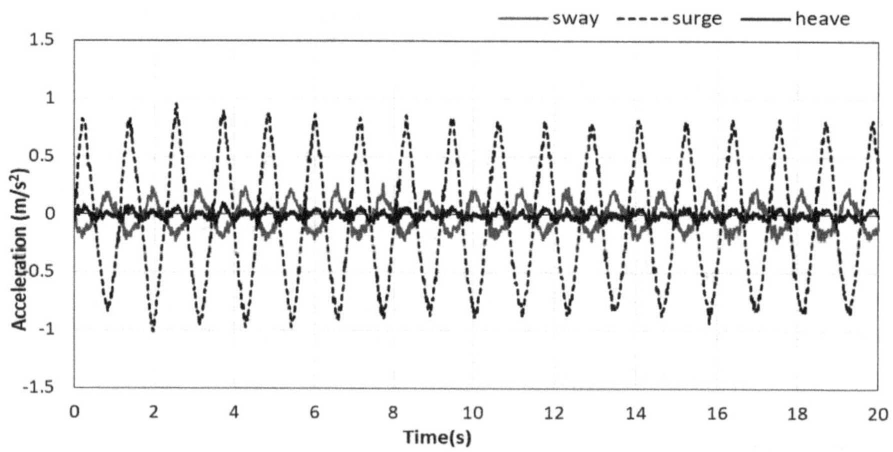

Figure 11. Translational accelerations in extreme design condition for modified TLP.

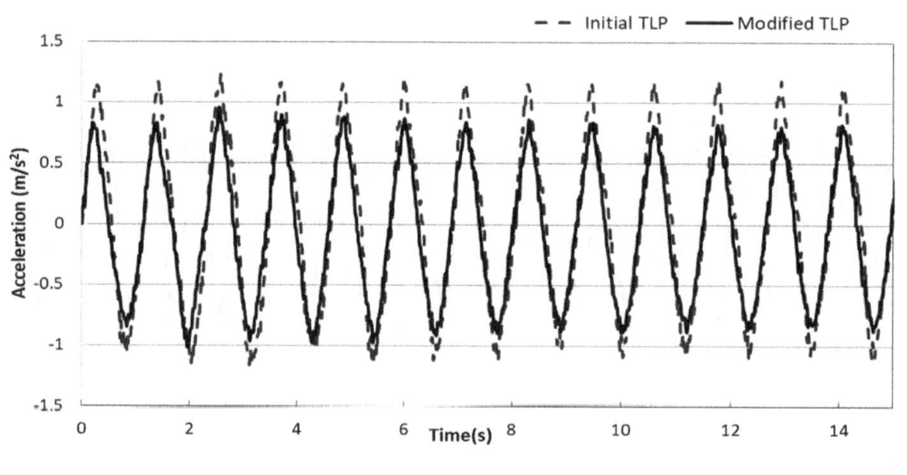

Figure 12. Surge acceleration comparison for both concepts in extreme design case.

As shown in Figure 13, the surge amplitude spectra derived from the time domain using the FFT, which defines a comparison between the surge acceleration in the extreme conditions of the two concepts, confirmed the mooring lines contribution in damping the surge motion even in extreme conditions. However, because the exhibited motion had a larger amplitude, the spectra appeared noisier than in the regular case, but the peaks were still well defined. Another aspect of concern is the loads applied to the structure in the extreme wave event. It was expected that the wave would have a significant impact on the structure in terms of loading, as it was proven that the added mass generated in the case of an extreme wave is higher owing to the high kinetic energy of the fluid around the structure, which causes an increase in both the body's motion and the water kinematics. Therefore, the loading on the system exerted a high tension on the tethers and the new mooring lines; however, this tension should remain within the safe range because an excess on the design value would result in the consequent destruction of the system. This means that the new model withstood the extreme design wave condition for a 100-year return period for the chosen site.

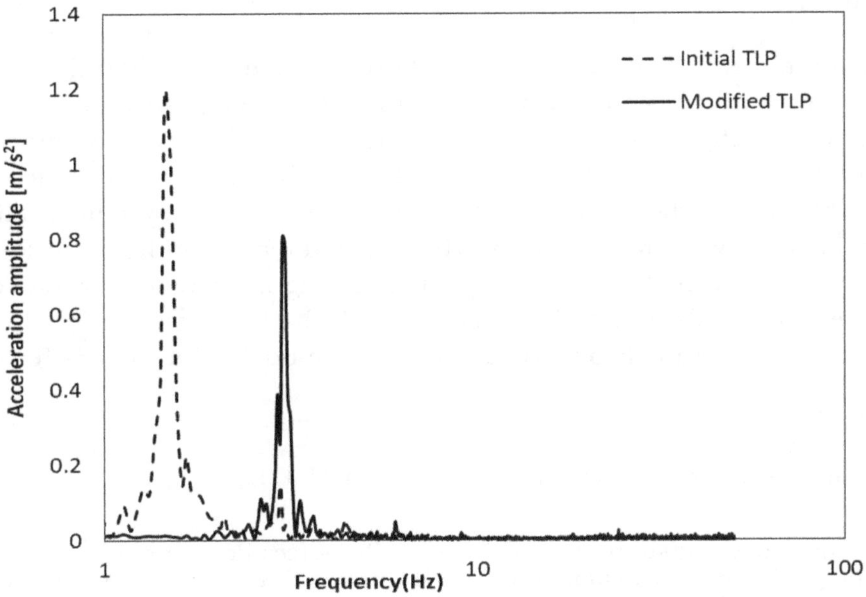

Figure 13. Initial and modified TLP's surge spectrum in extreme case.

The new findings in the behavior of the modified TLP under the onslaught of waves may play a fundamental role in providing better stability and thereby improve the overall performance of the system, especially in critical environmental loads. This is because the mooring lines tend to increase the stiffness, which contributes to the rigidity of the system, and the design life of offshore wind turbines may be enhanced as well.

5. Conclusions

A new tension leg platform conceptual design for offshore wind turbines was proposed with the objective of reducing the dominant surge motion of this kind of platform. The feasibility of the new model to operate under different wave conditions was also investigated. The design was proposed based on the NREL prototype and the wave conditions were chosen according to a report in the USA regarding the Northern Sea of California, where it is assumed that the new system will operate.

For the operational range of regular waves, the following can be stated:

- The modified concept model tends to decrease the velocity and acceleration of the overall system. A new method, known as the omega arithmetic method, which provides accurate and exact conversions, was introduced to integrate the displacement from the acceleration data.

- The RAO surge response proved that the two concepts avoid the natural wave frequency. This means that resonance is avoided, which is one of the main requirements for FOWT in order to operate in a safe environment. Furthermore, the water depth also has the same effect as the mooring lines, as it was proven that the motion amplitudes decrease with increasing water depth, which confirms that the new model is suitable for deep water.
- As regards extreme sea conditions for a return period of 100 years, the experimental results indicate that the motion amplitude will increase without exceeding the range of accelerations where the floating system is known to be able to operate safely. Further, the modified concept was verified to have smaller motion amplitude than the initial concept even in harsh environmental conditions.
- The loads applied on the structure by the waves were also measured using the Morison equation after matching all the conditions required for the use of the equation.
- The results obtained for the new model show that the addition of the mooring lines plays a positive role in reducing the wave loads compared to the initial model in the case of both regular and extreme waves.

These findings on the performance of the model under wave effect should be taken into account when constructing this kind of platform in the future. However, additional studies need to be conducted in order to investigate the coupled dynamic motion under all combined environmental loads combined with the tension on the tethers to enable optimization of the modified TLP system with additional lines. This would further improve the modified concept TLP system and help to effectively minimize the drawbacks to withstanding all environmental loads and increase the performance of offshore wind turbines. On top of these findings, the scaling issue would be one of the important considerations when the test data of scaled model match the full scale model. In particular, most measurements in the flume ertr made only on a small-scale model, which needs further investigation in the future.

Author Contributions: Supervision, J.S.; Writing—original draft, H.-C.L.

Acknowledgments: This work was supported by "Human Resources Program in Energy Technology" of the Korea Institute of Energy Technology Evaluation and Planning (KETEP), granted financial resource from the Ministry of Trade, Industry & Energy, Republic of Korea (No. 20164030201230). In addition, this work was supported by the National Research Foundation of Korea (NRF) grant funded by the Korea government (MSIP) (No. 2016R1A2B1013820).

Appendix A

Matlab code: omega arithmetic method for converting acceleration into velocity and displacement.

```
Fs=100; dt=1/Fs; et=15; t=0:dt:et;
y=xlsread('surge');
subplot(2,1,1); plot(t,y,'r');


xlabel('time(s)'); Y=fft(y); n=length(y); Amp=abs(Y)/n; NumuniquePts=ceil((n+1)/2);
freq=(0:NumuniquePts-1)*Fs/(n-1); freq2=-1*freq(end:-1:1);
if mod(n,2)==0
freq2(1)=[ ];
end
freq3=[freq,freq2];
freq3';
size(freq3)
```

```
if mod(n,2)==1
freq3(n+1)=[ ];
end
if mod(n,2)==0
freq3(n)=[ ]];
end
subplot(2,1,2); plot(freq,Amp(1:NumuniquePts),'b');
subplot(2,1,1); plot(freq,Amp(1:NumuniquePts),'b');
xlabel('frequency(hz)'); ylabel('amplitude');

B=reshape(Y,1,1501); G=B./(2*pi*freq3*1i);

G(1)=[1];
if mod(n,2)==0
G(n)=[1];
end
inversed=ifft(G);
plot(t,real(inversed),'b')
hold on
plot(t,y,'r')
title('omega arithmetic');
xlabel('time(s)');
ylabel('velocity blue,Accel red');
```

References

1. Jonkman, J.; Butterfield, S.; Musial, W.; Scott, G. *Definition of 5-MW Reference Wind Turbine for Offshore System Development*; Tech. Report NREL/TP-500-38060; NREL: Golden, CO, USA, 2009.
2. Wayman, E.N.; Sclavonous, P.D. Coupled Dynamic Modeling of Floating Wind Turbine Systems. In Proceedings of the Offshore Technology Conference, OTC, Houston, TX, USA, 1–4 May 2006.
3. Lee, S. Dynamic Response Analysis of Spar Buoy Floating Wind Turbine Systems. Master's Thesis, Massachusetts Institute of Technology, Cambridge, MA, USA, 2008.
4. Naqvi, S.K. Scale Model Experiments on Floating Offshore Wind Turbines. Master's Thesis, Worcester Polytechnic Institute, Worcester, MA, USA, 2012.
5. Hong, S.; Lee, I.; Park, S.H.; Lee, C. An Experimental Study of the Effect of Mooring systems on the Dynamics of a SPAR Buoy-type Floating Offshore Wind Turbine. *J. Naval Archit. Ocean Eng.* **2015**, *7*, 559–579. [CrossRef]
6. Withee, J.E. Fully Coupled Dynamic Analysis of a Floating Wind Turbine System. Ph.D. Thesis, Massachusetts Institute of Technology, Cambridge, MA, USA, 2004.
7. Murray, J.; Yang, C.K.; Wooseuk, Y. An Extended Tension Leg Platform Design for Post-Katrina Gulf of Mexico. In Proceedings of the International Offshore and Polar Engineering Conference, ISOPE, Osaka, Japan, 21–26 July 2009.
8. Karimirad, M.; Moan, T. Wave and Wind-Induced Dynamic Response of a Spar-Type Offshore Wind Turbine. *J. Waterw. Port Coast. Ocean Eng.* **2011**, *138*, 9–20. [CrossRef]
9. Ren, N.; Li, Y.; Ou, J. The Wind-Wave Tunnel of a Tension Leg Platform Type Floating Offshore Wind Turbine. *J. Renew. Sustain. Energy* **2012**, *631*, 1–17. [CrossRef]
10. Martinez, M.S.; Natarajan, A.; Henriksen, L.C. *Feasibility Study of a Semi Floating Spar Buoy Wind Turbine Anchored with a Spherical Joint to the Sea Floor*; Technical Report INWIND; EWEA: Frankfurt, Germany, 2013.
11. Hadjihosseini, A.; Peinke, J.; Hoffmann, N.P. Stochastic Analysis of Ocean Wave States with and without Rogue Waves. *New J. Phys.* **2014**, *16*, 053037. [CrossRef]
12. Hadjihosseini, A.; Wächter, M.; Hoffmann, N.P.; Peinke, J. Capturing Rogue Waves by Multi-point Statistics. *New J. Phys.* **2016**, *18*, 013017. [CrossRef]

13. Hadjihosseini, A.; Lind, P.G.; Mori, N.; Hoffman, N.P.; Peinke, J. Rogue waves and entropy consumption. *Front. Phys.* **2017**, *120*, 30008. [CrossRef]
14. Lind, P.G.; Herraez, I.; Wächter, M.; Peinke, J. Fatigue Load Estimation through a Simple Stochastic Model. *Energies* **2014**, *7*, 8279–8293. [CrossRef]
15. Lind, P.G.; Vera-Tudela, L.; Wächter, M.; Kühn, M.; Peinke, J. Normal Behaviour Models for Wind Turbine Vibrations: Comparison of Neural Networks and a Stochastic Approach. *Energies* **2017**, *10*, 1944. [CrossRef]
16. Marino, E.; Lugni, C.; Borri, C. The role of the nonlinear wave kinematics on the global responses of an OWT in parked and operating conditions. *J. Wind Eng. Ind. Aerodyn.* **2013**, *123*, 363–376. [CrossRef]
17. Marino, E.; Giusti, A.; Manuel, L. Offshore wind turbine fatigue loads: The influence of alternative wave modeling for different turbulent and mean winds. *Renew. Energy* **2017**, *102*, 157–169. [CrossRef]
18. Nielsen, A.W.; Schlutter, F.; Sorensen, J.V.T.; Bredmose, H. Wave loads on a monopole in 3D waves. In Proceedings of the ASME 2012 31st International Conference on Ocean, Offshore and Arctic Engineering, Rio de Janeiro, Brazil, 1–6 July 2012.
19. Paulsen, B.T.; Bredmose, H.; Bingham, H.B. An efficient domain decomposition strategy for wave loads on surface piercing circular cylinders. *Coast. Eng.* **2014**, *86*, 57–76. [CrossRef]
20. Bredmose, H.; Slabiak, P.; Sahlberg-Nielsen, L.; Schlutter, F. Dynamic Excitation of Monopiles by Steep and Breaking Waves: Experimental and Numerical Study. In Proceedings of the ASME 2013 32nd International Conference on Ocean, Offshore and Arctic Engineering, Nantes, France, 9–14 June 2013.
21. Patel, M.H.; Witz, J.A. *Compliant Offshore Structures*; Butterworth-Heinemann Ltd.: Oxford, UK, 2013; pp. 80–83.
22. Peeringa, J.M. *Wave Loads on Offshore Wind Turbines*; Tech. Report Duurzame Energie; ECN Publisher: Amsterdam, The Netherlands, 2004.
23. Journee, J.M.J.; Massie, W.W. Wave Forces on Slender Cylinders. In *Offshore Hydromechanics*, 1st ed.; Delft University of Technology: Delft, The Netherlands, 2001; pp. 469–498.
24. Sarpkaya, T.; Isaacson, M. *Mechanics of Wave Forces on Offshore Structures*; van Nostrand Reinhold: New York, NY, USA, 1981.
25. Burrows, R.; Tickell, R.G.; Hames, D.; Najafian, G. Morison Wave Forces Coefficients for Application to Random Seas. *J. Appl. Ocean Res.* **1997**, *19*, 183–199. [CrossRef]
26. Henderson, A.R. *Design Methods for Offshore Wind Turbines at Exposed Sites*; Tudelft: Delft, The Netherlands, 2003.
27. Dean, R.G.; Dalrymple, R.A. *Water Wave Mechanics for Engineers and Scientists*; World Scientific Publishing Co. Pte. Ltd.: Singapore, 1984.
28. Lim, H.C. Optimum Design of a Sloping-wall-type Wave Absorber Placed in a Sinusoidal Propagating Wave. *Ocean Eng.* **2014**, *88*, 588–597. [CrossRef]
29. Zhu, L.; Lim, H.C. Hydrodynamic Characteristics of a Separated Heave Plate Mounted at a Vertical Circular Cylinder. *Ocean Eng.* **2017**, *131*, 213–223. [CrossRef]
30. Barltrop, N.D.P. *Floating Structures: A Guide for Design and Analysis*; Oilfield Publications, Inc.: Surrey, UK, 1998.
31. Nematbakhsh, A. A Nonlinear Computational Model of Floating Wind Turbines. Ph.D. Thesis, Worcester Polytechnic Institute, Worcester, MA, USA, 2013.
32. Kim, H.J.; Hong, S.Y. The Shape Design and Analysis of Floating Offshore Wind Turbine Structures with Damper Structure and Shallow Draft. *J. Ocean Wind Energy* **2014**, *1*, 170–176.
33. Wolfram, J.; Naghipour, M. On the Estimation of Morison Force Coefficients and Their Predictive Accuracy for Very Rough Circular Cylinders. *Appl. Ocean Res.* **1999**, *21*, 311–328. [CrossRef]
34. Berg, J.C. *Extreme Ocean Wave Conditions for Northern California Wave Energy Conversion Device*; Tech. Report SAND2011-9304; Sandia Report: Livermore, CA, USA, 2011.

14

Investigation on Optimization Design of Offshore Wind Turbine Blades based on Particle Swarm Optimization

Yong Ma [1,2,3], Aiming Zhang [1], Lele Yang [1,3,*], Chao Hu [2] and Yue Bai [2]

1 School of Marine Engineering and Technology, Sun Yat-sen University, Guangzhou 518000, China; mayong3@mail.sysu.edu.cn (Y.M.); zhangaim5@mail2.sysu.edu.cn (A.Z.)

2 College of Shipbuilding Engineering, Harbin Engineering University, Harbin 150001, China; huchao@hrbeu.edu.cn (C.H.); mayong02@hrbeu.edu.cn (Y.B.)

3 Southern Marine Science and Engineering Guangdong Laboratory (Zhuhai), Zhuhai 519000, China

* Correspondence: yanglele@mail.sysu.edu.cn

Abstract: Offshore wind power has become an important trend in global renewable energy development. Based on a particle swarm optimization (PSO) algorithm and FAST program, a time-domain coupled calculation model for a floating wind turbine is established, and a combined optimization design method for the wind turbine's blade is developed in this paper. The influence of waves on the power of the floating wind turbine is studied in this paper. The results show that, with the increase of wave height, the power fluctuation of the wind turbine increases and the average power of the wind turbine decreases. With the increase of wave period, the power oscillation amplitude of the wind turbine increases, and the power of the wind turbine at equilibrium position decreases. The optimal design of the offshore floating wind turbine blade under different wind speeds is carried out. The results show that the optimum effect of the blades is more obvious at low and mid-low wind speeds than at rated wind speeds. Considering the actual wind direction distribution in the sea area, the maximum power of the wind turbine can be increased by 3.8% after weighted optimization, and the chord length and the twist angle of the blade are reduced.

Keywords: wind turbine; PSO algorithm; FAST; time-domain coupled model; blade optimization

---

1. Introduction

Climate change is showing abnormal effects all over the world. Environmental pollution is continuously aggravated, and energy demand is increasingly variable. Developing and utilizing clean and pollution-free renewable energy has become an effective way to alleviate energy shortages, reduce environmental pollution and improve climate conditions [1].

Wind energy is one of the most mature forms of renewable energy, which has promising large-scale development conditions and commercial development prospects. Wind energy includes onshore wind energy and offshore wind energy. In recent years, the development and utilization of onshore wind energy has encountered bottlenecks due to the limitation of land and wind resources. Compared with onshore wind power, offshore wind power has many advantages, such as higher wind speed, more stable wind direction, and smaller turbulence and wind shear. With the rapid development of wind power technology, the leading direction of wind power development in the future will focus on offshore wind power [2].

A large number of studies on fixed wind turbines have been carried out. Apparent models suitable for the wind turbines have been developed to predict the aerodynamic characteristics of the wind turbine [3,4]. In the engineering application, most design software for the wind turbine follows the dynamic stall model in the helicopter field, and modifies the working environment of the fixed wind turbine [5,6]. It can basically meet the design requirements of the fixed wind turbine. At present, fixed wind turbine technology has become relatively mature and accomplished commercialization in China and several European countries. However, the floating wind turbine can move freely in the ocean environment. It is always in an obvious oscillation dynamic state under the influence of waves. Continuous oscillation not only causes the dynamic response of the wind turbine, but also enhances the unsteady effect of aerodynamic load, which brings great difficulties to the design and evaluation of floating wind turbine. The technology is still in the primary stage of research and development [7].

The offshore floating wind turbine is a very complex system, which mainly consists of the top wind turbine generator set, the tower supporting the wind turbine, the floating platform infrastructure and the mooring positioning system. For the wind turbine generator set alone, blades with excellent aerodynamic performance are the core of wind turbines [8]. Blade design is a complex multi-objective optimization design process, which needs to meet a number of design indicators, such as aerodynamic performance, structural strength and economic characteristics [9]. There are three theoretical research methods: blade element momentum (BEM) method based on momentum conservation; generalized dynamics wake (GDW) based on vortex theory; CFD method by solving N-S equation directly. The BEM method can simply calculate the energy efficiency and the force of horizontal axis impeller. It is effective and fast for estimating the overall load of the impeller. It is suitable for wind turbines with equal blade lengths and no less than a 3 blade tip speed ratio. The BEM method is the most widely used and technologically mature aerodynamic performance prediction method for the wind turbine [10,11]. GDW method includes the dynamic wake effect, tip loss and skewed wake effect, and does not require additional modified model [12]. However, the method assumes that the induced velocity is small relative to the inflow velocity. Thus, it is suitable for the wind turbines with heavy load. When the wind speed is small, the wake appears to be in a turbulent state, and the method becomes unstable [13]. CFD method can simulate yaw, tower shadow, wind shear and other complex conditions. Through CFD calculation, abundant flow field information can be obtained [14,15]. However, due to the large dimension of the wind turbine and the complexity of blade shape, the CFD method requires a lot of computer resources and computing time. Besides, this method is affected by the turbulence model, transition model, discrete format, boundary conditions and other factors [16].

In the application research process of the wind turbine, numerous methods synthetically simulating wind turbine loads have been developed based on the above theoretical research methods. Some scholars have synthetically simulated the preliminary design of floating wind turbines by using the linear frequency domain method commonly used in offshore oil and gas exploitation. For example, Bulder et al. [17] used the linear frequency domain method to simulate a wind turbine with a three-dimensional floating body. The response amplitude operator and standard deviation of the six rigid-body motion modes of the floating platform were calculated and analyzed. Tracy [18] used a similar method to study the optimal design parameters of tension leg and mast-type wind turbine. However, these studies only introduce the aerodynamic and structural characteristics of the wind turbines by adding additional mass, additional damping and an additional recovery matrix, without simulating the real wind turbine. The linear frequency domain method can't simulate the non-linear dynamic characteristics and unsteady effects. Some scholars use other methods to study the floating wind turbine. For example, Henderson et al. [19] used a method called "state domain" to study the effect of platform motion on the fatigue load of the wind turbine. Combining aeroelastic load simulation software with hydrodynamic load simulation software, Withee [20] established the time domain dynamic model for floating wind turbine, and studied a tension leg wind turbine. Because they employed a simple hydrodynamic model, aerodynamic model or structural dynamic model, these studies still have some limitations and couldn't develop a better blade optimization design scheme.

Wind turbine blade design is also a complex multi-objective optimization design process. Strong interactions exist between the disciplines of aerodynamics, dynamics, structures, and economies. A number of previous studies have optimized blades with intelligent calculation methods, including a neural network algorithm, genetic algorithm and data mining algorithm [21–23]. Particle swarm algorithm is a modern heuristic algorithm, which has great advantages in computing speed and memory consumption, and has fewer parameters to adjust. Liao et al. [24] employed an improved PSO algorithm to optimize wind turbine blades. The comparison results of optimized blades and reference blades indicated that this method was feasible and practical. Combined with the improved PSO algorithm with FAST program, they pursued the minimum blade mass to reduce the cost of the wind turbine. The thickness and the location of the layers in spar caps were selected as the optimization variables [25].

In addition, the wind speed of 12 m/s is mainly used as the design reference for international mainstream wind turbines. But the wind field with high-quality wind resources is limited. The wind field with low wind speed occupies a larger proportion, and these wind fields are generally closer to the power grid [26]. Taking China as an example, only the wind speed of 9.5 m/s in Fujian sea area is relatively close to the average wind speed in Europe, while the average wind speed in other sea areas is generally less than 8m/s, and even the wind speed in many sea areas is less than 7 m/s.

Therefore, this paper combines the PSO algorithm and FAST V8 computational analysis software for the wind turbine to optimize the design of offshore wind turbine blades in sea areas with different wind speeds including low wind speed. The "aero-hydro-control-structure" coupled time-domain calculation model for floating wind turbine is developed. By using this model, the motion state of floating wind turbine under real working conditions is simulated. Aiming at getting the maximum output power of the wind turbine, the optimal design of the wind turbine blade can be searched in many combinations by changing the distribution of twist angle and chord length. In addition, the influence of sea state and wind conditions in the actual sea area on the optimal design of the wind turbine blades is investigated. The study can provide theoretical guidance for the optimal design of an offshore floating wind turbine blade.

2. Blade Optimization Based on PSO Algorithm and FAST Program

FAST is an open source software for simulating dynamic response of horizontal-axis wind turbines, which is developed by the United States Department of Energy's NREL. FAST was originally used as a program to calculate the aerodynamic performance of horizontal axis wind turbines based on BEM theory. Thereafter, the hydrodynamic module, coupling calculation module, noise module, elastic module, ice load module, servo system module and wind environment load module were gradually added, making FAST a floating wind turbine solver with multiple performance modules. Input files should be provided before running the FAST, including the basic operation parameters of the wind turbine, blade geometry parameters, wind spectrum file, tower data file, wave load file, and platform foundation file. When FAST runs, each module reads its corresponding input file, and the data interaction between each module is completed by a dynamic link library. All modules cooperate to achieve time-domain coupled calculation.

There are two different analysis modes in FAST V8. The first analysis mode is linearization, in which BEM is used to calculate the aerodynamic force; the second analysis mode is simulation, in which the time marching method is used to solve the nonlinear motion equation. This is also the method used in this study. In the simulation process, the aerodynamic response and the structural response of the wind turbine are processed in real time. FAST V8 provides the S-Function (System Function) function interface for the Simulink module library in MATLAB, as shown in Figure 1. The S-Function function is compiled by C++ language, and the optimization algorithm is embedded to carry out the optimization design of the wind turbine blades.

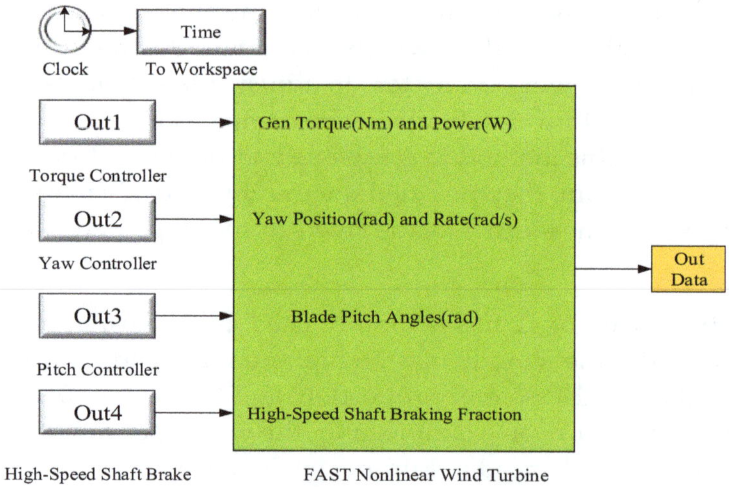

Figure 1. FAST Simulink S-function.

In this study, the NREL 5 MW wind turbine is used as the optimized parent form. The wind turbine is a three-blade horizontal axis wind turbine, which can be used as onshore or offshore wind turbine, and can be installed in deep-sea areas. It has gradually become a research example of large offshore wind turbines. The overall parameters of the wind turbine system are shown in the Table 1. PSO algorithm is used to optimize the design of the wind turbine blades. As shown in Figure 2, each blade of NREL 5 MW wind turbine has 19 control surfaces. After the airfoil of each control surface is selected, each control surface is determined by its twist angle and chord length. Figure 3 shows the blade layout. The control surfaces from the blade root to 4.1 m from the blade root are circular, and the overall shape is cylindrical. Therefore, the lift coefficient is 0, which makes no contribution to the aerodynamic performance of the whole blade. Therefore, only 16 control surfaces in the range of 4.1 m from the blade root to the blade tip need to be considered with a total of 16 × 2 parameters. Particle swarm space is set as 32 dimensions. For each particle, $y_1 \ldots y_{16}$ in $Y1 = (\mathrm{y}_1 \ldots \mathrm{y}_{32})$ represents the twist angle of 16 control surfaces of NREL 5 MW wind turbine blade in turn, and $y_{17} \ldots y_{32}$ represents the chord length of 16 control surfaces of a blade of NREL 5 MW wind turbine blade in turn.

Table 1. Overall parameters of NREL 5 MW wind turbine system.

Name	Parameter
Unit-power	5 MW
Blade layout	Upwind Direction, 3 blades
Control system	Variable speed, variable pitch
Drive system	High-speed multi-stage gearbox
Blade and hub diameter	126 m, 3 m
Hub center height	90 m
Cut-in, rated and cut-out wind speed	3 m/s, 11.4 m/s, 25 m/s
Cut-in and rated blade rotate speed	6.9 rpm, 12.1 rpm
Rated tip speed	80 m/s
Cantilever length, axis inclination angle, pre-deflection angle of the blade	5 m, 5°, 2.5°
Total system mass	8,130,315.388 kg
Total gravity center height (below hydrostatic surface)	77.3829 m

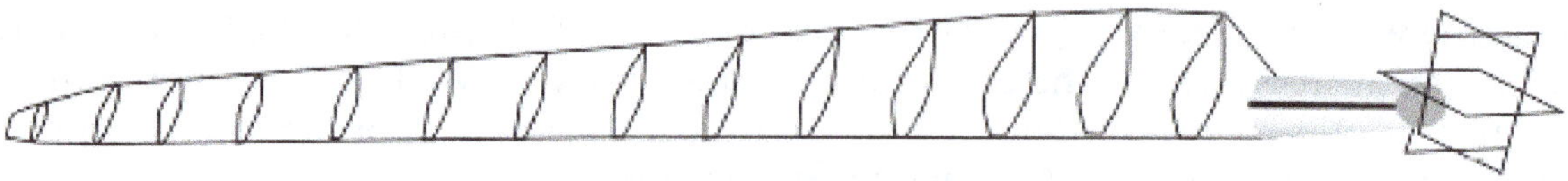

Figure 2. Blade structure of NREL 5 MW wind turbine.

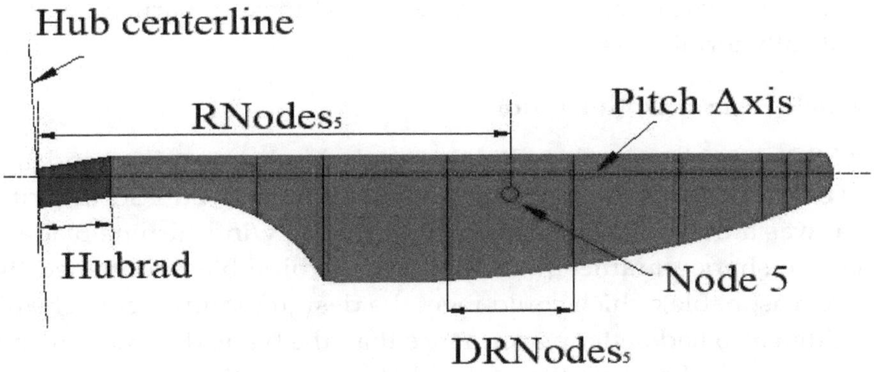

Figure 3. Blade layout.

After K iterations, the updating speed of the ith particle in the next iteration is as follows:

$$V_{\mathrm{i}} = (v_1, v_2, \ldots v_d, \ldots v_D) \tag{1}$$

where

$$\begin{aligned} v_{id}^{k+1} &= \omega v_{id}^{k+1} + C_1 r_1 \left(Pbest_{id}^{k} - x_{id}^{k}\right) \\ &+ C_2 r_2 \left(Zbest_{id} - x_{id}^{k}\right) \end{aligned} \tag{2}$$

In the next iteration, the position of the ith particle is:

$$X_i^{k+1} = X_i^{k+1} + V_i^{k+1} \tag{3}$$

where C_1 and C_2 are learning factors and non-negative constants, which can accelerate the convergence rate if appropriate selected; ω is inertia weight; r_1 and r_2 are random numbers distributed between [0,1].

In order to prevent premature convergence, which leads to local optimal solution of the algorithm, an improved PSO algorithm is adopted. After K iterations of the ith particle, the velocity update equation is changed to:

$$V_{id}(t+1) = \omega(t) V_{id}(t) + c_1(t) r_1(t) (P_{f_i d}(t) - X_{id}(t)) \tag{4}$$

Improved PSO algorithm can enable particles to learn from excellent particles in the crowd and enhance particle search ability. The aerodynamic profile of offshore wind turbine blades is determined by the optimization results of PSO algorithm. Under the combined action of wind and waves, the floating wind turbine power is in a fluctuating state. The wind turbine power at balance position in a period of time is taken as the average power during this period. The optimization objective is to select the wind turbine blade with the maximum power at balance position. The fitness of each particle of the particle swarm is selected as the wind turbine power at balance position generated by FAST. By changing the distribution of twist angle and chord length of the blade and studying the change rule of aerodynamic performance of the wind turbine, the optimal design of the wind turbine blade can be determined.

When the program runs, the whole particle swarm is initialized to generate the initial population. The PSO algorithm needs to set an upper bound (ub) and a lower bound (pb) to prevent the particles from deviating from the solution range in the search process. At the same time, the algorithm can reduce the impact of the optimization results on the blade structure, strength and control system. In this study, the upper and lower bounds of particle swarm are ±10% of the original parameters of NREL 5 MW wind turbine blade. When the twist angle and chord length of any one-dimensional interface of a particle in particle swarm exceed the limit, the program sets the exceeding values as ub and pb. After the program starts to run, FAST is automatically called to calculate the fitness of each particle in the iteration process. When the set number of iterations is reached or the relative error between the average fitness of particle swarm and the fitness of particle swarm is less than 0.5%, the program automatically terminates.

3. Validation of Particle Swarm Optimization

The improved PSO algorithm employed in this study has been verified and applied in the work of Liao et al. [24]. They compared the optimized results of improved PSO algorithm with those of Glauert method that was a traditional design method for the wind turbine blade. The comparison results showed that the shape parameters of the wind turbine blade obtained by improved PSO algorithm were more reasonable, which could meet the design requirement. The blade designed by improved PSO algorithm also had better performance than the blade designed by the Glauert method. In this study, in order to verify the feasibility of the blade optimization design method combining PSO algorithm with FAST program, the twist angle, chord length and power improvement of the wind turbine blades before and after optimization are analyzed by taking the NREL 5 MW wind turbine as the optimized parent form. During validity verification, the 5 MW wind turbine is considered as an offshore fixed wind turbine, and the foundation is anchored to the sea floor. The optimization of the wind turbine is carried out under steady wind. Therefore, in the main input file of FAST, CompMooring and CompHydro are both equal to 0. The six degrees of freedom of the platform in the ElastoDyn module is set to 0. The particle swarm size is 30. The number of iteration steps is set to 30. The time step is 0.0125 s, and the calculation time is 1000 s.

3.1. Optimization of Fixed Wind Turbine Blade at Rated Wind Speed

Comparing the twist angle and chord length of the optimized particle with that before optimization, as shown in Figures 4 and 5, the chord length of the optimized blade is slightly larger than that of the optimized blade after 10 m from the hub surface, and the twist angle is almost the same before and after optimization. From the graph, it can be seen that the distribution of the twist angle and chord length of the blade will jump after optimization by particle swarm optimization. This is because particle swarm optimization is only a theoretical calculation. In the calculation, the optimal twist angle and chord length of each control surface are considered only in theory, and the continuity of blade structure is not considered in the actual manufacturing process. Therefore, in the following blade optimization process, if there is a large jump in the distribution of twist angle and chord length, polynomial fitting is used to smooth the blade.

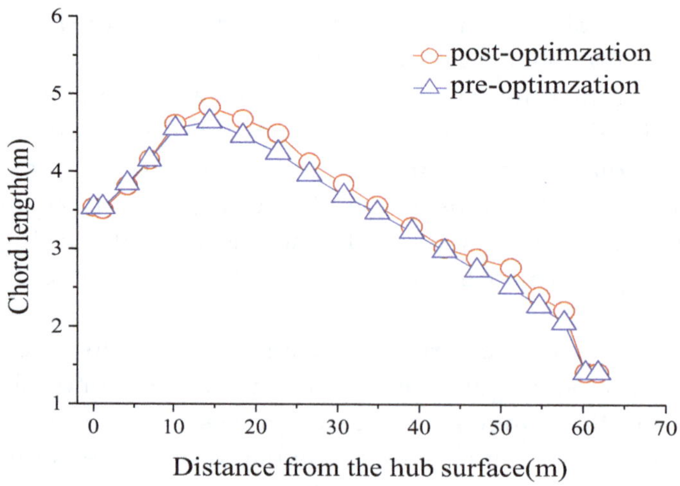

Figure 4. Distribution of the chord length for fixed wind turbine blade at rated wind speed before and after optimization.

The optimized power is 5037.3 kW, which is approximately equal to 5030 KW of the wind turbine before optimization, and the improved power is 0.145%. Under rated wind speed condition, the optimization effect of the wind turbine blade is not great. This is because the NREL 5 MW wind turbine is designed according to the rated wind speed of 11.4 m/s, and the twist angle and chord length of each control surface have been designed according to the optimal size.

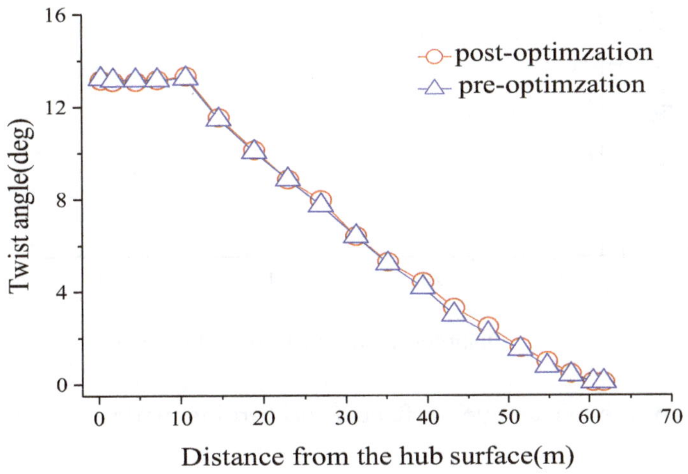

Figure 5. Distribution of the twist angle for fixed wind turbine blade at rated wind speed before and after optimization.

3.2. Optimization of Fixed Wind Turbine Blade at Low Wind Speed

In order to verify the PSO algorithm, the typical operating conditions of China's offshore waters with a wind speed of 6 m/s are selected. The time step is set to 0.0125 s and the time domain calculation period is 300 s. The time domain calculation result of NREL 5 MW wind turbine power is shown in Figure 6.

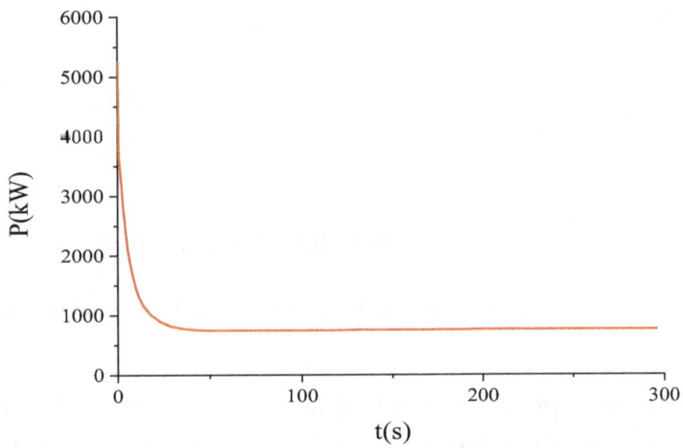

Figure 6. Time-domain calculation of fixed wind turbine power.

Figures 7 and 8 depict the distributions of the chord length and twist angle before and after optimization. After iterative screening of the PSO algorithm designed in this paper, the distribution of the chord length in the range of 10 m to 35 m from the hub surface of the wind turbine is less than that before optimization. The chord length of the control surface at 34.85 m is basically the same as that before optimization. The chord length in the range of 34.85 m from the hub surface to blade tip decreases further. The distribution of the twist angle in the range of 10.25 m to 38.95 m from the hub surface of the wind turbine is less than that before optimization. After 38.95 m, the distribution of twist angle after optimization is close to that before optimization.

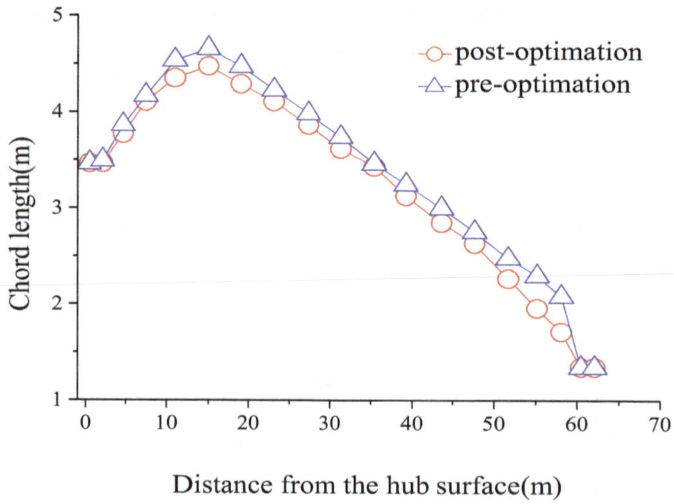

Figure 7. Distribution of the twist angle for fixed wind turbine blade at low wind speed before and after optimization.

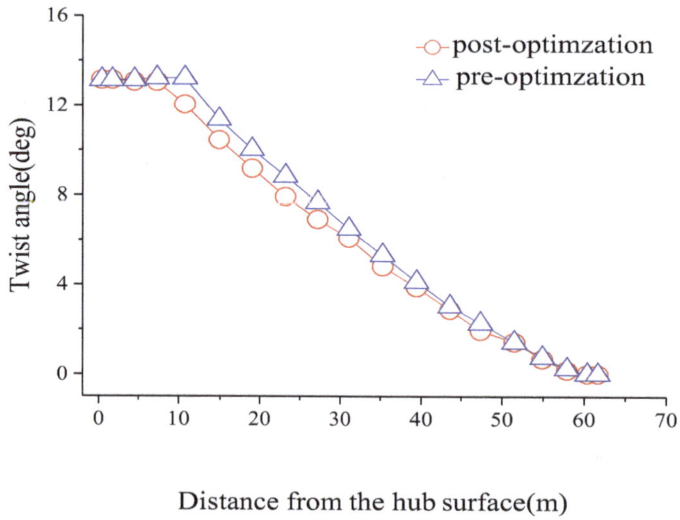

Figure 8. Distribution of the twist angle for fixed wind turbine blade at low wind speed before and after optimization.

The optimized wind turbine power is 760 kW, larger than 750 kW before optimization, and the power is increased by 1.4%. The enhancement effect is obviously larger than that at rated wind speed. In addition, when the wind turbine works, the wind turbine adjusts its position by yaw to adapt to the incoming wind. The phase angle between the impeller and the wind turbine blade changes periodically. Thus, the aerodynamic characteristics of the wind turbine blade need to have roughness. As can be seen in Figure 9, the optimized wind turbine power has been promoted obviously in the range of 5.5 m/s to 6.5 m/s of low wind speed. This indicates that the optimized wind turbine blades meet the roughness requirements of aerodynamic characteristics. Therefore, the blade optimization program based on PSO algorithm and FAST secondary development has an obvious effect on blade optimization at low wind speed. The sensitivity to wind speed is reduced, and the twist angle and chord length are also reduced. The ability of PSO algorithm to solve non-linear and global optimization problems is verified. It also proves the effectiveness of PSO algorithm in aerodynamic optimization of the wind turbine blades.

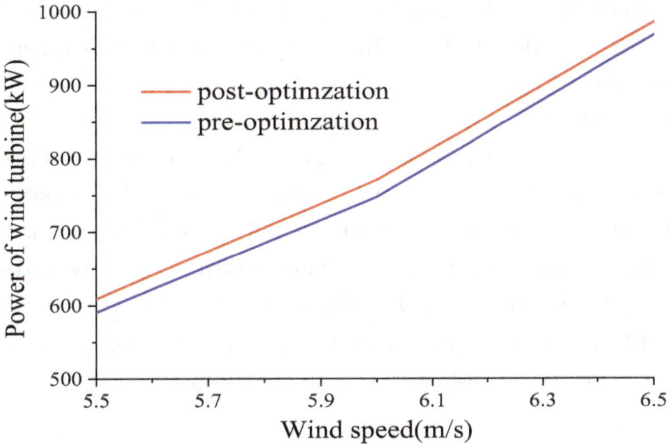

Figure 9. Wind turbine blade power at low wind speed before and after optimization.

4. Optimization of Offshore Wind Turbine Blades

The waves cause the motion of the foundation platform of floating wind turbines, which result in change of relative wind speed of the wind turbine. Accordingly, the aerodynamic performance of the wind turbine also changes, which is the main difference between floating wind turbine and fixed wind turbine. Therefore, for floating wind turbines installed in different sea areas, the influence of sea conditions should be taken into account in the design of the wind turbine blade. It has significant theoretical and practical meaning to develop optimization design method for the wind turbine blade that is suitable for different sea conditions.

In this paper, the optimization design of NREL 5 MW_OC3Hywind wind turbine blade is carried out by using the sea state parameters listed in Table 2. The incident wave is regular wave parallel to the hub axis.

Table 2. Sea state parameters.

Working Conditions	Wind Speed (m/s)	Significant Wave Height (m)	Peak Period (s)	Wave Direction (deg)
rated wind speed	11.4	6	12	0
mid-low wind speed	10	6	12	0
low wind speed	6	3	6	0

4.1. *Influence of Regular Wave Height and Period on Wind Turbine Power*

In order to study the effects of wave height and period of different regular waves on floating wind turbines, several typical wave heights and periods are selected respectively. The coupling model of 5 MW wind turbine-SPAR platform-mooring system is developed by FAST to calculate for different conditions. The wind turbine powers at wave crest, wave trough and equilibrium position are selected for analysis.

4.1.1. Effect of Wave Height on Wind Turbine Power

Figure 10 shows the change of the wind turbine power with a significant wave height, with the direction and period of the wave remaining unchanged. As the significant wave height increases from 2 m to 6 m, the wave peak power fluctuates greatly. The wave peak power shows a parabolic distribution in the range of 3 m to 4 m for significant wave height, increasing from 5460 kW to 5610 kW. The wave trough power increases slowly at first and then decreases sharply, from 3980 kW to 3770 kW. The wave peak power and wave trough power of the wind turbine fluctuate irregularly with wave height. This is due to that the floating foundation is fixed on the sea floor through anchor chains. When the floating platform is affected by wave force, the wind force is applied above the wind turbine.

The coupling effect of the two forces makes the phase angle between the wind turbine and the incoming wind change due to the platform shaking. The phase angle change is different for different wave height. Thus, the wind turbine power fluctuates irregularly. The power at the equilibrium position decreases rapidly in the range of 2 m to 5 m for significant wave height, from 5010 kW to 4790 kW. In the range of 5 m to 6 m, the power at the equilibrium position is stable at 4790 kW. The overall power fluctuation at the equilibrium position is only 4%, which is very small. This is because the Spar water line surface is very small and the draft depth is large, and thus the change of wave height has little influence on the dynamic response of the floating platform. Fixed onshore 5 MW wind turbine has a power of 5000 kW at 12 m/s wind speed. Floating wind turbine has a lower power at the equilibrium position. In conclusion, with the increase of incident wave height, the power fluctuation of the wind turbine increases, and the average power of the wind turbine decreases.

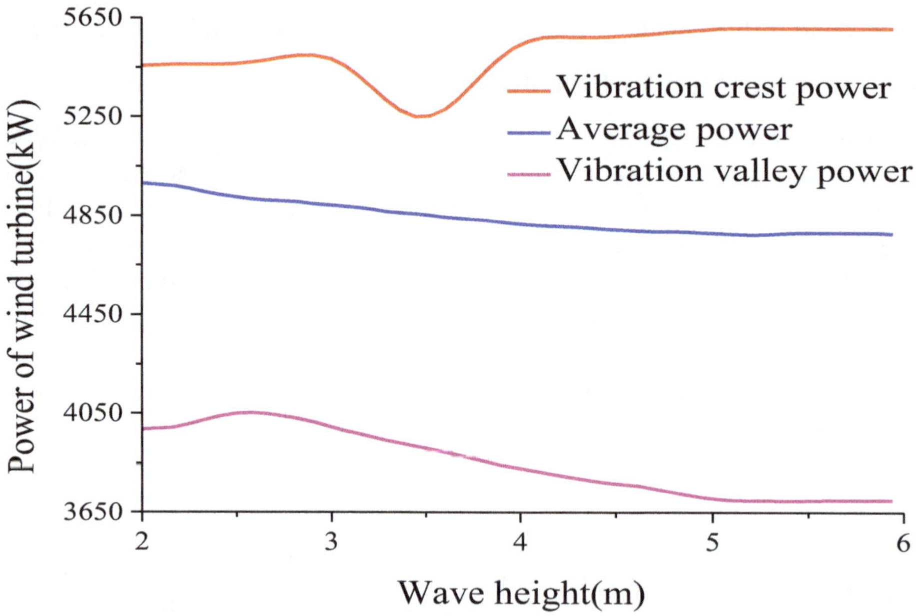

Figure 10. Comparison of the wind turbine power at different wave heights.

4.1.2. Effect of the Wave Period on the Wind Turbine Power

Figure 11 shows the change of the wind turbine power with the wave period, with the direction and significant wave height of the wave remaining unchanged. As the wave period increases from 1 s to 12 s, the wave peak power increases from 5410 kW to 5610 kW. The wave trough power decreases from 4680 kW to 3710 kW. The amplitude of power fluctuation increases gradually. When the incident wave period is 1 s, the wind turbine power at the equilibrium position is 4969 kW, which is the same as the power of 5 MW wind turbine on land at the same wind speed. It means that the influence of waves on the wind turbine power is small. But when the incident wave period is 12 s, the wind turbine power at the equilibrium position is only 4790 kW. The influence of waves on the wind turbine power becomes large. With the increase of the incident wave period, the wind turbine power at the equilibrium position decreases. NREL 5 MW wind turbine is a variable-speed and variable-pitch wind turbine. The control system can make the speed and pitch angle of the wind turbine correspond to the actual working conditions as much as possible. However, the large wave period results in a poor adaptive ability of the wind turbine. Then, the wind turbine cannot adjust its attitude to adapt to the incoming wind in time, resulting in a power reduction. In conclusion, the period of the wave is closely related to the wind turbine power. With the increase of incident wave period, the power oscillation amplitude of the wind turbine increases and wind turbine power at the equilibrium position decreases.

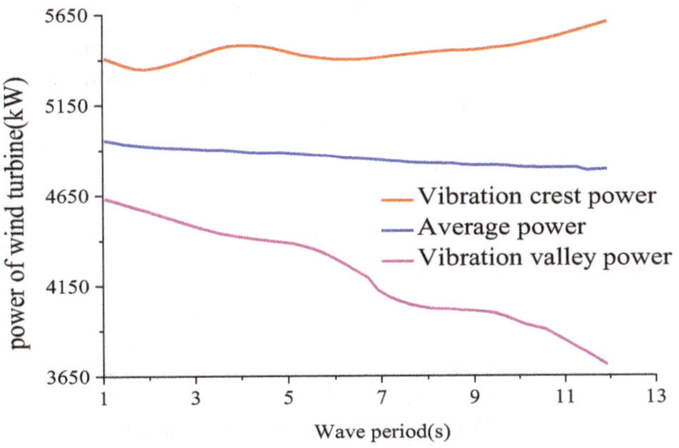

Figure 11. Comparison of the wind turbine power at different wave periods.

4.2. Optimization of the Wind Turbine Blade at Rated Wind Speed

At the rated wind speed, the PSO algorithm is used to optimize the floating wind turbine blades. The fitness output of the optimal particle for each iteration of PSO is shown in Figure 12. The evolution of the population can be observed. In 0-14 iterations, the particle swarm grows rapidly, and Zbest of the optimal particle increases from 4718 kW to 4761 kW. In 14–25 iterations, the growth rate of Zbest slows down, increasing from 4761 kW to 4766 kW. After 25 iterations, the optimal particle of particle swarm does not renew. The PSO algorithm stops automatically after 35 iterations, indicating that the difference between the average fitness and the best fitness of the population is less than 0.5%. The particle swarm as a whole approaches the optimal solution in the search area. Compared with the power before optimization, the power is increased by 1.017% after optimization by the PSO algorithm under this condition.

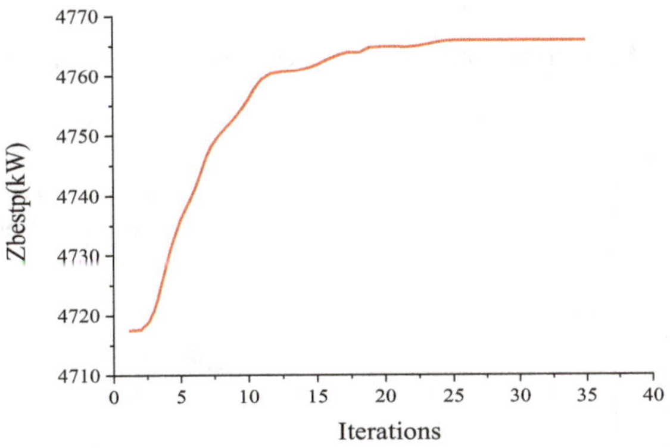

Figure 12. Evolution of particle swarm at rated wind speed.

The distributions of twist angle and chord length after optimization by the PSO algorithm are compared with those before optimization, as shown in Figures 13 and 14. The optimized twist angle is slightly larger than that before optimization in the range of 10.25 m to 47.15 m from hub surface. The optimized twist angle is almost close to that before optimization in the range of 47.15 m from the hub surface to the blade tip. The current optimization design only considers the best twist angle and chord length of each control surfaces by using blade element momentum, without considering the continuous process and actual operation of the wind turbine. Thus, some changes have taken place in the distribution of the chord length. The changes of the distribution of twist angle are more obvious. The chord length is smoothed by a spline curve in MATLAB. The distribution of chord length after smoothing by the spline curve is more reasonable, as shown in Figure 15. The continuity of blade structure is ensured by modification, which facilitates the manufacture of actual blades. The wind turbine power with smooth blades is 4751 kW, slightly less than that before smoothing. Compared with the power before optimization, the wind turbine power after smoothing is increased by 0.7%.

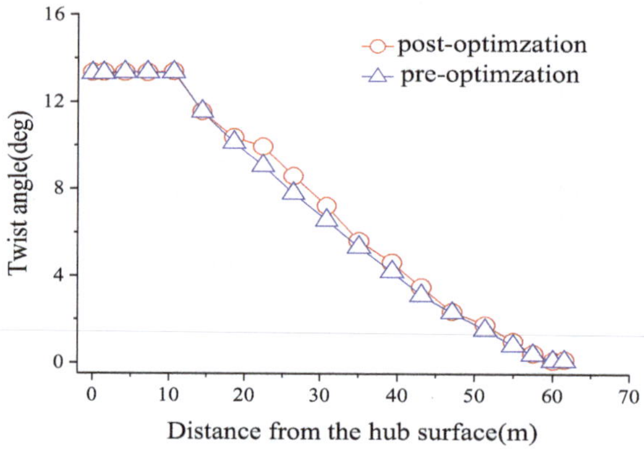

Figure 13. Distribution of the twist angle at rated wind speed before and after optimization.

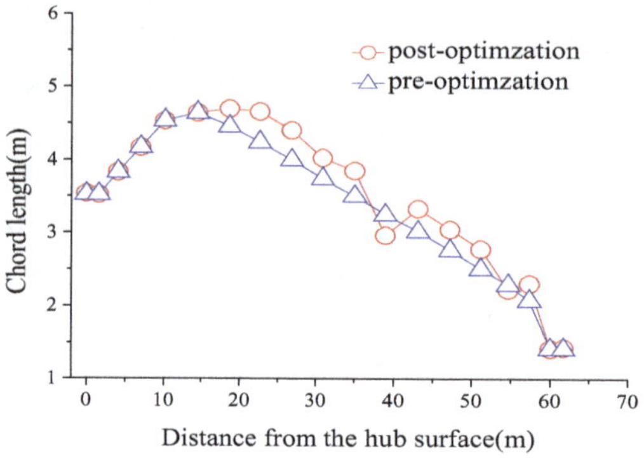

Figure 14. Distribution of the chord length at rated wind speed before and after optimization.

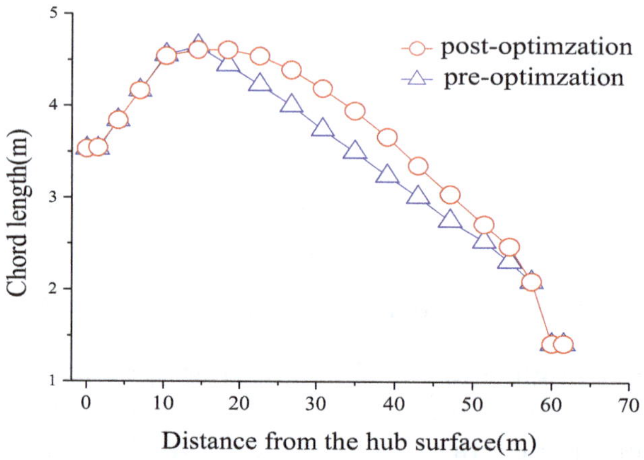

Figure 15. Distribution of the chord length at rated wind speed before and after smoothing.

4.3. Optimization of the Wind Turbine Blade at Mid-Low Wind Speed

At mid-low wind speeds, the fitness output of the optimal particle for each iteration of PSO can be observed in Figure 16. In 0~34 iterations, each iteration produces a better individual (Zbest). After 34 iterations, each iteration does not produce a better individual, but the particle swarm as a whole approaches the optimal solution in the search area. The PSO algorithm stops automatically after 50 iterations, indicating that the difference between the average fitness and the best fitness of the population is less than 0.5%.

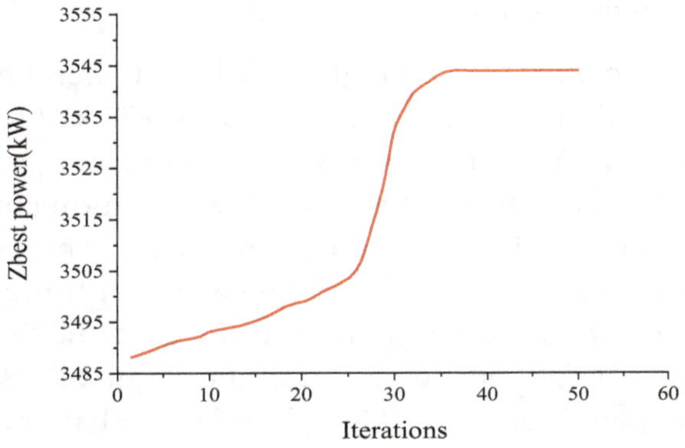

Figure 16. Evolution of particle swarm at mid-low wind speed.

Figures 17 and 18 are the distributions of twist angle and chord length after optimization at mid-low wind speed. The change of the twist angle and chord length in 15 control surfaces of floating wind turbine blades can be observed. In the range of 0 m to 10 m from the hub surface, the first four control surfaces are cylindrical, which are the connecting parts between blades and hub. Thus, there is no change in the twist angle and chord length. In the range of 10 m to 60 m from the hub surface, the distribution of twist angle after optimization is slightly larger than that before optimization, and the chord length is slightly larger than that before optimization. Except for the cylindrical control surface, the chord length in each control surface of the wind turbine after optimization is smaller than that before optimization. Therefore, the area of the wind turbine blade is reduced, and the material of the wind turbine blade is saved, which can save the cost. At mid-low wind speed, the power of the wind turbine at equilibrium position is 3485 kW before optimization. After optimization, the power of the wind turbine at equilibrium position is increased by 1.7%, namely 3544 kW.

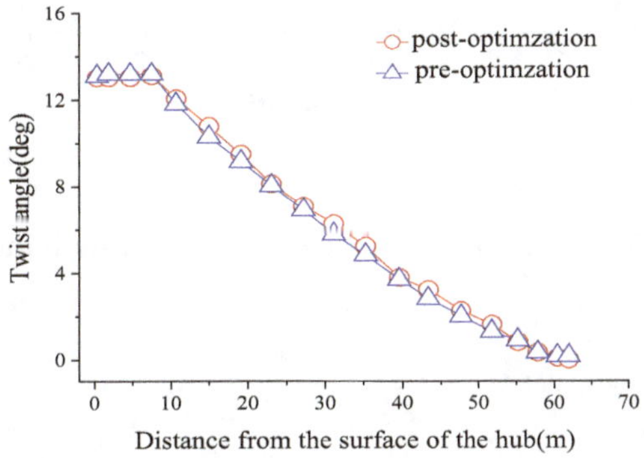

Figure 17. Distribution of the twist angle at mid-low wind speed before and after optimization.

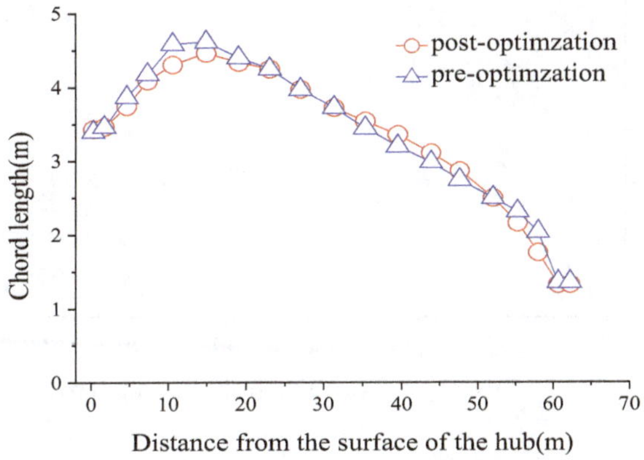

Figure 18. Distribution of the chord length at mid-low wind speed before and after optimization.

4.4. Optimization of the Wind Turbine Blade at Low Wind Speed

Population evolution can be observed from Figure 19. In 0~11 iterations, the particle swarm grows rapidly, and the Zbest of population particles increases from 777 kW to 797 kW. After 11 iterations, the optimal particle is no longer updated and no new extremum emerges. When the algorithm runs to 26 iterations, the particle swarm algorithm achieves the termination condition. The particle swarm as a whole approaches the optimal solution in the search area. The power of the wind turbine is increased by 2.5% through the optimization of particle swarm optimization. Figures 20 and 21 are the distributions of twist angle and chord length after optimization. The optimized twist angle and chord length are less than those before optimization in the range of 6.83 m from the hub surface to blade tip. The difference of twist angle and chord length before and after optimization is the greatest in the control surface 6.83 m from the hub surface. Along the direction of blade tip, the twist angle and chord length are gradually close to those before and after optimization. In summary, when the blade is optimized at low wind speed, the power of the wind turbine get promoted, while the manufacturing difficulty and cost of the blades are reduced.

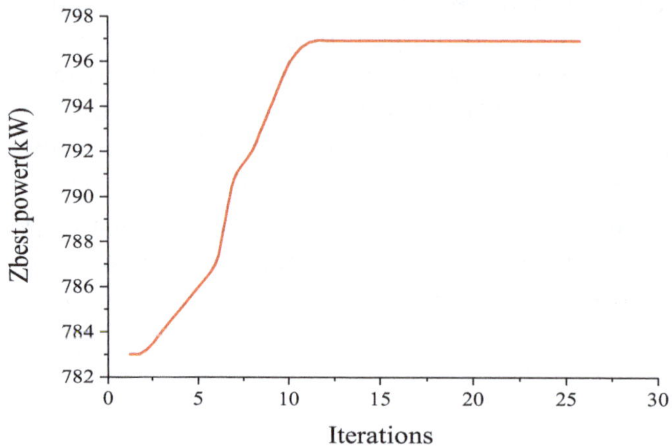

Figure 19. Evolution of particle swarm at low wind speed.

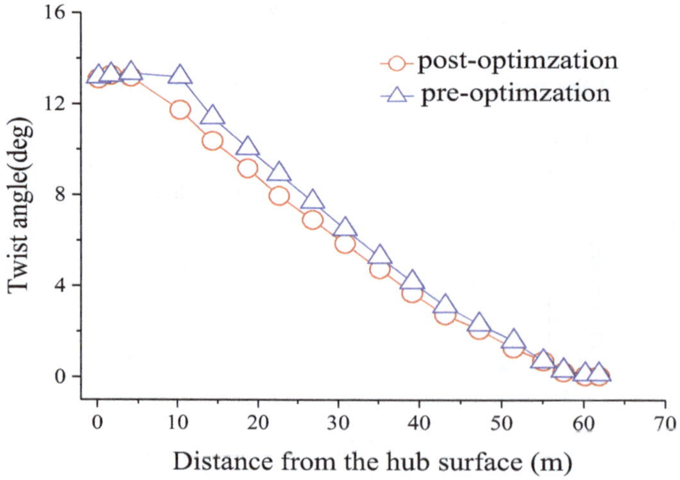

Figure 20. Distribution of the twist angle at low wind speed before and after optimization.

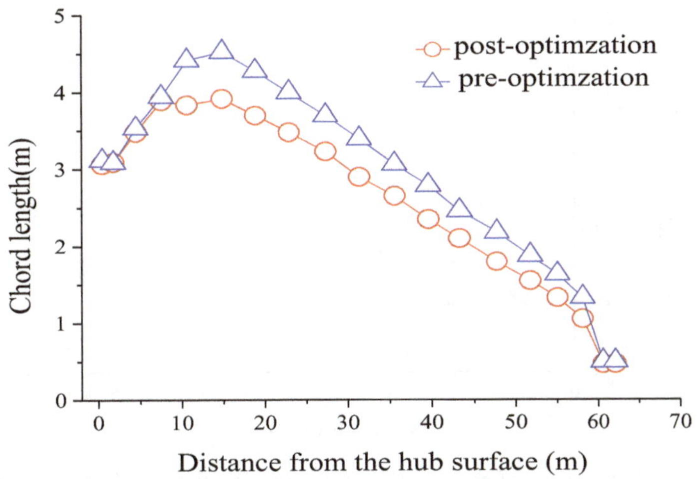

Figure 21. Distribution of the chord length at low wind speed before and after optimization.

4.5. Optimization of the wind turbine Blade under Actual Sea Conditions

The marine meteorological environment of site section should be considered in the offshore wind turbine design. In this paper, the location of the wind turbines is assumed to be in the China's Bohai Bay. According to the statistics of wind wave characteristics in Bohai Bay within one year, the average wind speed is 7 m/s, and the average significant wave height is 1.3 m. The wave period is 3.25 s. The frequency of the wind wave direction is 27% in the north direction, 22% in the northwest direction, and 8.3% in the northeast to northwest direction.

In the optimization process, the generated power of the wind turbines in eight wind wave directions (N, NW, W, SW, S, SE, E, NE) is calculated by FAST software. The particle fitness of the particle population is determined by the weighted average of the wind wave direction frequency. The expression is as follows:

$$\begin{aligned} Fitness = \; & (P(\mathrm{N}) \times 0.27) + (P(\mathrm{NW}) \times 0.23) \\ & +(P(\mathrm{W}) \times 0.083 + (P(\mathrm{SW}) \times 0.083) \\ & + (P(\mathrm{S}) \times 0.083) + (P(\mathrm{SE}) \times 0.083) \\ & +(P(\mathrm{E}) \times 0.083) + (P(\mathrm{NE}) \times 0.083) \end{aligned} \quad (5)$$

The wave spectrum is chosen as Jonswap spectrum. Under the rated wind speed, the floating wind turbine blade is optimized by PSO algorithm. As shown in Figure 22, particle swarm grows rapidly in 0~13 iterations, and the Zbest of particle swarm increases from 1180 kW to 1225 kW. After 13 iterations, the optimal particle of particle swarm does not renew. When it reaches 24 iterations, particle swarm stops automatically, indicating that the difference between the average fitness and optimal fitness of the population is less than 0.5%. The algorithm achieves the termination condition, and the particle swarm as a whole approaches the optimal solution in the search area. Under this condition, the power of the wind turbine is increased by 3.8% through the optimization of particle swarm optimization.

As shown in Figures 23 and 24, the optimized torsion angle is slightly less than that before optimization in the range of 10.25 m to 47.15 m from the hub surface. The optimized torsion angle is almost close to that before optimization in the range of 47.15 m to 60 m from the hub surface. The chord length in each control surface of optimized blade is less than that before optimization. The overall optimization effect is obvious.

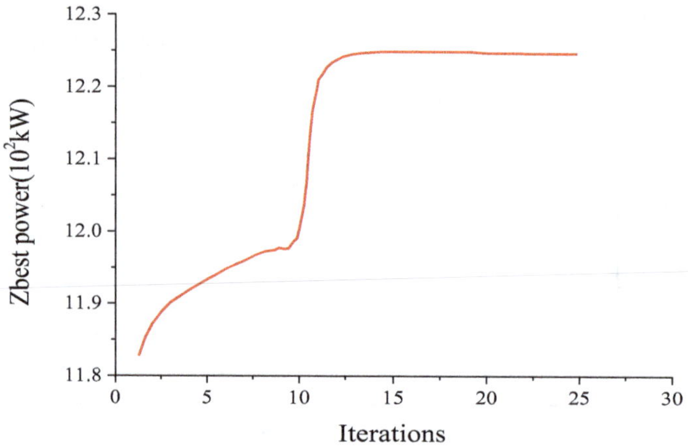

Figure 22. Evolution of the particle swarm under actual sea conditions.

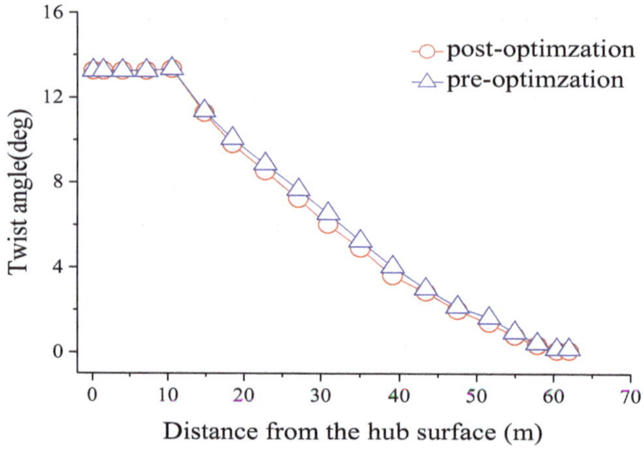

Figure 23. Distribution of the twist angle under actual sea conditions before and after optimization.

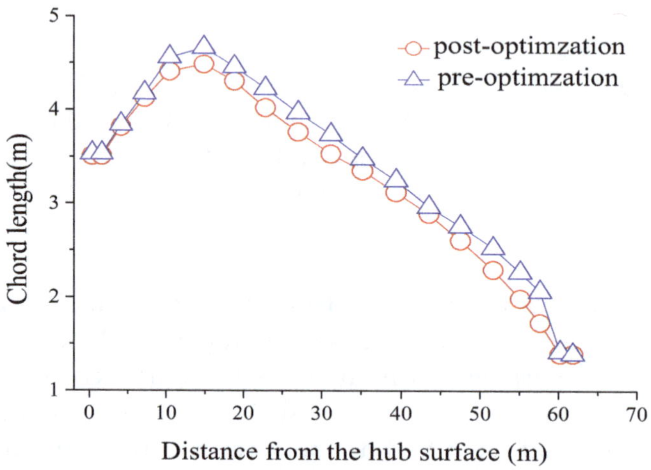

Figure 24. Distribution of the chord length under actual sea conditions before and after optimization.

5. Conclusions

Based on PSO algorithm and Fast program, a combined optimization design method for a wind turbine blade is developed in this paper. The optimization results of the NREL 5 MW fixed wind turbine show that this optimization method can not only improve the power of the wind turbine, but also reduce the manufacturing cost of the wind turbine blade. The feasibility of PSO is verified. The influence of wave on the power of floating wind turbine is studied in this paper. The results show that, with the increase of wave height, the power fluctuation of the wind turbine increases and the average power of the wind turbine decreases. With the increase of the wave period, the power oscillation amplitude of the wind turbine increases, and the power of the wind turbine at equilibrium position decreases. The optimal design of the offshore floating wind turbine blade under different wind speeds is carried out. The results show that the power of the optimized wind turbine blade increases less under rated wind speeds, and the optimization effect of the blade is more obvious at mid-low and low wind speeds. The Bohai Strait is chosen as the operation site of the wind turbine, and the wind wave direction is weighted. After optimization, the maximum power of the wind turbine can be increased by 3.8% after weighted optimization, and the chord length and the twist angle of the blade are reduced. The results of this study can provide theoretical guidance for the optimal design of a offshore floating wind turbine blade.

Author Contributions: Conceptualization, L.Y. and Y.M.; Methodology, L.Y.; Software, L.Y. and A.Z.; Validation, L.Y. and Y.M.; Investigation, L.Y.; Resources, Y.M.; Data Curation, C.H.; Writing-Original Draft Preparation, L.Y.; Writing-Review & Editing, A.Z. and Y.B.; Visualization, A.Z.; Supervision, C.H.; Project Administration, Y.M.; Funding Acquisition, Y.M.

Acknowledgments: Authors gratefully acknowledge the Institute of Marine Engineering Equipment and Technology in Sun Yat-sen University.

References

1. Oh, K.-Y.; Nam, W.; Ryu, M.S.; Kim, J.Y.; Epureanu, B.I. A review of foundations of offshore wind energy convertors: Current status and future perspectives. *Renew. Sustain. Energy Rev.* **2018**, *88*, 16–36. [CrossRef]
2. Fontana, C.M.; Hallowell, S.T.; Arwade, S.R.; DeGroot, D.J.; Melissa, E.L.; Charles, P.A.; Brian, D.; Andrew, T.M.; Senol, O. Multiline anchor force dynamics in floating offshore wind turbines. *Wind Energy* **2018**, *21*, 1177–1190. [CrossRef]
3. Madsen, H.A.; Rasmussen, F. A near wake model for trailing vorticity compared with the blade element momentum theory. *Wind Energy* **2010**, *7*, 325–341. [CrossRef]
4. Leishman, J.G. Challenges in modelling the unsteady aerodynamics of wind turbines. *Wind Energy* **2002**, *5*, 85–132. [CrossRef]
5. Butterfield, C.P.; Simms, D.; Scott, G.; Hansen, A.C. Dynamic stall on wind turbine blades. In Proceeding of the Wind power '91Conerence and Exposition, Palm Springs, CA, USA, 24–27 September 1991; pp. 393–401.
6. Pierce, K.G. Wind Turbine Load Prediction Using the Beddoes-Leishman Model for Unsteady Aerodynamics and Dynamic Stall. Master's Thesis, University of Utah, Salt Lake, UT, USA, 1996.
7. Liang, L.; Yuanchuan, L.; Zhiming, Y.; Gao, Y. Wind field effect on the power generation and aerodynamic performance of offshore floating wind turbines. *Energy* **2018**, *157*, 379–390.
8. Li, X.; Yang, K.; Liao, C.; Bai, J.; Zhang, L.; Xu, J. Overall design optimization of dedicated outboard airfoils for horizontal axis wind turbine blades. *Wind Energy* **2018**, *21*, 320–337. [CrossRef]
9. Chehouri, A.; Younes, R.; Ilinca, A.; Perron, J. Review of performance optimization techniques applied to wind turbines. *Appl. Energy.* **2015**, *142*, 361–388. [CrossRef]
10. Pratumnopharat, P.; Leung, P.S. Validation of various windmill brake state models used by blade element momentum calculation. *Renew. Energy.* **2011**, *36*, 3222–3227. [CrossRef]
11. Bai, C.J.; Wang, W.C.; Chen, P.W.; Chong, W.T. System integration of the horizontal-axis wind turbine: The design of turbine blades with an axial-flux permanent magnet generator. *Energies* **2014**, *7*, 7773–7793. [CrossRef]
12. Gaonkar, G.H.; Peters, D. Review of dynamic inflow modeling for rotorcraft flight dynamics. *Vertica* **1988**, *12*, 213–242. [CrossRef]
13. Laino, D.; Hansen, A.C. Current efforts toward improved aerodynamic modeling using the AeroDyn subroutines ASME wind energy symposium. In Proceeding of the Aerospace Sciences Meeting and Exhibit, Reno, NV, USA, 5–8 January 2004; pp. 329–338.
14. Shahizare, B.; Nazri Bin Nik Ghazali, N.; Chong, W.; Saeed Tabatabaeikia, S.; Izadyar, N. Investigation of the optimal omni-direction-guide-vane design for vertical axis wind turbines based on unsteady flow CFD simulation. *Energies* **2016**, *9*, 146. [CrossRef]

15. Gebhardt, C.G.; Preidikman, S.; Massa, J.C. Numerical simulations of the aerodynamic behavior of large horizontal-axis wind turbines. *Int. J. Hydrogen Energy* **2010**, *35*, 6005–6011. [CrossRef]
16. Benjanirat, S. Computational studies of the horizontal axis wind turbines in high wind speed condition using advanced turbulence models. Ph.D. Thesis, Georgia Institute of Technology, Atlanta, GA, USA, 2006.
17. Bulder, B.H.; van Hees, M.T.; Henderson, A.; Huijsmans, R.H.M.; Pierik, J.T.G.; Snijders, E.J.B.; Wijnants, G.H.; Wolf, M.J. *Study to Feasibility of and Boundary Conditions for Floating Offshore Wind Turbines*; Technische Universiteit Delft: Delft, Netherlands, 2002.
18. Tracy, C.C.H. Parametric Design of Floating Wind Turbines. Ph.D. Thesis, Massachusetts Institute of Technology, Boston, MA, USA, 2007.
19. Henderson, A.R.; Patel, M.H. On the modelling of a floating offshore wind turbine. *Wind Energy* **2003**, *6*, 53–86. [CrossRef]
20. Withee, J.E. Fully coupled dynamic analysis of a floating wind turbine system. Ph.D. Thesis, Massachusetts Institute of Technology, Cambridge, MA, USA, 2004.
21. Han, B.; Zhou, L.; Zhang, Z. LIDAR-assisted radial basis function neural network optimization for wind turbines. *IEEJ Trans. Electr. Electr.* **2018**, *13*, 195–200. [CrossRef]
22. Lee, J.; Hajela, P. Parallel genetic algorithm implementation in multidisciplinary rotor blade design. *J. Aircr.* **1996**, *33*, 962–969. [CrossRef]
23. Kusiak, A.; Zheng, H. Optimization of wind turbine energy and power factor with an evolutionary computation algorithm. *Energy* **2010**, *35*, 1324–1332. [CrossRef]
24. Liao, C.C.; Xi, G.; Xu, J.Z. An improved PSO algorithm for solution of constraint optimization problem and its application. *J. Eng. Thermophys.-Rus.* **2009**, *24*, 256–260. (In Chinese)
25. Liao, C.C.; Zhao, X.L.; Xu, J.Z. Blade layers optimization of wind turbines using FAST and improved PSO Algorithm. *Renew. Energy* **2012**, *42*, 227–233. [CrossRef]
26. Yang, H.; Chen, J.; Pang, X. Wind turbine optimization for minimum cost of Energy in low wind speed areas considering blade length and hub height. *Appl. Sci.* **2018**, *8*, 1202. [CrossRef]

Permissions

All chapters in this book were first published in MDPI; hereby published with permission under the Creative Commons Attribution License or equivalent. Every chapter published in this book has been scrutinized by our experts. Their significance has been extensively debated. The topics covered herein carry significant findings which will fuel the growth of the discipline. They may even be implemented as practical applications or may be referred to as a beginning point for another development.

The contributors of this book come from diverse backgrounds, making this book a truly international effort. This book will bring forth new frontiers with its revolutionizing research information and detailed analysis of the nascent developments around the world.

We would like to thank all the contributing authors for lending their expertise to make the book truly unique. They have played a crucial role in the development of this book. Without their invaluable contributions this book wouldn't have been possible. They have made vital efforts to compile up to date information on the varied aspects of this subject to make this book a valuable addition to the collection of many professionals and students.

This book was conceptualized with the vision of imparting up-to-date information and advanced data in this field. To ensure the same, a matchless editorial board was set up. Every individual on the board went through rigorous rounds of assessment to prove their worth. After which they invested a large part of their time researching and compiling the most relevant data for our readers.

The editorial board has been involved in producing this book since its inception. They have spent rigorous hours researching and exploring the diverse topics which have resulted in the successful publishing of this book. They have passed on their knowledge of decades through this book. To expedite this challenging task, the publisher supported the team at every step. A small team of assistant editors was also appointed to further simplify the editing procedure and attain best results for the readers.

Apart from the editorial board, the designing team has also invested a significant amount of their time in understanding the subject and creating the most relevant covers. They scrutinized every image to scout for the most suitable representation of the subject and create an appropriate cover for the book.

The publishing team has been an ardent support to the editorial, designing and production team. Their endless efforts to recruit the best for this project, has resulted in the accomplishment of this book. They are a veteran in the field of academics and their pool of knowledge is as vast as their experience in printing. Their expertise and guidance has proved useful at every step. Their uncompromising quality standards have made this book an exceptional effort. Their encouragement from time to time has been an inspiration for everyone.

The publisher and the editorial board hope that this book will prove to be a valuable piece of knowledge for researchers, students, practitioners and scholars across the globe.

List of Contributors

Sabra Ahyaten and Jalal El Bahaoui
Department of Physics, Faculty of science Tetouan, Abdelmalek Essaadi University, 93 030 Tetouan, Morocco

Ko Matias Adrian Kosasih, Hideyuki Niizato and Shigeki Okubo
Hitachi Zosen Corporation, Osaka 551-0022, Japan

Hideyuki Suzuki
Department of Systems Innovation, Graduate School of Engineering, The University of Tokyo, Tokyo 113-8656, Japan

Yuan Song
Department of Advanced Mechanical Engineering, Kangwon National University, Chuncheon 24341, Gangwon, Korea

Insu Paek
Department of Integrated Energy & Infra System, Kangwon National University, Chuncheon 24341, Gangwon, Korea
Department of Mechatronics Engineering, Kangwon National University, Chuncheon 24341, Gangwon, Korea

Minhui Tong
Logistics Engineering College, Shanghai Maritime University, Shanghai 201306, China
Department of Mechanical Engineering, University of Maryland, Baltimore County, Baltimore, MD 21250, USA

Weidong Zhu, Meilin Yu, Kan Liu and Gang Li
Department of Mechanical Engineering, University of Maryland, Baltimore County, Baltimore, MD 21250, USA

Xiang Zhao
Department of Mechanical Engineering, University of Maryland, Baltimore County, Baltimore, MD 21250, USA
School of Civil Engineering and Geomatics, Southwest Petroleum University, Chengdu 610599, China

Francesco Natili , Francesco Castellani, Davide Astolfi and Matteo Becchetti
Department of Engineering, University of Perugia, Via G. Duranti 93, 06125 Perugia, Italy

Georgios Gasparis, Wai Hou Lio and Fanzhong Meng
Department of Wind Energy, Technical University of Denmark, 4000 Roskilde, Denmark

Miguel-Ángel Manso-Callejo, Calimanut-Ionut Cira , Ramón Alcarria and José-Juan Arranz-Justel
Departamento de Ingeniería Topográfica y Cartografía, E.T.S.I. en Topografía, Geodesia y Cartografía, Universidad Politécnica de Madrid, 28031 Madrid, Spain

Davide Astolfi and Francesco Castellani
Department of Engineering, University of Perugia, Via G. Duranti 93, 06125 Perugia, Italy

Xiaobing Kong, Lele Ma, Xiangjie Liu and Qian Wu
The State Key Laboratory of Alternate Electrical Power System with Renewable Energy Sources, North China Electric Power University, Beijing 102206, China

Mohamed Abdelkarim Abdelbaky
Electrical Power and Machines Engineering Department, Faculty of Engineering, Cairo University, Giza 12613, Egypt

Lu Ma, Xiaodong Wang, Jian Zhu and Shun Kang
Key Laboratory of Power Station Energy Transfer Conversion and System, Ministry of Education, North China Electric Power University, Beijing 102206, China

Iñigo Aramendia and Unai Fernandez-Gamiz
Nuclear Engineering and Fluid Mechanics Department, University of the Basque Country UPV/EHU, Nieves Cano 12, 01006 Vitoria-Gasteiz, Spain

Ekaitz Zulueta, Aitor Saenz-Aguirre and Daniel Teso-Fz-Betoño
Automatic Control and System Engineering Department, University of the Basque Country UPV/EHU, Nieves Cano 12, 01006 Vitoria-Gasteiz, Spain

Zhenzhou Shao, Ying Wu, Li Li, Shuang Han and Yongqian Liu
State Key Laboratory of Alternate Electrical Power System with Renewable Energy Sources, North China Electric Power University, Changping District, Beijing 102206, China

Juhun Song and Hee-Chang Lim
School of Mechanical Engineering, Pusan National University, 2, Busandaehak-ro 63 beon-gil, Geumjeong-gu, Busan 46241, Korea

Aiming Zhang
School of Marine Engineering and Technology, Sun Yat-sen University, Guangzhou 518000, China

Chao Hu and Yue Bai
College of Shipbuilding Engineering, Harbin Engineering University, Harbin 150001, China

Yong Ma
School of Marine Engineering and Technology, Sun Yat-sen University, Guangzhou 518000, China
College of Shipbuilding Engineering, Harbin Engineering University, Harbin 150001, China
Southern Marine Science and Engineering Guangdong Laboratory (Zhuhai), Zhuhai 519000, China

Lele Yang
School of Marine Engineering and Technology, Sun Yat-sen University, Guangzhou 518000, China
Southern Marine Science and Engineering Guangdong Laboratory (Zhuhai), Zhuhai 519000, China

Index

www.ingramcontent.com/pod-product-compliance
Lightning Source LLC
LaVergne TN
LVHW082138200926
848484LV00008B/2073